Science Theory for Engineers and Physicists

This book is a concise hands-on guide that provides students a useful tool for a systematic approach to scientific inquiry.

Science Theory for Engineers and Physicists provides a basic understanding of "the method" as a question and to understand how a reliable answer that can be validated is constructed. It covers the basic terminology and methods which are exemplified for ease of understanding for the reader. Testing of hypotheses and quantitative understanding of uncertainty and uncertainty propagation are also illustrated.

The book is intended to be used with experimental work as a final or thesis project for undergraduate or graduate students who have not had prior experience in science theory or data handling and uncertainty.

Science Theory for Engineers and Physicists

Anders E. W. Jarfors

CRC Press
Taylor & Francis Group
Boca Raton London New York

CRC Press is an imprint of the
Taylor & Francis Group, an **informa** business

First edition published 2025
by CRC Press
2385 NW Executive Center Drive, Suite 320, Boca Raton FL 33431

and by CRC Press
4 Park Square, Milton Park, Abingdon, Oxon, OX14 4RN

CRC Press is an imprint of Taylor & Francis Group, LLC

ISBN: 978-1-032-84671-2 (hbk)
ISBN: 978-1-032-85479-3 (pbk)
ISBN: 978-1-003-51837-2 (ebk)

DOI: 10.1201/9781003518372

Typeset in Times
by Newgen Publishing UK

This short handbook and guide is dedicated to my wife Anna-Lena for being my guiding light and having patience with all my follies!

I love you more than life itself!

Contents

Preface

In the book *Reading the Book of Nature: An Introduction to the Philosophy of Science*, by Peter Kosso, it is stated that to understand nature, you need to read nature, but it is written in an unfamiliar language using an unknown alphabet and must as such be interpreted. This interpretation is to be made by a scientist. The ability to interpret requires both knowledge and a process or method.

The understanding of what a scientific approach is and when you have proven a relationship, or not, has gone lost in may engineering educations. The importance of what a significant digit means likewise. The entry of computers en-mass, together with machine learning, and an increasingly black box situation, where the numbers popping out from a computer are taken verbatim. There is a need for a sanity check and for support in what is a scientific base, experimental uncertainty and an understanding of the difference between noise, systematic errors and proof.

This book aims to give a short hands-on guide to the theory of science to provide students in engineering fields and physicists with a tool for a systematic approach to inquiry. Basic terminology and methods are introduced and exemplified for ease of understanding. Testing of hypotheses and quantitative understanding of uncertainty and uncertainty propagation are also illustrated.

The book is intended to be used with experimental work as a final or thesis project for undergraduate or graduate students who have not previously had science theory or data handling and uncertainty. It should serve as a short handbook and guide to the topic—a tool to improve the systematic quality of the work.

Anders E. W. Jarfors
Jönköping 2024-05-28

Acknowledgements

This booklet is created to be used in teaching as supplementary material to final-year project works. The author acknowledges Professor Petter Leisner for inspiring and creating the possibility to write this book. Furthermore, it was part of the Light Weight Design for Sustainable Mobility or LiDeSum project and the partners in the project, Professor Jernej Klemenc, with the University of Ljubljana, who took the initiative to the project, and Professor Manuel Lagache with the University of Savoie Mont Blanc as a project partner. The author also wishes to thank the colleagues at the Materials and Manufacturing Department for providing feedback and inspirational comments, especially Dr Vasilios Fourlakidis for finding the sources of inspiration and helping me reach the required number of words!

Thank you!

Author

Anders E. W. Jarfors is a Swedish researcher in materials and manufacturing. He earned an MSc and a PhD at the Royal Institute of Technology in Sweden. He has worked at the Royal Institute of Technology in Stockholm and the Swedish Institute of Metals Research (formerly KIMAB and now RISE and Swerim). He has also worked at the Singapore Institute of Manufacturing Technology and MMI Holding Pte Ltd in Singapore. He was also an adjunct associate professor at Nanyang Technological University. He has held guest and visiting professorships in China with the China Academy of Machinery Sciences and Technology and at Nanyang Technological University, Singapore. He is currently a professor of materials and manufacturing at Jönköping University and a visiting professor of sustainable manufacturing at Cranfield University.

Professor Jarfors has a long list of empirical work in materials science and manufacturing processes ranging from metal casting and sheet metal forming to powder metallurgy and spray forming. His primary research interests are in metal casting and automotive components. He is an active teacher who recognized a need for a simple description of the foundations of the scientific approach. This is commonly seen in the approach and at the start of final-year projects where there are gaps in approach and the ability to understand what is proven and what is suspected.

Introduction

1

1.1 WHAT IS SCIENCE AND A SCIENTIFIC METHOD?

The two words research and science are intimately connected and often used with more or less the same meaning. Is the meaning of a researcher and a scientist the same? Is there a difference, and what is what?

The Oxford Dictionary states several meanings, but one is:

Science: A particular area of knowledge or study; a recognized branch of learning; spec. (in the Middle Ages) each of the seven subjects forming the trivium (grammar, logic, and rhetoric) and quadrivium (arithmetic, geometry, music, and astronomy)

Research: Systematic investigation or inquiry aimed at contributing to knowledge of a theory, topic, etc., by careful consideration, observation, or study of a subject. In later use also: original critical or scientific investigation carried out under the auspices of an academic or other institution.

This means that science is the topic of study, and the research is the action of investigation. Science theory is thus more related to the actual topic of study and the actual constituents of the subject area. Research methodology is how you conduct your research, and the scientific method is how the methodology is made in relation to the rules and relationships of the specific research area. The rules of the science area will dictate specific rules or systems for the study, meaning that there is a systematic approach, and the research will impose study methods. The theory of science will then guide a researcher or scientist on how to work with scientific questions using a systematic methodology for the research made.

DOI:10.1201/9781003518372-1

It should also be noted that ever since the Middle Ages, the initial branches of science were grouped into two different main types of science:

- trivium
- quadrivium

It originates from how the seven sciences were taught where trivium (grammar, logic, and rhetoric) was the foundation to move on to the second stage, quadrivium (arithmetic, geometry, music, and astronomy). To some extent, the trivium was the tool required for the application in quadrivium. Schooling is very different today, and the number of sciences has exploded, especially in the applied and engineering sciences. To some extent, though, the sciences still have two groups with

- a priori science
- empirical science

The a priori science consists of logic and rhetoric. It should be noted that arithmetic, geometry, and mathematics, in general, are now more associated with the a priori sciences. The empirical sciences are any study of our physical environment and most commonly in the form of experimentation in many of the old and new sciences such as mechanical engineering, engineering physics, metallurgy, materials science, chemistry, astronomy, and so forth. Interestingly enough, with the increased simulation capability and the use of machine learning and metamodels, the use of computational experimentation is entering into science. This is highly a priori based, and the output is virtual, but it is close enough that it is possible to use as a description of reality, and by varying input, it is possible to emulate input and functional uncertainties of processes. It is a virtual analogy to reality.

1.2 THE ROLE OF SCIENCE AND A SCIENTIST

To start, it must be understood what a scientist or a researcher is. The Oxford Dictionary states the following:

Scientist: A person who conducts scientific research or investigation; an expert in or student of science, esp. one or more of the natural or physical sciences.

Researcher: A person who researches; an investigator, inquirer.

These two definitions clearly show that a scientist needs a scientific approach and method to do research in a scientific field and may be called a researcher. On the other hand, a researcher is not necessarily a scientist, as not all research is scientific. An example here is that a journalist may be doing research on an event, or an actor can do research for a movie, but this may not necessarily be scientific. The current book focuses on scientific research and methods. This is also why this treatise is on science theory and not research methodology.

The hallmark of a scientist is an inquisitive mind. An inquisitive mind tends to ask questions. It is man's (and scientist's) nature to ask questions. Questions are asked for many reasons, but in science, the questions are asked to learn.

Asking questions to learn something that holds over time is a different matter. The answers cannot be arbitrary, nor should they be biased. Moreover, it should be possible to validate the answers.

The first matter to establish is what is science and what is not science. Contrary to what many think, a basic foundation of science is that it is falsifiable, meaning that it should be possible to be proven wrong in one or another way, at least in principle. This statement is essential since all science does not always include the possibility to observe (logic, mathematics, and so-called a priori sciences). This distinction is critical to separate dogmatism from science. This is also the essence of a scientific truth. It can change and evolve.

Another critical matter is that of pseudoscience, fringe science, or so-called "cargo cult science". This is more difficult as some researchers genuinely believe they are doing science, and their work appears scientific. The difference is that it lacks what can be seen as utter honesty that allows a rigorous evaluation of their work. The lack of transparency or incomplete data sets or reporting and explanation of the methods and models used will result in a risk of entering the work into the "cargo cult science" compartment. Here also lies the concept of bias, or more appropriately confirmation bias, where only material supporting the claim is in, and there is no argument with arguments contradicting the outcome that need falsification before acceptance of a new idea.

The purpose of science is thus to determine the truth or even the ultimate truth. Some matters may be difficult to observe or visualise, leading to a suspicion that there are questions that science cannot answer, which may be atoms, electrons, reciprocal spaces, or null spaces spanned by eigenvectors. As the understanding of both science and technology evolves, this frontier is moving with science, and more and more elements are directly observable, and with that there is also a change of the truth, allowing the falsification of old truths.

One critical function of science is establishing facts and truths explaining the world around us. Other streams also state that science and scientific theories should be judged on usefulness. From this perspective, it is not critical whether the theories are true or not as long as they are functional.

Example :

> The law of gravity is a well-accepted law with a well-known equation. This is a reasonably accepted fact. This law is likely to be more accurately described using neural networks. These neural networks are likely capable of an even more accurate description of gravity. However, neural networks will not give an equation, and it is not easy to explain what the law is. This is an example of a truly functional approach, but it is more or less impossible to pinpoint if it is right or wrong as there is no explicit visible outcome. Both approaches do result in an acceptable outcome and are acceptable from a scientific standpoint. They do, however, require two different acceptances of what science is.

Modern society stands on the pillars of empiricism, where a number of truths have been developed and used in our daily lives. Empirical studies are often based on a need, followed by an investigation to get the answer. This answer is often used to build something. This is development work. Science also includes that the observation should hold not only to a single case but allow some degree of generalisation. This does not necessarily lead to a direct application but drives a new understanding or knowledge. This knowledge should then be possible to use not only for one application but for many. A sufficient bulk of new knowledge lays the foundation for an improved society. As a scientist, you build the foundation for tomorrow's society.

There are many aspects of being a society builder, including ethics, which is not a core part of this treatise, but the reader is advised to study and use an ethical approach to the work being done. Another aspect is that as a scientist generates new knowledge, he or she has the duty to ensure that this new knowledge is used in society. A scientist must, therefore, teach. Students are the vehicles of new knowledge, and being a scientist is synonymous with being a teacher.

1.3 THE DIFFERENCE BETWEEN DIFFERENT SCIENCES

The current treatise deals with the theory of science related to the natural sciences. There are some significant differences, for instance, in social sciences. These differences arise from the fundamental difference in the nature of the problems investigated. The purpose of all sciences is to build knowledge to predict outcomes. No science or society benefits from inaccurate, not precise, or biased models.

In natural science, taking physics as an example, the target is to explain the most straightforward description of reality, including a small number of components and the use of idealisation/simplification of often isolated systems. This has proven successful in understanding natural phenomena such as gravity, electricity, condensed matter, and so on. This has been a very successful concept based on the ability to simplify and then to generalise with the understanding of what the simplifications and delimitation or isolation of the problem mean for the generalisation.

On the other hand, social scientists describe human interaction, meaning that a number of complex components of a study has become immense. The ability to quantify with accuracy becomes dramatically reduced by this condition. The human element is far more complex than the atoms' behaviour.

The level of complexity in social science compared to natural sciences makes generalisation more difficult. Natural sciences and the laws developed gain general acceptance as they tend to stand over time. The level of complexity, the drift of attitudes, and the mixture of people in societies make generalisation more complex, making the laws' validity and observation difficult to last permanently. This also means that within social sciences, the laws and relationships struggle to last over time, and new paradigms develop over time, often portraying the previous laws and regulations as lies or seriously flawed. One example is the different political beliefs: liberalism–capitalism–marxism.

Another difference is how explanations are being made. In social sciences, almost without exception, all explanations are phenomenological descriptions of the events being studied. In natural sciences, this is not an explanation but rather an observation.

1.4 POLITICAL BIAS IN SCIENCE

The basic task of science is to systematically explain, without bias, observation and events taking place in the area being studied. The hypotheses and theories being developed should be true or false, independent of your own opinions and ethics. The task of science is to be non-political. In many cases, for physical sciences, your opinions do not affect the actual physics. Gravity does not care about your opinion. The climate does not care if your beliefs are capitalistic or Marxist. It will proceed anyway as a response to your actions.

In social science, it is not necessarily a situation that abides by a repeatable law but could be a situation where different roles are assigned. The political risk here is when science is used to force the assigned roles as a law, maintaining prejudice or bias. One example is that their biology gives the role of women; therefore, it would be unnatural to interfere. A standard view within sociology is that humankind should be freed from socially assigned roles, making the discoveries of sociology a demanding foundation for political action.

Social sciences and politics have a closer tie, but funding systems and appointments in leading institutions also have a significant political influence on the natural sciences. All researchers should know this and understand the ethics and consequences of bias in observation and conclusions. The role of the funding system or sponsor of the research is critical, and to maintain utter honesty, sponsoring and funding information should be publicised openly. This is important as it allows for a judgement of the conclusions and the risk of judgement. Today, it is important that the funding agent or sponsor does not influence the research design, evaluation, or conclusions. This may often be the case, but what topics are being funded themselves is biased. This is not unimportant in the definition of use-inspired research with a dedicated purpose and drive towards faster implementation than other approaches. This is common in research areas such as engineering and medicine. In order to promote the construct of tomorrow's society, public funding for research is vital as a complement to the use of inspired research. It is essential to avoid the fact that all research is company-funded and that commercial interest does not solely control the research arena. There must be room for curiosity-driven research as well.

1.5 THE PURPOSE OF THIS BOOK

Engineering science and engineering physics are terms intimately connected to various combinations of engineering topics such as physics or engineering combined with aerospace, biology, chemistry, electronics/electrical, mathematics, mechanical, materials, and so on. Critical to the success of the engineering sciences is the use of a scientific method as a rigorous foundation to allow successful ways to apply, design, and develop new engineering solutions for the good of humankind.

This book aims to provide a basic understanding of what is required for the work to be scientific and to understand the boundaries of science. To do this, science history and the empirical approach are introduced so that a systematic development of the construct of a study can be made. The whole concept and method of how to pose a question and understand how to derive a reliable answer that can be validated from empirical observation is explained. There is not just one route; a whole toolbox exists. This also includes some basic understanding of statistics, meaning understanding when something is proven and with what confidence. In this, there is also the understanding of error and uncertainty and how it is propagated in models and explanations.

Science and the Boundaries of Science

2

The origin of science can be seen as the study of the most repeatable laws of nature, such as gravity, the location of the stars, heat, pressure, and so forth. Giving quantitative and objective arguments and analysing the hypotheses made were challenging without measurement tools. This is very clear from the fact that it is hard to distinguish between an objective observation and a subjective one (= objective + processed through human senses). This also strongly implies that no empirical science is more robust and capable than its capability to measure. Empiricism, the development of metrology, and all other characterisation methods and equipment are essential for developing the progress of science and humankind.

Natural science is dedicated to what is observable, in contrast to, for instance, philosophy, which is more based on intrinsic human reasoning and its values and belief systems. As the ability to measure increases, the natural sciences are moving into the areas of the other sciences, where the discovery of the Higgs particle is one such prediction. The a priori sciences, such as mathematics, venture into the boundaries between the natural sciences and science, such as philosophy.

This is an interaction between these fields of science, as many of the human needs for the natural sciences have an intrinsic driver. This means that the boundaries of the natural sciences are not always given by the ability to measure, but rather that the direction of what to research has a drive based on ethics, morals, and trends in society or more or less political influence.

Nevertheless, due to the cardinal importance of technology development in the field of measurement, it is essential that these should be repeatable and operator-independent, which naturally is a significant challenge. Repeating studies and having more than one person take measurements do not duplicate results; instead, it is necessary for building knowledge. In many cases of hypothesis testing, it is only possible to do this with the existence of the right

DOI:10.1201/9781003518372-2

instrument or the appropriate experimental setup. This is an essential part of the deduction of the test to be designed.

Technological development also forces the reality of the truth to be relative and not constant. This is a natural consequence of expanding our knowledge; what is a truth is changing. The most essential quality of truth is that it is verifiable and testable, thereby also changeable!

In striving for the ultimate truth, it is possible to hypothesise that when the ultimate truth is reached, it will not change and that as this point is reached, the falsification resistance of truth is increasing. The same can be said about the ability to continue purposeful knowledge creation. These points would be boundaries of science explanations. This argument is based on the fact that:

- A bad explanation is easy to vary.
- The search for hard-to-vary explanation drives all progress.
- The absolute truth consists of hard-to-vary assertions about reality, which then would make up the boundaries of science for the physical world.

Critical to all societies is the concept of sustained creation of knowledge driving wealth and health. Essential to this are elements such as the concept of optimism, where problems are inevitable, and solutions will always exist. This is only possible if the right knowledge is sought out and acquired, continuously moving the boundaries of science and technology, making the boundaries of science non-absolute.

Laws

3

3.1 EMPIRICAL SCIENCE

Early science started with the answer to the question "Why?" and should be able to point back to an accepted law, discussed below. This works for more straightforward problems, and there must be a match between the nature of the observation and the nature of the explanation. There are issues in the actual events that have statistical variations and in the ability to observe which has uncertainty. These matters are discussed in detail later in this book. A better understanding comes from the fact that the scientific explanation must be statistically relevant to the outcome being studied and explained.

Commonly, the first step to a scientific observation is that something is repeatable and happens similarly or nearly similar at each occurrence. This is like Newton's apple always falling to the ground. This is an empirical observation or is based on experience. This is also referred to as an a posteriori science approach.

An actual law is something that happens regularly and is repeatable as an observation.

Examples of actual laws are

- If you drop a glass on the floor, it breaks.
- If it is humid, there may be morning dew on the grass.

There are also the laws of nature that come with a necessity.
Examples of laws of nature are

- If you increase the pressure of a closed gas container, the temperature increases.
- The general gas law is $pV = nRT$ p=pressure, V=volume, n=amount, R=gas constant, T=absolute temperature.

DOI:10.1201/9781003518372-3

The laws of nature can be

- qualitative or
- quantitative

A qualitative law is the observation that temperature increases if you increase the pressure of a fixed amount and volume of gas. A quantitative law is like the general gas law, where it is possible to specify precisely how much the temperature increases given a specific pressure increment.

In the empirical cycle, de Groot's approach is the customarily accepted cycle (Figure 3.1). This consists of the following steps:

- Observation: The first step is observing a phenomenon and enquiring about its causes. This is similar to a pre-study to understand what matters and what the question is about.
- Induction: A hypothesis regarding generalised explanations of the phenomenon is formulated from the initial observation.
- Deduction: A formulation of the experiments that should test the hypothesis and how data is collected.
- Testing: The actual experimental technique and procedures used to test and collect data.
- Evaluation: This is the interpretation of the results and the formulation of the theory. This is commonly made using an abductive argument that presents the result of the experiments as the most reasonable explanation for the phenomenon.

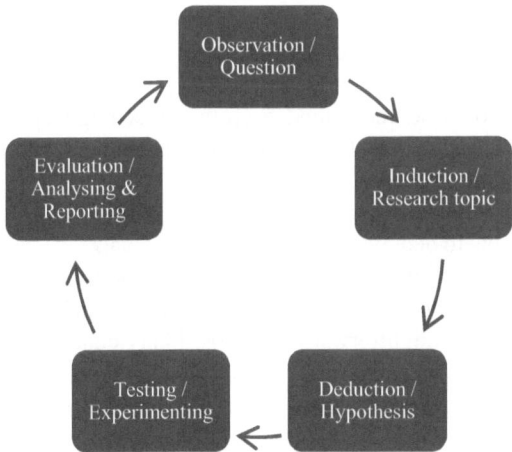

FIGURE 3.1 Illustration of A. D. de Groot's approach in empirical science.

Some new terms require explanation. Based on the first initial observation, the hypothesis is made by induction. Inductive reasoning refers to any method of reasoning in which broad generalisations of principles are derived from a body of observations.
Example:

> Induction can be illustrated by the following: There are 20 black or white balls in a box. A sample of five balls is taken, and one is white. A generalised induction would then suggest that there are four white balls in the box.

Induction is the process of inferring based on an observation that 99% of swans are white. If another swan arrives at the pond, it is possible to induce that this swan is white. There is still a 1% risk that this conclusion is wrong.

Deductive reasoning is used in de Groot's approach to design a test. Deductive reasoning is the mental process of drawing deductive inferences. An inference is deductively valid if its conclusion follows logically from its premises, that is, the premises cannot be true and the conclusion is false. A common error at this stage is that you look at, for instance, a casting process.
Example:

> Deduction can be illustrated by inferring the general principle that all persons in the city can ride a bike and the fact that my Red Lapierre is a bike. It follows that all people can ride my Red Lapierre.

In a casting process, there are a number of things that you can vary, and then you set up a plan to vary the "casting parameters". This may not be sufficient to test or prove your hypothesis. The inductive formulation of the hypothesis and the deductive design of the experiment should be considered.

The testing used has a few elements. The approach must reveal what is sought, which should be understood from the preceding steps. The available equipment for the evaluation should exist, or an agreement with parties having access should be made before the tests.

Firstly, the evaluation phase commonly requires using equipment and understanding how to deal with and treat measured data.

Once data collection is made, the actual reasoning of the analysis can be made using abduction. Abductive reasoning is a form of logical inference that seeks the most straightforward and most likely conclusion from a set of observations. Unlike deductive reasoning, abductive reasoning yields a plausible conclusion but does not definitively verify it. Abductive conclusions do not eliminate uncertainty or doubt, expressed in retreat terms such as "best available" or "most likely". While inductive reasoning draws general conclusions that apply to many situations, abductive conclusions are confined to particular observations. Occam's razor is another term for this, seeking the simplest explanation as the most likely explanation.

Deductive reasoning leading up to an experimental design often leads to what should be varied and the most beneficial factors from which to collect data. The experimental approach to proving a law is made in several different manners. There are three dominant approaches used throughout history. These three approaches are:

- The concurrence approach
- The differential approach
- The variation approach

3.1.1 The Concurrence Approach

The concurrence approach is suitable when understanding which factors are correlated. If you, for instance, make three different experiments and identify what factors are present, it is possible to see which are correlated. In experiment 1, A, B, C, and D were present. In experiments 2 and 3, A, B, E, D and A, C, F, D were present, respectively. Since only A and D were present, it is safe to conclude that A and D were correlated.
Example:

Three metal samples containing the following elements were analysed using optical emission spectroscopy.

- Sample 1: Fe, Mn, Co, Cu
- Sample 2: Fe, Ni, Cu
- Sample 3: Cu, Zn, Ni

All samples had a spectral line with a wavelength of 3248Å. Since Cu is present in all samples, it is safe to conclude that this line correlates with Cu's presence.

3.1.2 The Differential Approach

The differential approach is suitable for understanding if a specific factor correlates with another factor. If you, for instance, have a case where B, C, E, and F were present. If you, by the introduction of A, also find D, it is safe to conclude that D depends on A.
Example:

Three metal samples containing the following elements were analysed using optical emission spectroscopy.

- Sample 1: Fe, Mn, Co, Cu
- Sample 2: Fe, Ni, Cu
- Sample 3: Cu, Zn, Ni

All samples had a spectral line with a wavelength of 3248Å. Since Cu is present in all samples, it is safe to conclude that this line correlates with Cu's presence.

Example:

Suppose you place a compass needle next to a straight copper wire and allow it to attain equilibrium determined by the earth's magnetic field. Inducing a current through the copper wire will make the compass change orientation. If no other changes are made, it can be concluded that the compass needle orientation change depends on the electrical current in the copper wire.

3.1.3 The Variation Approach

In the variation approach, you can start when you have A, B, C, and D present and then vary the intensity of A. If you, by varying the intensity, A also observe that D varies in intensity and that B and C are unaffected, then it can be concluded that A and D are correlated. If D, on the other hand, does not change, then it is safe to conclude that D does not depend on A.

Returning to the example of the compass needle above, you change the experiment so that instead of introducing the current in the copper wire, you vary the amount of current passing through the wire. This becomes a different study. If the needle orientation change varies with the current level, then the current and the magnetic needle orientation are correlated entities. It is important to note that this is not the same conclusion as the differential approach.

3.2 A PRIORI SCIENCE

A priori science, compared to a posteriori science, is built on logical rules that do not necessarily correspond to something you can measure or perceive with

your sensory systems. A priori means independent on experience. The foundation is that you start with some non-defined basic terms and then deduce other terms. The foundation is based on so-called axioms that are given and regarded as obvious or safe. These are used to prove other elements, so-called theorems. The proving of theorems is made in a deductive manner.
Example

> Euclidean geometry is a deduced system based on a number of axioms, such as

> - A straight pass through two points, at least;
> - Through a point outside a straight line goes only a straight line parallel to the former (Euclid's Parallel Axiom).

An example of a theorem or law based on these is for an example the first congruence case:

- If two sides and the angle between these sides are similar to another triangle, these triangles are congruent.

Example

> It is easy to create a non-Euclidean geometry. If you start with an Euclidean space and bind it by a circle (Figure 3.2), letting a straight line, ST, be a Corda to the bounding circle, then the concept of distance can be introduced using the following definition. The distance AB is equally long as A1B1 provided that

$$\frac{AT}{BT} \cdot \frac{BS}{AS} = \frac{A1T1}{B1T1} \cdot \frac{B1S}{A1S1}$$

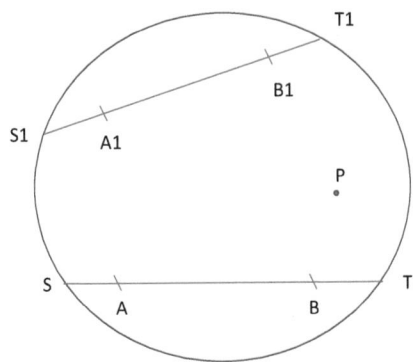

FIGURE 3.2 Deduction of a non-euclidean space.

The distance AB is shorter than A1B1 if the left-hand side is smaller than the right-hand side. Figure 3.2 shows that it is possible to draw an infinite number of straight lines that do not cut the straight-line ST from an arbitrary point P outside the line ST. Thus, the space construction is a non-Euclidean space that will not fulfil Euclid's Parallel Axiom.

The rest of this treatise will only deal with empirical approaches and not any type of a priori approaches.

Definitions

4

4.1 STIPULATING DEFINITIONS

Once the concept of science is well understood, a number of critical tools are required to execute scientific work and analysis. The first set of tools explained in the previous chapter was related to designs on how to investigate and how to design investigations using different methods for both empirical and a priori science. As stated in the last chapter, this treatise is all about empirical sciences, and even though much of what is discussed may also be used for a priori science, the target and framing of the discussion are made for empirical sciences where physical observation and measurement are the core capabilities and means to reach the sought outcome of the study.

A stipulative definition defines the meaning of an existing or even a new term that is narrowed down in meaning for a specific purpose. For existing terms, this definition may or may not contradict the lexical definition of the term. Consequently, a stipulative definition does not have the property of being "correct" or "incorrect". At most, this can differ from other definitions. The only important property is that it is defined for a specific purpose in a specific context. A stipulating definition states how a term will be used. These can be either more of an analytical type or functioning more like a dictionary.

Example

- Force is the product of an object's mass times its acceleration (Analytical type)
- An acid is an electrolyte that releases hydrogen ions (Analytical type)
- Metallography = the science of the physics and chemistry of metals (Dictionary type)

In-text definitions are often used to introduce abbreviations to be later used in the text.

Example

DOI:10.1201/9781003518372-4

- Introduction at first use … the use of Hot-Isostatic-Pressing (HIP)…. after introduction and continued use … HIP was used to close porosity …

The most important form of definition is the analytical or descriptive definition. The analytical descriptive definition can describe the accepted meaning of a known concept. In the description, the terms used should already be known in the same manner as the abbreviations.

4.2 OPERATIONAL DEFINITIONS

Stipulative definitions are one approach to defining what is meant by the terms and concepts used in the work's description. These are necessary but not always sufficient; therefore, operational definitions have been defined and are commonly used. Operational definitions differ from the stipulating definitions in that an operational definition determines a term or a concept by stipulating the type of result that should follow a certain action.
Example

- The length of a line is measured using a Vernier Calliper, which measures the distance from the starting point to the endpoint.

This type of definition is important since some units are constructed out of others.
Example

- The SI unit Newton is constructed in a way that it is kg m/s^2 in the founding units in the SI system.
- In the older CGS system, capacitance has the unit cm.

In the world of definitions, unit constructions can have odd outcomes. Capacitance in the CGS system cannot be measured using a Vernier Calliper as there is no distance. As a consequence, as with capacitance, not all objects have all properties, but rather, all objects of a certain kind have certain properties.
The fact that there is an intricate relationship between the object properties and the measured quantity also affects how the characterisation is made. One such example is Mass. Utilising gravity, the mass of a body can be measured using a scale. This mass is referred to as weight. In the absence of gravity, since force is the product of mass and acceleration, the mass can be determined by measuring force and acceleration. Mass is given by dividing force with accelerations. This mass is referred to as inertial mass.

Hypotheses

5

5.1 HYPOTHESES AND REQUIREMENTS OF THEM

A scientific approach entails many different elements, starting with the careful observation to try to understand an event, followed by the induction of the nature of the observation, leading to the deduction of the test. In order to define and test an idea, the deduction of the test commonly leads to a first draft of a hypothesis. The definition of a test and a hypothesis go hand in hand. A hypothesis's final version is usually not made until after the test.

A hypothesis is what could be stated as a proposed explanation for a phenomenon. The whole idea behind science is the ability to falsify an idea, making an essential property of a hypothesis testable. The experiment's design and the hypothesis's definition are intimately related, as the design of the tests must be made so that the hypothesis is testable or falsifiable.

Often, the words theory and hypothesis are synonyms in science. Science involves a process where a rigorously made paper or thesis offers satisfactory explanations for a theory. At this first stage, it can be seen as a working hypothesis. It is not until a hypothesis has been tested by others independently and accepted that the provisionally accepted working hypothesis is an accepted hypothesis or theory.

A proven true hypothesis may gain general acceptance, leading to it becoming a Law of Nature. If the hypothesis is part of a system of hypotheses that describe a complex problem and are all consistent, then this becomes a theory. The hypothesis should serve as an explanation and allow for model development. Consequently, the boundaries between Hypothesis, Explanation, modelling laws, and theory are diffuse.

DOI:10.1201/9781003518372-5

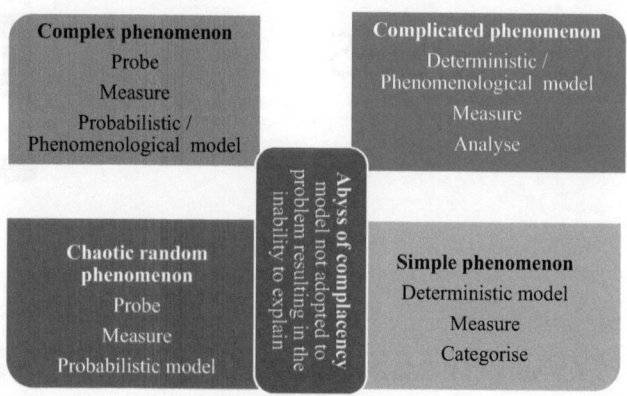

FIGURE 5.1   A Cynefin framework for empirical science.

Based on the observation, the first step in de Groot's empirical approach (Figure 5.1), it is possible to move on to the Induction step. Based on the observations of factors A, B, C, and D ..., some basic understanding of their relationship has been achieved. The induction work means that a first guess on how the factors are related is made. This initial guess is not random but more of an educated guess. The initial guess is also referred to as a hypothesis.

Formulating a hypothesis may not be difficult, but considering the following requirements on how it should be used is more complicated. The requirement of a hypothesis contains the following essential elements:

- The hypothesis must be possible to test.
- It must be possible to conclude that it is true or false.
- The elements required to test the hypothesis must be measurable and/ or observable.

The hypothesis and its testability are prerequisites for giving it some explanatory value or content. The testability of the hypothesis is its single most important property. Without this property, a hypothesis lacks value or interest.

The simplest hypothesis is that the hypothesis can be tested and validated through measurement, for instance, with Ohm's law that if you put a potential over a resistance, it predicts a current. This strength is also why Ohm's law has moved from a Hypothesis to a Law of Nature.

Another interesting case is a bridging hypothesis that joins previous models and theories. A bridging hypothesis aims to bind other hypotheses and laws together. An excellent example is from physics, which deals with whether light is an electromagnetic wave or a particle stream. De Broglies' experiment showed that both interpretations were correct, and as such, the previous interpretations became special cases of De Broglies' hypothesis and experimental validation of his hypothesis.

The testability of a hypothesis is an essential matter. Equally important is that it is fully acceptable to question a hypothesis. The knowledge bases are changing, and the technology and the capability to measure and observe are evolving, leading to the possibility that old truths may change.

There are no laws capable of predicting something with 100% certainty. This can be seen by, for instance, Heisenberg's uncertainty relation, clearly stating that it is impossible to state location and speed simultaneously. The inaccuracy, on the other hand, can be very small. Laws that can predict an outcome with a near 100% outcome are referred to as deterministic laws. Deterministic laws are also seen as necessary laws. Laws with a more significant stochastic element are probabilistic and can be seen as factual laws. The boundary between deterministic laws and probabilistic laws is diffuse.

5.2 EXPLANATIONS

The primary purpose of the hypothesis is to explain a phenomenon. This also means that if you can predict an outcome, then you can also explain the outcome. Since understanding and the ability to characterise phenomena are constantly evolving, predicting and explaining a phenomenon contributes with a step towards the most accurate explanation. It is in this step that the novelty requirement of research lies. Research without novelty or preparation and copying does not contribute to the advancement of humankind, which is the primary purpose of a researcher.

The actual choice of explanation is an open discussion with no particular rule that applies and is a case-by-case decision. One way to choose is to apply Occam's razor principle and choose the simplest explanation that explains what all the other explanations do.

Looking at the differences between hypotheses, explanations, and models, we can see that these concepts have no exact boundaries, and their boundaries are diffuse.

Phenomenological explanations are another class where observing behaviours directly inspires the model or hypothesis. This means that the model is not derived but instead mimics the physics of the phenomenon investigated. The most extreme form of this is found in multivariate data analysis and regression modelling, which is evolving into machine learning, where physics does not matter, but all is about replicating events and including a large number of inputs. This then also requires training to fit reality as well as possible.

5.3 MODELLING

Formulating a hypothesis or theory can often be complemented with an inter-pretation that expresses the hypothesis or theory in terms of already accepted knowledge. This interpretation can also effectively represent the hypothesis itself and, as such, become the hypothesis. This is what is commonly referred to as a model. In the modelling approach, it is essential not to over-simplify since there is a risk that the models will not be able to replicate the actual phe-nomenon in a more general manner or explain rendering the model useless. This can be structured in a so-called Cynefin framework (Figure 5.1).

The Cynefin framework is a conceptual framework to aid decision-making. The term Cynefin originates from the Welsh language and means habitat. The Cynefin framework defines five different habitats or domains to make sense, initially to make sense in management decisions for intellectual capital management. The Cynefin framework is a tool to help test and match the nature of the problem with the nature of the engineered model or hypoth-esis. It is critical to the ability to both design the observation and to match the nature of the hypothesis and the nature of the problem. This is essential to allow high-quality explanations and reasonable grounds for generalisations moving beyond the capability of describing a single problem.

The Cynefin framework reminds us that it is easy to understand for a simple system and that deterministic modelling is a direct approach. This can be, for instance, a relatively well-known phenomenon and often what is needed in physics and engineering physical properties.
Example:

Thermal conductivity of a material or a combination of materials often leads to measuring physical properties or fitting coefficients for studies in the simple phenomenon area.

More complicated will require a more profound analysis where adjust-ment or development of partly new models is required.
Example:

Heat transfer in a multiphase material, including partial transparency and view factors. Here, this is currently an area where the physics is known. However, it has been complicated to characterise, leading to combinations of deterministic and phenomenological laws to adequately describe heat conduction, radiation transmission reflection and scattering to be combined.

Complex systems include more variables and sometimes more than one output. This is often not only a specific phenomenon being included but also a whole system.

Example:

Warping of a cast component:

There are inherent variations in the filling that cannot be measured nor fully quantified due to their stochastic nature. The problems included heat transfer, phase change stress, and the solidifying body and tooling interaction. In addition, weather and slight variations in the shot cycle affect the die release agent evaporation, affecting the heat transfer. This means that all the physical effects are well known but not fully quantifiable, and therefore, a combination of phenomenological models and probabilistic models, also sometimes supported by deterministic models, are needed to describe this phenomenon:

A chaotic problem is even more challenging to understand and model. This often includes studies that would be hard to follow during their occurrence, and all investigations need to be made post-mortem. This is a common situation in metals casting research where in-situ investigations are complicated, and post-mortem methods such as CT X-ray scanning and metallographic techniques are used to identify the outcome, followed by an interpretation of what is found, leading to a constructed model.

Example:

Porosity in casting can be shrinkage or gas porosity. Gas porosity can be the result of physical entrainment of the gas during the mould-filling process or the result of chemical reactions. To predict shrinkage, porosity is relatively accurate, but entrainment posterity is driven by the waviness of the fill front, which is not accurately observable and challenging to predict. Probabilistic modelling can be used, or simpler qualitative models can represent only the risk of excessive porosity.

5.3.1 Analogies

In terms of models, several different models can be used. Not all models are mathematical or based on reasoning, but they can be physical models replicating events on different scales. An analogy can be used for reasoning.

Example:

> Electric current and water have similarities, and the reasoning related to what flows in a pipe can be seen as an analogy to electrical current.

Example:

> Analogies can also be physical models of an actual phenomenon. In metal casting, it is difficult to see what happens due to the opaque nature of the metal. Water, on the other hand, is transparent. Given the understanding of the physics of the flow, it is possible to scale the events in metal flow to water flow as an analogy and thereby measure properties in the water model as a physical model or analogy to the liquid metal.

It should be noted here that the range of model validity needs to be well understood for interpretation.

5.3.2 Models

The most characteristic description of a model is that it describes reality as if it would appear just as the model. It must be understood that the model is not reality and will never get all reality's properties!

It is common to develop this on a simplified ideal case and then test it on other cases to see how generalisable it is or if it is only valid for a particular case. The fact is that some models and laws of nature, such as Newton's second law, F=ma, describe an idealisation of reality. A simple mathematical relationship is just what it is: a simpler and more manageable relationship serving as a tool to understand. This is a challenge as Artificial Intelligence and Machine Learning enter the scene. Computational power gives a new ability to phenomenologically and statistically replicate reality, including more complex scenarios. The difficulty in understanding and learning from this is just as challenging as understanding the flaws induced by idealisation.

In all cases, it must be possible to understand under what conditions the model is valid and what it can predict without significant errors or other quantitative controversy.

5.4 TEST OF A HYPOTHESIS

Testing of a hypothesis can be done in several manners. However, the workflow of a hypothesis test follows a typical schedule, as shown in Figure 5.2.

FIGURE 5.2 Workflow for hypothesis testing.

In the design of a hypothesis test, there is always a risk of confirmation bias. Confirmation bias commonly originates from entertaining a single hypothesis. Strong inference emphasises the need to entertain multiple alternative hypotheses to avoid this. This is also true to avoid artefacts.

5.4.1 Null Hypothesis Specification

The null hypothesis is based on the accepted fact or the matter being investigated. Null originates from the fact that it is nullifiable, meaning disproved. This also means the null hypothesis will be proven correct until something else disproves it. Experimental work and science set out to disprove the null hypothesis. The null hypothesis is commonly identified as H_0

As before, the null hypothesis can be qualitative or quantitative, or, as a theory, a combination of both sometimes makes it necessary to have more than one test and several alternative hypotheses.

5.4.2 Alternative Hypothesis Specification

The alternative hypothesis is the so-called other opinion, commonly denoted as H_1.

Example

> H_0: The earth is flat
> H_1: It is not flat

Example

> H_0: Diffusion obeys a parabolic law
> H_1: It does not obey a parabolic law

Example

> H_0: Group A is different from Group B
> H_1: Group A is not different from Group B

5.4.3 Hypothesis Test Approach

In defining the alternative hypothesis, it is important to understand what type of test is undertaken. There are two types of tests.

- One-tailed tests
- Two-tailed tests

From a purely statistical significance testing point of view, the one-tailed and two-tailed tests are different ways to calculate the statistical significance of a parameter based on a data set. A two-tailed test is desired when a value is greater or smaller than a specific range of values. Consequently, a one-tailed test investigates a specific event, and the outcome is a quantification greater or smaller than a specific value. Furthermore, this has the consequence that a two-tailed test is direction-independent, and the actual question is whether or not the outcome is the same. A one-tailed test has a direction, often referred to as a right-hand or left-hand one-tailed test, depending on whether the test relates to an inclusive query or excluding query with significance. The one-tailed test is mainly found in the natural and engineering sciences due to the quantitative nature of the models, which often have mathematical characteristics, and their fit to data would be quantitatively identified.

There is always a possibility to get the wrong outcome depending on the hypothesis and the actual conditions or outcome of the reality targeted to be predicted or described. There are two types of errors (see Table 5.1):

- Type 1 error: The false positive
- Type 2 error: The false negative

The likelihood of errors partly depends on the hypothesis or how well it explains reality. The other part is the measurements and the data quality, which is equally important. Last but not least are the statistics of the hypothesis testing and the criteria for the test's significance. Statistics commonly only deal with precision and not accuracy, which is one shortcoming, but it is an effective manner of understanding repeatability, which, on the other hand, is one of the best guides to identifying a law of nature.

TABLE 5.1   Errors in hypothesis testing that may occur

	H_0 is true	H_0 IS FALSE
H_0 accepted	OK	Type 2 false negative
H_0 rejected	Type 1 false positive	OK

5.4.4 Hypothesis Rejection Criteria Specification

The two manners in which a hypothesis can be tested in the current treatise will be presented. This is by far not the whole picture. In engineering and physics, the experiment's efforts drive everyone towards the idea that the samples should be independent. In the experimental planning, the sequence of experiments should be randomised to avoid, for instance, all tests being made at a lower temperature than the previous so that systematic thermal history can influence the outcome and quality of the models.

The two methods to be dealt with are well suited for independent samples with two, three, or more samples. The two methods to be described are:

- Chi-square test
- Analysis of variance (ANOVA)

These will be described in detail below.

5.4.5 The Chi-squared Test

The chi-squared test is a statistical hypothesis test used to examine whether two categorical variables are independent in influencing the test statistic. The chi-square test is applicable to a large number of problems. The chi-squared test is commonly used by scientists and researchers for the following situations:

- Perform a goodness-of-fit test,
- Analyse the significance of the association between two attributes, and
- Investigate the homogeneity or the significance of population variance.

For a goodness of fit test, this means that calculating the chi-squared value and this is below a certain tabulated value given a predetermined confidence level, the fit is considered to be good. A good fit implies that the divergence between the observed and expected values is due to sampling fluctuations. On the other hand, if the value is greater than the calculated chi-squared value, the fit is not considered good. This test is based on the concurrence approach described earlier.

For the variables' independence test, the chi-squared test enables us to investigate whether or not two attributes are associated. An example would be a new die lubrication for a high-pressure die casting die improves the rejection rate. The chi-squared test will assist in deciding this. A suitable start is to assume that the die lubrication and rejection rate are independent as a null hypothesis, implying that the new die lubricant is ineffective in improving the rejection rate. The first step is calculating the expected rejection rate and then the chi-squared value. Suppose the calculated chi-squared value is less than the tabulated value at the decided confidence level, the null hypothesis means that the new lubricant and the rejection rate are independent or not associated, meaning that the new die lubricant is ineffective in affecting the rejection rate. On the other hand, if the calculated chi-squared value is greater than the tabulated value, the null hypothesis would not hold, and the rejection rate and the new die lubricant would be associated. This suggests that the new die lubricant is effective and should be used. This approach is different approach to an investigation and allows a statistical decision to a situation that may not be so black and white as the example with the Cu spectrum line used as an example earlier.

In the tests, however, the magnitude of the effect is not considered, which requires a different approach and investigation targeting a variation approach in the experimental approach.

Understanding the differences in the application above makes it possible to look more closely at how the chi-square test is executed. In this test's standard applications, all observations are divided into mutually exclusive classes. Choosing the null hypothesis that the classes are all the same, the test statistic computed from the observations follows a chi-squared distribution. Moreover, a right-tailed approach is appropriate. This means that the right side of the distribution is used. This approach aims to evaluate how likely the observed data would be assumed under the assumption that the null hypothesis that all values are the same is true. The chi-square test works in a way where the distribution of the test statistic approaches the chi-square distribution asymptotically. The chi-squared test is thus better for more extensive samples than smaller data sets.

The first step in all hypothesis testing is to decide on the goodness of fit or confidence for rejection. A 95% confidence interval or confidence level is usually typical. This means there is only a 5% risk that the opposite event would occur due to chance. The relationship between confidence levels and standard devotion is shown in Table 5.2.

In this context, the degree of freedom is a central concept. Since the degrees of freedom in a chi-squared test is to distinguish between two variables that may be dependent, each variable of the sample set, say consisting of MN samples, results in that the degrees of freedom will be $(M-1)(N-1)$. The chi-square table for different degrees of freedom and confidence levels is shown in the Appendix.

TABLE 5.2 Standard deviations and confidence intervals

STANDARD DEVIATIONS	CONFIDENCE LEVEL (%)
$\pm s$	68.2
$\pm 2s$	95.4
$\pm 3s$	99.6
$\pm 4s$	99.99
$\pm 5s$	99.9999
$\pm 6s$	99.9999998
$\pm ns$	$100\left(1 - erf\left(\dfrac{n}{\sqrt{2}}\right)\right)$

TABLE 5.3 Porosity rejection in different foundries

DEFECT LOCATION/ FOUNDRY	A	B	C	D	TOTAL
1	90	60	104	95	349
2	30	50	51	20	151
3	30	40	45	35	150
Total	150	150	200	150	650

Example:

In metallography, porosities and microstructure can be studied. The location of porosity has been investigated in 4 foundries with the following outcome (Table 5.3).

The question now is if Foundry A is overperforming or underperforming. The following hypothesis can be formulated.

H_0: The foundries are performing similarly.

H_1: The foundries are performing differently.

The required confidence is that it should be within a 95% certainty or for the chi-square test $\alpha = 0.05$.

Taking the location that was the most critical—location 1—the overall expected value based on the outcome of all foundries would be the fraction in location 1 and the overall porosity level times that porosity rejection level in foundry A.

The overall porosity fraction for location 1 is $\dfrac{349}{650}$ which results in the expected outcome for Foundry A being $150\dfrac{349}{650} \approx 80.54$. The expected outcome for Foundry A in location 2 would be $150\dfrac{151}{650}$, and so on. The chi-square distribution value sought is the sum of all the cells in the table (excluding the summation cells)

$$Y = \sum_i \frac{\left(y_i - y_{exp}\right)^2}{y_{exp}} \equiv \frac{(90 - 80.54)^2}{80.54} + \frac{(30 - 34.85)^2}{34.85} + \cdots$$

$$+ \frac{(20 - 80.54)^2}{34.85} + \frac{(35 - 34.62)^2}{34.62} = 24.57$$

The degrees of freedom can be calculated as $(number\ of\ columns - 1)(number\ of\ rows - 1)$ which, in the current situation, is $(3-1)(4-1) = 6$. Using that chi-square table in the Appendix with $\alpha = 0.05$ $df = 0.05$ gives the critical value as $Y_{crit} = 12.592$. Now since $Y > Y_{crit}$, the hypothesis must be rejected as they do not perform similarly.

5.4.6 The Analysis of Variance or ANOVA Approach

The ANOVA approach is handy for comparing more than two populations, such as in comparing the yield of production line output from several factories, the tolerance variations from several production lines, or the performance of a number of casting machines in a foundry. Doing this and considering all possible combinations of the populations would be very time consuming. The ANOVA approach allows us to investigate the differences among the means of all the populations concurrently. It is possible to state that the essence of ANOVA is that the total variation in a data set of data is broken down into two types of causes. These two types are as follows:

- Attributed to chance only and
- Attributed to specified causes

In model building and especially heuristic linear regression modelling, the ANOVA approach allows the determination of factors that are hypothesised to influence the dependent variable.

ANOVA is not just one method but a family of techniques. ANOVA is based on the total variance, and different variables are partitioned into components attributed to different sources of variation. ANOVA provides a statistical test of whether two or more population means are equal, meaning that ANOVA is used to test the difference between two or more means. ANOVA allows the analysis of comparative experiments. The purpose is to analyse the differences in outcomes. The general principle is to assess the statistical significance of a test based on two different variances through the ratio of two variances. The variance ratio is independent of matters such as multiplication by a factor (increase spread in data or random error increase) or adding a constant (shifting the data, meaning systematic errors). The ANOVA statistical significance result is thus independent of constant bias and scaling errors, making it powerful and highly useful in engineering sciences. The ANOVA analysis approach uses something that is called the F-test and not the chi-squared test to assess the significance. The F-test is a statistical test designed to compare the ratio of variances between multiple samples. It is commonly used for comparing regression models fitted using least squares to a data set allowing the decision and identification of which model best fits the data set. This procedure is described in detail in Appendix and is found together with the F-test critical values for the F-distribution.

Example

Repeating the analysis above by using ANOVA. Starting with the sum of squares within SSW

$$SSW = SS_A + SS_B + \cdots + SS_D$$

To execute this, we need the averages of all the columns :

$$\bar{y}_A = \frac{1}{3}(90 + 90 + 30) = 50$$

This must then be made for the other columns as well. The next step is to calculate the sum of squares for each column :

$$SS_A = (90-50)^2 + (30-50)^2 + (30-50)^2 = 2400$$

The result is that SSW becomes

$$SSW = 2400 + 200 + 2108.7 + 3150 = 7858.7$$

The next step is to calculate the sum of squares between the columns, which is made overall individual cells as

$$SSB = \sum_{i=1}^{RC}\left(y_{Factor\ i} - \bar{y}_0\right)^2$$

This requires us to calculate $\bar{y}_0$ that is the average of the average column deviations.

$$\bar{y}_0 = \frac{1}{4}(50+50+66.7+50) = 54.2$$

This then makes SSB to be

$$SSB = (90-54.2)^2 + (30-54.2)^2 + \ldots + (20-54.2)^2 + (45-54.2)^2$$
$$= 8483.7$$

The degrees of freedom required for the SSW and SSB are given by

$$df_W = (R-1)C = (3-1)4 = 8;\ df_B = (C-1) = (4-1) = 3$$

/	dfB=1	2	3	4
dfW=1	161.44	199.50	215.71	224.58
2	18.518	19.000	19.164	19.247
3	10.1280	9.5521	9.2766	9.1172
4	7.7086	6.9443	6.5914	6.3882
5	6.6079	5.7861	5.4095	5.1922
6	5.9874	5.1433	4.7571	4.5337
7	5.5914	4.7374	4.3403	4.1203
8	5.3177	4.4590	4.0662	3.8379
9	5.1174	4.2565	3.8625	3.6331
10	4.9646	4.1028	3.7083	3.4780

FIGURE 5.3   Illustration of the use of the F-test table.

The mean sum of squares is then possible to calculate for both cases :

$$MS_W = \frac{SSW}{df_W} = \frac{7858.7}{8} = 982.3; \; MS_B = \frac{SSB}{df_B} = \frac{8483.7}{3} = 2827.9$$

The F-value that is sought is given by

$$F = \frac{MS_B}{MS_W} = \frac{2827.9}{982.3} = 2.878$$

Using the degrees of freedom and $\alpha = 0.05$, there is only a 5% probability that this difference occurs by chance using the tables in Appendix section (Figure 5.3).

This then leads to

$$F = 2.878 < 4.0662 = F_{Crit}$$

Furthermore, the conclusion is that the difference between the columns is statistically significant and will not occur by chance. Hence, the hypothesis that they perform the same must be rejected.

Design of Experiment

6

In the design of an experiment, the experiment must be adapted to the research question so that an answer is possible, often meaning that the null hypothesis is falsifiable and that this can be established in a quantifiable manner. This should be facilitated using De Groot's approach in the observation and induction phases. The actual experimentation can then be defined in the manner that was defined in Chapter 3. Three different ways to approach a problem were defined, and three different manners were given:

- The concurrence approach
- The differential approach
- The variation approach

The concurrence approach and the differential approach are more defined to see if there is an interaction or correlation between different variables and responses. The variation approach is very common today, and most engineering and physics problems aim to quantify relationships to build a model and use this to optimise different scenarios or situations. This last method actually requires some thinking and consideration in experimental planning. There are some standard ways to analyse a system with a limited number of variables. As the system's complexity grows, the difficulty, time, and effort to make experiments grow exponentially, and other approaches are required.

In the design of an experiment, the experimental window must be determined, and to do this, it is critical first to establish which are the independent variables and which are the dependent variables, functions, or responses.

Following this, establishing a wide enough experimental window for the independent variables is critical. The discussion below will deal with choosing the experimental settings within this window.

We can follow the tailor's advice to measure three times and cut once in all empirical activities! This means it is not enough to plan and run the

DOI:10.1201/9781003518372-6

experiments once for each point. There are three critical principles in the design of experiments (DoE), and these are as follows:

- Randomise your experiments to ensure that there are no "history effects". An example is running a casting experiment and increasing the tool temperature between each setting. This may affect the machine temperature distribution, which would be different by cooling the tool between each experiment altering tolerances and clearances in the equipment.
- Replicate at least some experiments to ensure that the experimental uncertainty is testable. Common practice is to use at least five pints as replicates in a more extensive series of experiments. A small series (< 10 experiments) of at least three experiments at each point is a good practise.
- Lack-of-Fit tests where there is a small deviation from the experimental settings, so these points are almost identical to the actual points. This is to be used to understand the model quality.

6.1 THE FULL FACTORIAL PLANNING

The full factorial is a simple way of defining experiments, and it has been used for a long time since it is easy to plan, explain, and use. We must plan the experiments accordingly depending on the nature of the expected relationship, for instance, linear or non-linear behaviour. If the behaviour is non-linear, we cannot plan the experiments as a linear system, as the Cynefin framework suggests.

For a linear system, it must be remembered that a linear system can be represented by a straight line, meaning that each variable requires two points to pass through to resolve this system. Take a response U that depends on the variables x and y. The description for a linear set up would be to be able to describe the following equation:

$$U(x, y) = a + bx + cy$$

The most straightforward plan to generate experiments for this would be to have two values for x and two values for y to have a straight-line representation for each dependence. This means that the plan should be as in Table 6.1.

TABLE 6.1   Full factorial experimental design requiring four experiments for linear behaviours

VARIABLES	y_1	y_2
x_1	$U(x_1, y_1)$	$U(x_1, y_2)$
x_2	$U(x_2, y_1)$	$U(x_2, y_2)$

TABLE 6.2   Full factorial experimental design requiring four experiments for non-linear behaviours

VARIABLES	y_1	y_2	y_3
x_1	$U(x_1, y_1)$	$U(x_1, y_2)$	$U(x_1, y_3)$
x_2	$U(x_2, y_1)$	$U(x_2, y_2)$	$U(x_2, y_3)$
x_3	$U(x_3, y_1)$	$U(x_3, y_2)$	$U(x_3, y_3)$

Two variables with a linear dependence require a minimum of four experiments in the plan. It is possible to show that for linear systems, the number of experiments required is 2^k experiments where k is the number of variables, and the number 2 originates from the fact that the planning is based on a linear system with two levels of variables. In empirical science, duplication of experiments is a good practice to ensure reasonable control of experimental and measurement uncertainty. This is not necessary for computational experiments.

Expanding the problem to a more common situation, which is non-linear behaviour, the situation is changed, and the equation to deal with then multiplies. The equation to model then becomes

$$U(x, y) = a + bx + cy + exy + fx^2 + gy^2$$

The experimental plan then becomes, as shown in Table 6.2.

Two variables with a quadratic (non-linear) dependence require a minimum of nine experiments in the plan. It is possible to show that for a quadratic system, the number of experiments required is 3^k experiments where k is the number of variables, and the number 3 originates from the fact that the planning is based on a quadratic system with three levels of variables.

It is not difficult to understand that experiments in non-linear systems will rapidly grow in an effort to investigate, and it is uncommon to see full factorial experiments with more than two variables, as adding a third variable would result in 27 experiments. There are several ways to deal with order reduction, including full factoring and mixed approaches, with parts being linear and parts being non-linear. The rise of both multivariate data analysis and machine learning has changed the scene somewhat, and this has also changed the way experiments are being planned.

6.2 RESPONSE SURFACE METHODS

Multivariate data analysis has a strong correlation to the area called design of experiments or DoE. This is based on rationalising the amount of experimenting without significantly reducing the fidelity of the experimental resolution. Machine learning is the next generation, where the amount of data is not necessarily planned, but you take all the data and group the data to facilitate efficient analysis. Both require some computational capacity where machine learning is heavy, and multivariate data analysis in conjunction with DoE is significantly more straightforward. This chapter will introduce the DoE concept with the target of achieving engineering-type optimisation.

Starting with the reduction concept to minimise fidelity loss, we return to the non-linear system with two variables and nine experiments. This was necessary as three levels for each variable were required. If, instead of the nine experiments, one chose to do the experimental planning as outlined in Table 6.3.

In essence, there are three levels for each variable. It is just that the experimental resolution is a bit sparse along the edges but reasonably good along the diagonal. This type of plan would result in larger uncertainty near the outer edges of the experimental window but would give a reasonable expression in the centre of the experimental window. In all experiments, it is important not to choose a small experimental window as all regression modelling is well suited for interpolation and not for extrapolation. This makes the edges of the experimental window more uncertain, so any optimum or sought-after solutions should be more central than peripheral.

TABLE 6.3 Hypothetical full factorial experimental design requiring four experiments for non-linear behaviours with a reduction of the number of experiments

VARIABLES	y_1	y_2	y_3
x_1	$U(x_1, y_1)$		$U(x_1, y_3)$
x_2		$U(x_2, y_2)$	
x_3	$U(x_3, y_1)$		$U(x_3, y_3)$

The idea presented above that it is possible to reduce the number of experiments without significant loss of fidelity has given rise to a research field called design of experiments, which has many different methods. In this treatise, the focus will be on so-called response surface methods, particularly optimal designs that are well suited for optimisation.

6.2.1 D-Optimal Design

The optimal design family of an approach with flexibility are that the experimental points are not predetermined as for other response surface methods. The strength of the optimal designs is that they work with unequal variable ranges, meaning that all the high and low values of the variable do not need to have the same difference. It is also possible to implement multi-component constraints. There are tools to do these designs as these designs require a numerical approach.

In these software, the design depends on the chosen model for the different responses and variables. The system also plans for replicates to understand the experimental uncertainty and points referred to as lack-of-fit points that test the model's sensitivity and are part of the model quality evaluation. Common is to add five replicates and five that lack fit. As mentioned above, without a lack-of-fit test, there may be no indication that an inadequate model fits the data.

Taking the non-linear experimental plan and expressing it as a matrix system gives

$$U = Xc$$

Or

$$
\begin{bmatrix} U_1(x_1, y_1) \\ \vdots \\ U_9(x_3, y_1) \end{bmatrix} = \begin{bmatrix} 1 & x_1 & y_1 & x_1 y_1 & x_1^2 & y_1^2 \\ \vdots & \vdots & \vdots & \vdots & \vdots & \vdots \\ 1 & x_9 & y_9 & x_9 y_9 & x_9^2 & y_9^2 \end{bmatrix} \begin{bmatrix} a \\ b \\ c \\ d \\ e \\ f \\ g \end{bmatrix}
$$

where x_1 to x_9 are the potential experimental variable settings for x and y_1 to y_9 are the variable settings for y for the experiments to be planned.

The D-optimal aims at maximising the amount of information for each experiment. This is the same as reducing the linear dependence of each experiment, which is in a sense similar to maximising the determinant of the matrix X. Now, most of these systems are overdetermined, and therefore, the approach will be to maximise the determinant of $X^T X$.

Most commonly, the variables are expressed from −1, the low value, and 1, the high value, and 0, the mid value. This normalisation is made for x and y, illustrated by y as

$$y_{CODE} = \frac{2y - \left(y_{max} - y_{min}\right)}{\left|y_{max} - y_{min}\right|}$$

With this, the full factorial plan for the nine experiments, excluding replicates and lack-of-fit points, is illustrated in Table 6.4.

On the other hand, the D-optimal design would not generate nine experiments but requires only six, which is not so far from the thought experiment at the beginning of this section. In that we used a central point, which would be $x = y = 0$. In the experiment, it is seen that there is no central point but rather four points slightly off the centre point, with values −0.13 and 0.4 for x and 0.13 and −0.39 for y, as shown in Table 6.5. Adding experimental points and moving away from "origo" or the experimental centrepoint, as schematically illustrated in Table 6.3, improves the data fidelity. Furthermore, it still significantly reduces experimental effort compared to the full factorial approach.

These designs when combined with ANOVA analysis for model building—choosing independent parameters and conducting regression analysis—is a very powerful tool for DoE aiming and performing optimisation. This

TABLE 6.4 Experimental parameters in coded form for a full factorial for a quadratic design

X	Y
1	1
1	0
1	−1
0	1
0	0
0	−1
−1	1
−1	0
−1	−1

TABLE 6.5 Experimental parameters in coded form for a D-optimal design for a quadratic design

X	Y
−1	−1
−1	1
−0.13	0.13
0.4	−1
1	1
1	−0.39

approach allows an efficient building of experimental data representation to better understand how the variables and responses are connected and quantitatively related. It may not be physically correct, but in a heuristic manner, it is a correct representation as a metamodel of the experiments.

6.2.2 S-Optimal Design

As mentioned, several different types of optimal designs have the common idea to maximise the information content of each experiment. In empirical studies with experimental uncertainty, the D-optimal model was chosen as a good approach, and it was recommended to add the so-called lack-of-fit point, which is at a point chosen near some strategic point, commonly five by default, in many tools. These are to judge the quality of the fitted model. The other recommendation was to add at least five replicates to understand the experimental uncertainty. These are strategies well suited for physical experimenting. Today, computational experimentation is also entering the scene, and since each computation will result in the same outcome, more or less replication points are unnecessary. In computational experimentation, it is also advisable to try to fill the whole space more evenly in an effort to reduce the error variation originating from the experimental designs. A method well suited for this is the S-optimal design, aiming at an equidistant space-filling approach.

The S-optimal design targets to maximise a quantity measuring the mutual column orthogonality of X and the determinant of the information matrix. The quality of orthogonality is given by

$$X^T X = XX^T = I$$

TABLE 6.6 Experimental parameters in coded form for an S-optimal design for a quadratic design

X	Y
−1	−0.33
0	−1
1	−1
0.33	1
−0.33	1
0.5	0

From this, it follows that

$$X^T = X^{-1}$$

The consequence for an orthogonal matrix is furthermore that

$$det\left(X^T X\right) = \det\left(I\right) = 1$$

Achieving an equidistant space, meaning that the distance between the points should be as large as possible and keeping the target to maintain the orthogonality as well as fulfilling the conditions above, leads to the objective to maximise a function $f\left(X\right)$ defined as

$$\max_{x}\left(f\left(X\right)\right) = \max_{x} \frac{\sum_{i,j=1,i\neq j}^{N}\left(\min\left|x_i - x_j\right|\right)}{N}$$

The corresponding space-filling solution for the sample problem, neglecting lack-of-fit and replication, results in Table 6.6.

Treatment of Measured Data

7

7.1 WHAT IS AND WHAT IS NOT A MEASUREMENT?

In general, measurement is a quantification of an entity. It must generate measurement data of some kind—a quantification. One of the most common measurements is the measure of length such as a metre (m), where a tape measure can be used as a tool. Using a tape measure, you might realise that you are no longer 2.00m tall but rather 1.99m. The uncertainty discussion is how to assess whether or not this is true.

Some matters appear to be a measurement, but they are not. One example is standing next to your brother and seeing the tallest one. This is not a measurement as it only generates a value: yes or no, you are not taller. This is a comparison; comparisons are not measurements as they do not generate a measurement value.

7.2 THE CRITICAL NATURE OF MEASUREMENT

The strength of empirical sciences and their importance to the progression of society can be deduced from the ability to quantify events. Quantifying events means that, for instance, the direction of change following an action of interference is not sufficient, but the core capability is to predict how far or long

DOI:10.1201/9781003518372-7

together with the direction is what brings value. This value must be accurate, predictable, and, foremost, measurable as part of the data collection in the test and the validation of a theory.

A systematic and careful collection of measurement data defines the critical difference between pseudoscience and actual science. Scientific measurements can be tabulated in tables or graphically represented as graphs and maps. Commonly, the data obtained from measurements are discrete, and statistical manipulation, such as correlation and regression analysis, is required to understand the nature of the data. Rarely do all points lie perfectly aligned to a plot, and there is almost always some level of deviation. This is in the nature of empirical science. There are errors originating from measurement tools, uncontrolled variables, and the human element involved in measurement and experiment execution.

The measurements might be made in a well-controlled laboratory setting with meticulously controlled parameters. This may be made in an industrial setting where production standards and inherent process variations influence the experiment's input and output parameters. Some scientific studies are made with the inability to vary or manipulate the objects studied, which is a common element in astronomy and astrophysics. Stars are difficult to manipulate or even impossible!

Common to all measurements is that they all require instrumentation. This instrumentation often includes specialised instruments such as Vernier callipers, laser interferometers, thermometers, spectroscopes, particle accelerators, and so on. Particle accelerators are an extreme version of instruments that require many operators, and each one is unique, meaning that benchmarking or Round-Robin testing is difficult, which is common for measurements to ensure that where and by whom does not influence the measurement. Reliable measurement is critical not only to science but also to the progression of society. It is not easy to venture or develop beyond what is measurable. This ties directly back to the boundaries of science, where the understanding that the search for hard-to-vary explanation drives all progress. If you cannot measure, you cannot explain. Therefore, the absolute truth consists of hard-to-vary assertions about reality which then would make up the boundaries of science for the physical world. The boundary of science is driven by the ability to measure; therefore, significant effort goes into the ability to measure, as this is a necessary foundation for societal progress. Examples are rulers, laser interferometers, optical emission spectrometers, computer tomographic scanners, biomarkers concentration analysers for heart disease, cancer, and so forth. All methods lead to a quantifiable outcome having a critical element delivered to humankind, allowing societal progress.

7.3 ACCURACY, PRECISION, AND ERROR

Measurement quality is a necessary and critical understanding and ability. However, what is a good quality measurement? To understand measurement quality, there are terms used such as good precision, high accuracy, low uncertainty, and a small error in the measurement, but what are all these terms?

The first thing about measurements is understanding the difference between precision, accuracy, uncertainty, and error.

- Accuracy is the closeness of agreement between a measured value and an actual or accepted value. Sometimes, this is referred to as inaccuracy. These terms are qualitative. It is important to understand that this is not the same as uncertainty, as uncertainty is quantitative and is associated with a number and a ± sign.
- Precision measures how well a result can be determined (without referencing a theoretical or actual value). It is the degree of consistency and agreement among independent measurements of the same quantity as the reliability or reproducibility of the result.
- Measurement error is the amount of inaccuracy.
- The uncertainty estimate associated with a measurement should account for the measurements' accuracy and precision.

It should also be noted that none of these entities are related to the concept of tolerance. Tolerances are acceptance limits for the measured value, commonly including measurement error or uncertainty.

Firstly, to understand the difference between accuracy and precision, refer to Figure 7.1 for a visual representation.

This means that the measured value, y, is related to the best estimate, y_{Best} and uncertainty, δy, as

$$y = y_{Best} \pm \delta y$$

Precision is often reported as the relative uncertainty related to the uncertainty δy and the measured value y_{Meas}, given as:

$$Relative\,uncertainty = \left| \frac{\delta y}{y_{Meas}} \right|$$

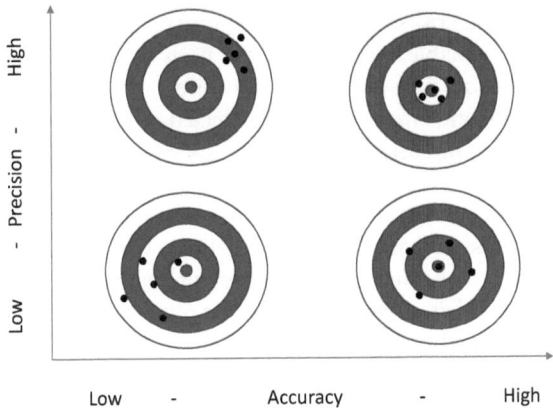

FIGURE 7.1 Illustration of the difference between precision and accuracy.

Accuracy is often reported as the relative error, which is based on the measured value y_{Meas} and the actual or expected value y_{Expect}

$$Relative\,error = \frac{y_{Meas} - y_{Expect}}{y_{Expect}}$$

7.3.1 Types of Error

Some errors can occur in measurement and sample-taking in science. Understanding their nature is a critical element to manage and control them from occurring or to be able to handle, correct, and report these errors. The measurement error is part of understanding and deciding if two measurements are the same or different. It must be understood that the measurements can represent the same outcome even if they do not have the same identical values. The first step is to classify the errors, and this first step is often made in such a way that the errors are classified into two different categories:

- Random errors
  - o Stochastic fluctuations can be handled and evaluated through statistical analysis. Most commonly, these errors are managed through averaging over many observations and described through standard deviation and standard error.

- Systematic errors
  - o Reproducible inaccuracies for each source cause measurement deviation in a source-specific direction. Unlike random errors, systematic errors cannot be detected or reduced by increasing the number of observations.

Error sources can have many different origins, and each can give rise to a random error, a systematic error, or sometimes both. This understanding is essential as some can be managed in the measurement process, and some need a change in the number of measurements that require post-processing, such as statistical analysis and manipulations. The most common sources of measurement errors are:

- Incomplete definition (can be both systematic and random)
  - o It is critical to specify the measurement condition completely and carefully define what should be considered to avoid errors using a specific measurement technique.
- Failure to account for a factor (systematic)
  - o It is often caused by a flawed design of the experiment where significant factors are missed. The headwind effect can influence fuel consumption when driving in one direction. Systematic errors can be corrected once they are quantified from the specific source of the errors.
- Environmental factors (can be both systematic or random)
  - o Examples are errors induced by the immediate working environment, and precautions can be taken to minimise effects from vibrations, drafts, temperature variation, and equipment electronic interference causing noise.
- Instrument resolution (random)
  - o Instruments all have a finite precision, causing limitations in the resolution of resolving small measurement differences.
- Calibration (systematic)
  - o If calibration is possible for the measurement, calibrate it! This is a necessity for measurement traceability and credibility. If a calibration standard is not available, the accuracy of the instrument should be checked by comparing it with another instrument in a Round-Robin test.
- Zero offset (systematic)
  - o Always check the zero reading first for all measurements that can be re-zeroed. Always re-zero the instrument if possible or record the zero offset to allow a post-measurement correction.

- Physical variations (random)
  - o Always repeat measurements when possible to ensure no variation or at least a known variation of the sample investigated. This may reveal variations that might remain undetected. Repeated measurements may establish an average value and an error estimate if there is a variation of, for instance, a diameter.
- Parallax (systematic or random)
  - o This is for non-contact measurements and is caused by the distance between the measuring scale and the indicator used to measure. Its origin is the lack of alignment.
- Instrument drift (systematic)
  - o For more extended duration measurements, this issue may become significant. Due to an instrument's thermal and mechanical stability, an inevitable measurement drift may occur over time.
- Lag time and hysteresis (systematic)
  - o Instruments such as a scanning electron, microscopes, and energy-dispersive spectroscopy require time to reach equilibrium. Taking a reading before thermal equilibrium will result in a measurement that is too high or low. The same is valid for simpler instruments such as thermometers or thermocouples. The effect is measurement hysteresis originating from a memory effect. One manner to avoid this is to understand the physics of your measurement. Similarly, randomising your experiments will remove certain types of errors originating from your experiments being made after heating or cooling if a temperature change is made to your sample.
- Operator errors (can be systematic and random)
  - o Common causes are carelessness, poor technique, bias, fatigue, and stress on the experimenter's part. It should be noted that operator errors are not part of the measurement uncertainty, and operator errors should be discarded as flawed and defective measurements are not valid.

7.4 UNCERTAINTY ESTIMATION

The estimation of uncertainty in a result has many perspectives. The first is understanding the uncertainty of measuring a single entity, such as a length.

The first guess of the value sought in a measurement is to use statistical means, and then, as a start, the value is the mean value, $\bar{y}$, of a number of measurements, N.

$$\bar{y} = \frac{1}{N} \sum_{i=1}^{N} y_i$$

In ordinary situations, 10 measurements is a good practice and is commonly considered sufficient. There are two ways to describe an error:

- Average deviation, d
- Standard deviation, s

It should be noted that the standard deviation is slightly greater than the average deviations, which originated in how these were calculated.

The average deviation, $\bar{d}$, is calculated as

$$\bar{d} = \frac{1}{N} \sum_{i=1}^{N} |y_i - \bar{y}_i|$$

Whereas the standard deviation is calculated as

$$s = \sqrt{\frac{1}{N-1} \sum_{i=1}^{N} \delta y_i^2}$$

It should be noted that uncertainty is commonly used as $\delta y_i = |y_i - \bar{y}_i|$

An important distinction is the use of the concept of the standard error, which is the standard deviation of the mean which is defined as

$$\sigma_{\bar{y}} = C \frac{s}{\sqrt{N}}$$

The factor C depends on the confidence interval. So, if it is based on the standard deviation $C=1$ and corresponds to the standard error, taking this to a 95% confidence level, it needs to be greater, and the factor should be 1.96, as illustrated in Figure 7.2. This means that to be certain all scenarios are captured and that no measurements are excluded. The error grows as more extreme values are included, and the only remedy is to increase the number of measurements on each occasion.

This concept can be extended, and for other confidence intervals, the value of C is changing and is also often referred to as a coverage factor as it signifies the coverage of the whole population.

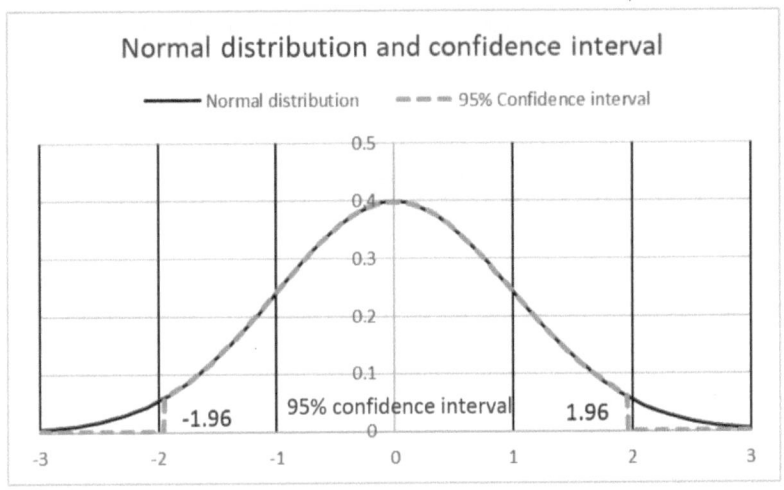

FIGURE 7.2 Illustration of the Gaussian or normal distribution and the 95% confidence interval.

- $C = 1$ for a confidence level of 68%
- $C = 1.96$ for a confidence level of 95%
- $C = 2.58$ for a confidence level of 99%
- $C = 3$ for a confidence level of 99.7%

7.5 ERROR IN MEASUREMENTS AND SIGNIFICANT DIGITS

The most common mistake in presenting data is using too many digits and not realising that the values in excess lack meaning and will be different if you repeat the same measurements without changing anything.

The number of digits to use is easily understood if scientific notation is used in the representation of data. The standard convention is then the following:

- A relative uncertainty of about 10% to 100% → 1 significant digits
- A relative uncertainty of about 1% to 10% → 2 significant digits
- A relative uncertainty of about 0.1% to 1% → 3 significant digits
- A relative uncertainty of about 0.01% to 0.1% → 4 significant digits
- And yes, it goes on …

In this practice, showing significant digits plus the first uncertain digit is acceptable.

7.6 UNCERTAINTY PROPAGATION

In many cases, some entities are determined that depend on more than one measurement. This means that the error does not depend on one measurement error but rather on several different measurements. The propagation of uncertainty depends on the different variables and how they are related to each other. The relationship can be described as a measured value, which is the function of these two variables.

$$y = f(x,t)$$

The error propagation in the measurement can be expressed by first taking the partial derivative to get the individual influence of the factors,

$$\delta y = \left(\frac{\partial f}{\partial x}\right)\delta x + \left(\frac{\partial f}{\partial t}\right)\delta t$$

The square of this is the law of uncertainty propagation.

$$\delta y^2 = \left(\frac{\partial f}{\partial x}\right)^2 \delta x^2 + \left(\frac{\partial f}{\partial t}\right)^2 \delta t^2 + 2\left(\frac{\partial f}{\partial x}\right)\left(\frac{\partial f}{\partial t}\right)\delta x \delta t$$

If x and t are uncorrelated, we get the error.

$$\sigma_f = \sqrt{\left(\frac{\partial f}{\partial x}\right)^2 \sigma_x^2 + \left(\frac{\partial f}{\partial t}\right)^2 \sigma_t^2}$$

where

$$\sigma_x = C\frac{s_x}{\sqrt{N}}; \sigma_t = C\frac{s_t}{\sqrt{N}}$$

Example:

Uncertainty propagation measuring area and perimeter using a ruler. The shape considered is a rectangle with the sides x and y. The ruler results in the same characteristics as the standard deviation, and the measurement for each element is made at the same time, meaning that

$$\sigma_x = \sigma_y = \sigma = C\frac{s}{\sqrt{N}}$$

The area is defined by

$$f(x,y) = xy$$

The error for the area then becomes

$$\sigma_f = \sqrt{\left(\left(\frac{\partial f}{\partial x}\right)^2 + \left(\frac{\partial f}{\partial y}\right)^2\right)\sigma^2} = \sqrt{\left((y)^2 + (x)^2\right)\left(C\frac{s}{\sqrt{N}}\right)^2}$$

and the perimeter is defined as

$$g(x,y) = 2(x+y)$$

The error for the perimeter then becomes

$$\sigma_g = \sqrt{\left(\left(\frac{\partial g}{\partial x}\right)^2 + \left(\frac{\partial g}{\partial y}\right)^2\right)\sigma^2} = \sqrt{\left((2)^2 + (2)^2\right)\left(C\frac{s}{\sqrt{N}}\right)^2}$$

Even if the measurements are the same, the behaviour of the error changes for the perimeter and is not the same as for the area due to the nature of the use of the measurements and the error propagation.

7.7 HOW DO WE KNOW THAT THERE IS A SIGNIFICANT DIFFERENCE BETWEEN MEASUREMENTS

In order to be significant, the difference means the values, including the uncertainty, must be separate from each other. This means that the average value of two sets of measurements and their uncertainties cannot overlap and be different.

$$\bar{y}_1 \pm \delta y_1, \ \bar{y}_2 \pm \delta y_2$$

This can be judged by taking the difference between the two measurements and dividing it by the average uncertainty of the measurements. If this difference is greater than 2, then the error bars will be separate and non-overlapping. These conditions can be expressed as:

$$\frac{\left|\bar{y}_1 - \bar{y}_2\right|}{\left(\delta y_1 + \delta y_2\right)/2} > 2$$

or simplified

$$\frac{\left|\bar{y}_1 - \bar{y}_2\right|}{\delta y_1 + \delta y_2} > 1$$

If this condition is fulfilled, the two measurements are different, and there is a proven difference between these two points. This allows a distinct conclusion that the difference is proven.

On the other hand, the averages are different, and the condition is not fulfilled. This means there is no significant difference proven, and there is no possibility to conclude that there is a difference. Based on the confidence interval used, the points likely reflect the same value with that confidence. If the error is a 95% confidence interval and they overlap, it is at least a $(0.95)(0.95) = 0.9025$ or 90% chance that these measurements are the same. It is possible if there is a Law of Nature that explains a difference, it is a tendency according to that Law of Nature that your results follow that specific Law of Nature, but it cannot be used to confirm the law itself.

7.8 COMPARING DATA AND NORMALISATION

In the discussion of the design of experiments, a coded factor scale ranging from −1 to 1 was introduced. This form is a common way to make the variables of the same magnitude independent of the units the factors are written in.

Several more methods are used for normalising the data as their relative contribution to a phenomenon can be better assessed as they all have the same dimensionless magnitude.

There are many reasons for normalising variables, but the most common reason is to allow objective comparison of importance for variables with different dimensions having several orders of magnitude in difference. This can be made to understand the factors' relative contribution better. Depending on the normalising scale, there is also a possibility of understanding variations and distributions of factors with different magnitudes to trace errors and deviations. Some examples will be discussed below.

Another important matter is that using normalised values in numerical simulations allows for improved stability and reduced floating number issues that may arise with large and small magnitude variables, reducing computational accuracy and precision.

7.8.1 Min-Max Feature Scaling

The so-called min-max feature scaling is similar to the coded factor scaling but ranges from 0 to 1 instead of from −1 to 1. The feature scaling is made by:

$$y_{Feature} = \frac{y - y_{min}}{y_{max} - y_{min}}$$

It is furthermore possible to extend into the use for data in different classes such, for instance, that are used looking at populations of different elements and then breaking down these to subgroups between any arbitrary point, y_0, and, y_1, as

$$y_{Feature} = y_0 + \left(\frac{y - y_{min}}{y_{max} - y_{min}} \right)(y_1 - y_0)$$

The min-max scaling gives an understanding of the variations within an experimental window and has specific applications normalising experiments and outputs in empirical sciences. There is, however, no information on the quality of the distribution of the values. This will be discussed in the following sections.

7.8.2 The Raw Score and the Z-score

Understanding the average value $\bar{y}$ and the standard deviation s_y of a complete population allows the deviation from the mean for a value y to be normalised by the standard deviation. This feature scaling is made as follows:

$$y_{Raw} = \frac{y - \bar{y}}{s_y}$$

If the difference, $y - \bar{y}$, is greater than s_y, then something can be said about the confidence level of the deviation if it is a random and normal-distributed deviation. This is good for model building in regression modelling, where the intent is to have a normal distribution of the residuals for having a model with model-building independent residual. A random deviation is just expressing this.

If the complete population data is unknown, then the data of what was sampled from the population is used instead. This measure is then referred to as not the Raw score but the Z-score. This is then expressed as

$$y_Z = \frac{y - \bar{y}_{Sample}}{s_{y, Sample}}$$

It might be seen as splitting hair, having both the Raw score and the Z-score. This is perhaps true because, for a sufficient sample, there should be little or no difference between the sample and the population. In the essence of cost-efficiency, samples may be borderline sufficient, and therefore, the distinction is motivated by a stringent and rigorous approach. It should be noted that this is not always made, and as such, a caution is issued. In both cases, the numerator and denominator of the normalising for the Raw- and the Z-score have the same units and cancel out through division, making both the Raw score and the Z-score dimensionless quantities.

7.8.3 Studentised Residuals and the Cook's Distance

A particular case is related to regression modelling where a linear regression is made using, for instance, the least squares method or some other fitting or optimisation approach to match the experimental input, x_i, and experimental response, y_i, to a mathematical model, which is here a polynomial. Taking the first-order polynomial as an example, this would be expressed as:

$$y = a + bx$$

The fitting objective is to minimise the residual, ε_i, for each fitting point. This means that, in reality, there is an error, ε_i.

$$y_i = a + bx_i + \varepsilon_i$$

The error ε_i is not an actual error but rather a fitting residual, $\hat{\varepsilon}_i$. In the ideal case, for a successful least-square fitting, the sum of the fitting residuals, $\hat{\varepsilon}_i$, is zero, meaning that normal statistics is ineffective as

$$\sum_i^N \hat{\varepsilon}_i = 0$$

and

$$\sum_i^N \hat{\varepsilon}_i x_i = 0$$

A workaround for this is to use the residual as an estimate of the variance of the fit. There are two approaches used for this one, with an internal frame of reference as

$$s_{i,\,int}^2 = \frac{1}{N-m} \sum_{i=1}^N \hat{\varepsilon}_i^2$$

And one with what is referred to as an external frame of reference:

$$s_{i,ext}^2 = \frac{1}{N-m-1} \sum_{\substack{j=1 \\ j \neq i}}^N \hat{\varepsilon}_i^2$$

Returning to the experimental design matrix for linear regression, it is written as

$$y = \begin{bmatrix} y_1(x_1) \\ \vdots \\ y_9(x_N) \end{bmatrix} = \begin{bmatrix} 1 & x_1 \\ \vdots & \vdots \\ 1 & x_N \end{bmatrix} \begin{bmatrix} a \\ b \end{bmatrix} = Xc$$

The least-square solution for finding the vector consisting of the sought values a and b is

$$\hat{c} = \left(X^\mathrm{T}X\right)^{-1} X^\mathrm{T} y$$

The vector, $\hat{c}$, is the uncorrected value, including the potential errors. The actual prediction of the responses can then be expressed as

$$\hat{y} = X\hat{c} = X\left(X^\mathrm{T}X\right)^{-1} X^\mathrm{T} y$$

The term created by manipulating the experimental design matrix X is called the hat matrix, and H is defined by.

$$H = X\left(X^\mathrm{T}X\right)^{-1} X^\mathrm{T}$$

The hat matrix diagonal is a matter of interest, as each diagonal term of the row h_{ii} is representing a distance in the projection in space made in the least-square fitting.

The variance of the residual $\mathrm{var}\left(\hat{\varepsilon}_i\right)$ is then

$$\mathrm{var}\left(\hat{\varepsilon}_i\right) = s_{i,\,int}^2 \left(1 - h_{ii}\right)$$

Or

$$\mathrm{var}\left(\hat{\varepsilon}_i\right) = s_{i,ext}^2 \left(1 - h_{ii}\right)$$

depending on an internal or external frame of reference. The term $\left(1 - h_{ii}\right)$ is referred to as the leverage. In essence, this means explicitly that

$$\mathrm{var}\left(\hat{\varepsilon}_i\right) = s_{i,\,int}^2 \left(1 - \frac{1}{n} + \frac{\left(x_i - \bar{x}\right)^2}{\sum_{i=1}^{N}\left(x_i - \bar{x}\right)^2} h_{ii}\right)$$

or

$$\mathrm{var}\left(\hat{\varepsilon}_i\right) = s_{i,\,ext}^2 \left(1 - \frac{1}{n} + \frac{\left(x_i - \bar{x}\right)^2}{\sum_{i=1}^{N}\left(x_i - \bar{x}\right)^2} h_{ii}\right)$$

depending on the use of an external or internal frame of reference. The normalisation of the calculated residuals can now be made as

$$t_i = \frac{\hat{\varepsilon}_i}{\sqrt{\text{var}\left(\hat{\varepsilon}\right)}}$$

This is referred to as a studentised residual. The measure is referred to as externally or internally studentised residual depending on which frame of reference is used. A well-devised model should render normal-distributed studentised residuals.

The leverage can be used to devise a measure based on the internally studentised residuals called the Cook's distance used to find so-called outliers, meaning statistically deviating experiments that hamper the construction of the model building. The Cook's distance is constructed using the so-called min-max feature scaling, similar to the coded factor scaling but ranges from 0 to 1 instead of −1 to 1. The standard score is made by:

$$D_i = t_i \left(\frac{h_{ii}}{1 - h_{ii}} \right) \frac{1}{p}$$

The new factor p is the number of independent factors in the model. This means that for a liner model, it is 2, and for a second-order model, 3, even if the variable is x only as it is the rank of the experimental design matrix or $Rank(X)$.

Appendix

A.1 THE CHI-SQUARE TEST (PEARSON'S VARIANT)

The chi-squared distribution, with k degrees of freedom, is the distribution of the sum of the squares of k independent standard normal random variables. The chi-squared distribution is built on the gamma distribution. The chi-square test is a test of the goodness of fit. It is set out by considering n random observations divided into k classes with a number of observations x_i in each class. It also states that the null hypothesis is that the likelihood that an observation ends up in a class i is p_i.

$$\sum_{i=1}^{k} p_i = 1$$

This also means that the expected value, m_i, has the following properties.

$$\sum_{i=1}^{k} m_i = n \sum_{i=1}^{k} p_i = n$$

The central assumption is that as $n \rightarrow \infty$, the limiting distribution of the difference is given by

$$Y = \sum_{i=1}^{k} \frac{\left(x_i - m_i\right)^2}{m_i} = n \sum_{i=1}^{k} \frac{\left(x_i\right)^2}{m_i} - n$$

It is least likely that Y would follow the chi-square distribution. Figure A.1 shows an example of the chi-square distribution.

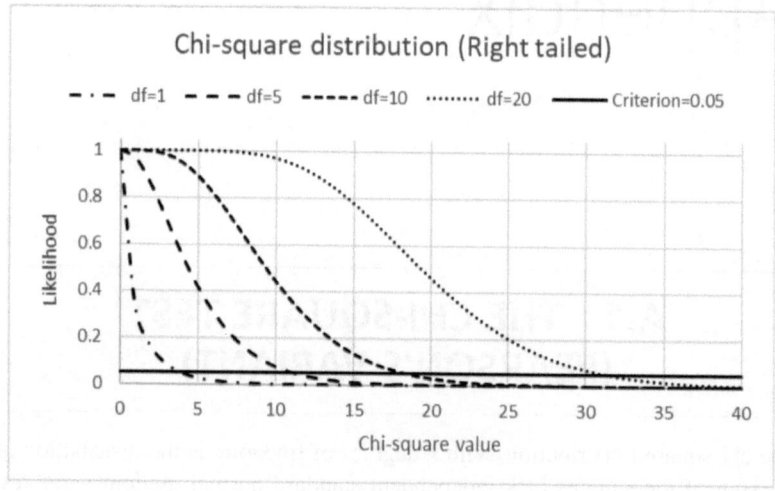

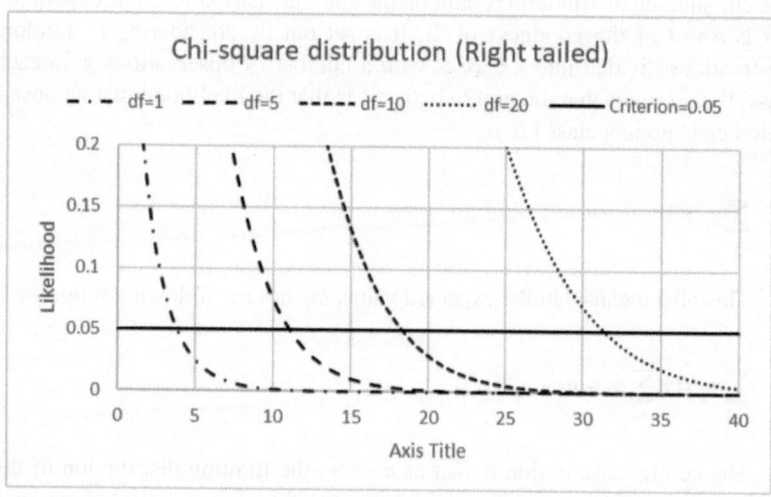

FIGURE A.1 The chi-square distribution illustrating (a) the overall appearance and (b) a close-up with the straight line illustrating the mark where 5% remain outside of the area to the right.

A.2   THE CHI-SQUARE TABLE

TABLE A.1   Critical values of a right-tailed chi-square as different degrees of freedom and significance levels α

DEGREES OF FREEDOM (DOF)	0.99	0.975	0.95	0.9	0.1	0.05	0.025	0.01
1	-	0.001	0.004	0.016	2.706	3.841	5.024	6.635
2	0.020	0.051	0.103	0.211	4.605	5.991	7.378	9.210
3	0.115	0.216	0.352	0.584	6.251	7.815	9.348	11.345
4	0.297	0.484	0.711	1.064	7.779	9.488	11.143	13.277
5	0.554	0.831	1.145	1.610	9.236	11.070	12.833	15.086
6	0.872	1.237	1.635	2.204	10.645	12.592	14.449	16.812
7	1.239	1.690	2.167	2.833	12.017	14.067	16.013	18.475
8	1.646	2.180	2.733	3.490	13.362	15.507	17.535	20.090
9	2.088	2.700	3.325	4.168	14.684	16.919	19.023	21.666
10	2.558	3.247	3.940	4.865	15.987	18.307	20.483	23.209
11	3.053	3.816	4.575	5.578	17.275	19.675	21.920	24.725
12	3.571	4.404	5.226	6.304	18.549	21.026	23.337	26.217
13	4.107	5.009	5.892	7.042	19.812	22.362	24.736	27.688
14	4.660	5.629	6.571	7.790	21.064	23.685	26.119	29.141
15	5.229	6.262	7.261	8.547	22.307	24.996	27.488	30.578
16	5.812	6.908	7.962	9.312	23.542	26.296	28.845	32.000
17	6.408	7.564	8.672	10.085	24.769	27.587	30.191	33.409
18	7.015	8.231	9.390	10.865	25.989	28.869	31.526	34.805
19	7.633	8.907	10.117	11.651	27.204	30.144	32.852	36.191
20	8.260	9.591	10.851	12.443	28.412	31.410	34.170	37.566
21	8.897	10.283	11.591	13.240	29.615	32.671	35.479	38.932
22	9.542	10.982	12.338	14.041	30.813	33.924	36.781	40.289
23	10.196	11.689	13.091	14.848	32.007	35.172	38.076	41.638
24	10.856	12.401	13.848	15.659	33.196	36.415	39.364	42.980
25	11.524	13.120	14.611	16.473	34.382	37.652	40.646	44.314
26	12.198	13.844	15.379	17.292	35.563	38.885	41.923	45.642
27	12.879	14.573	16.151	18.114	36.741	40.113	43.195	46.963
28	13.565	15.308	16.928	18.939	37.916	41.337	44.461	48.278
29	14.256	16.047	17.708	19.768	39.087	42.557	45.722	49.588
30	14.953	16.791	18.493	20.599	40.256	43.773	46.979	50.892

DEGREES OF FREEDOM (DOF)	0.99	0.975	0.95	0.9	0.1	0.05	0.025	0.01
40	22.164	24.433	26.509	29.051	51.805	55.758	59.342	63.691
50	29.707	32.357	34.764	37.689	63.167	67.505	71.420	76.154
60	37.485	40.482	43.188	46.459	74.397	79.082	83.298	88.379
70	45.442	48.758	51.739	55.329	85.527	90.531	95.023	100.425
80	53.540	57.153	60.391	64.278	96.578	101.879	106.629	112.329
100	61.754	65.647	69.126	73.291	107.565	113.145	118.136	124.116
1000	70.065	74.222	77.929	82.358	118.498	124.342	129.561	135.807

A.3   THE F-TEST

The F-test is a generic manner of testing a hypothesis. It is possible to compare the variances of two samples. For more than two samples, the ratio of variances is compared. For the F-test to be valid, a number of conditions need to be fulfilled, especially for use together with the Analysis of Variance (ANOVA). The prerequisites to be fulfilled are

• The distributions should be normally distributed.
• The variance should be homogeneous.
• The errors should be independent of all other factors.
• The samples should be random.

The fact that the F-test is sensitive to non-normality—meaning that there is a deviation from the normal distribution—is a significant shortcoming. Other tests have been developed for such cases. Although this is a more advanced topic, those interested can search for Levene's, Bartlett's, and Brown-Forsythe's tests for self-study purposes.

An example of difficulties can come from error propagation. In the case of measuring the length of the two sides, the measurements themselves have no problem. In this case, it was also clear that the error for the perimeter was independent of the length measurements and only dependent on the uncertainty of the length measurements. For the area, this was different. The error was dependent on both length measurements, and the error increased with the actual measurements. This would then create a problem when treating the area with the F-test. There are workarounds, but they are not part of the current treatise.

For the comparison of two samples under two conditions resulting in one set of measurements, the average is $\bar{x}$ and the standard deviation is s_x, meaning that it has a variance of s_x^2. For the other set of data, the average is $\bar{y}$ and the standard deviation is s_y, meaning that the variance is s_y^2. The F-value is then given by

$$F = \frac{s_x^2}{s_y^2}$$

In evaluating the F-test using the F-test table, the degrees of freedom are important based on the number of measurements or samples used. If the x values were made up of N samples and the y values were made up of M samples, the F-distribution was then made for the degrees of freedom (DOF) of $(N-1)$ and $(M-1)$. The level of desired significance will then be possible to be read from the table found in Section A.5.

A.4 ANALYSIS OF VARIANCE (ANOVA)

Analysis of variance or ANOVA separates the different contributions between different contributing factors. Often, a study is made of factors and observations as R rows and C columns (Table A.2).

The ANOVA procedure relies on the sum of squares as a characteristic but uses different groupings for the analysis. The first concept is the sum of squares within or SSW (W indicates within). This starts by calculating the column average.

TABLE A.2 Table of factors and experiments for ANOVA explanation

OBSERVATION/FACTOR	FACTOR 1	FACTOR 2	...	FACTOR C
Observation 1	y_1			
Observation 2	y_2			
...				
Observation R	y_R			
Column average	$\bar{y}_{Factor 1}$	$\bar{y}_{Factor 2}$		$\bar{y}_{Factor C}$

$$\overline{y}_{Factor\,1} = \frac{1}{R}\sum_{i=1}^{R} y_i$$

Furthermore, it is repeated for all columns.

$$SSW = SS_{Factor\,1} + SS_{Factor\,2} + \cdots + SS_{Factor\,C}$$

where

$$SS_{Factor\,1} = \sum_{i=1}^{R}\left(y_i - \overline{y}_{Factor\,1}\right)^2$$

This is repeated for all columns to allow the sum of SSW.

The next step is calculating the sum of squares between the columns SSB (B indicated between). This starts with the average of the column averages.

$$\overline{y}_0 = \frac{1}{C}\sum_{i=1}^{C} \overline{y}_{Factor\,i}$$

This SSB is then calculated by removing the average $\overline{y}_0$ from all the observations, for example, all cells in the table from 1 to RC.

$$SSB = \sum_{i=1}^{RC}\left(y_{Factor\,i} - \overline{y}_0\right)^2$$

The DOF differ for the within- and between-column assessments. For SSW, the DOF is df_W and for SSB, it is df_B

$$df_W = (R-1)C$$

$$df_B = (C-1)$$

This then results in the mean squares for the two cases, which can be described as

$$MS_W = \frac{SSW}{df_W}$$

$$MS_B = \frac{SSB}{df_B}$$

The two different means of the sum of squares are then used to create the hypothesis test following the F-test statistic as

$$F = \frac{MS_B}{MS_W}$$

This is used to test the hypothesis with their DOF for which critical values are found as shown in Table A.3.

A.5   THE F-TEST TABLE

TABLE A.3   The F-test table for $\alpha = 0.05$

/	DFB=1	2	3	4	5	6	7	8	9
dfW=1	161.44	199.50	215.71	224.58	230.16	233.99	236.77	238.88	240.54
2	18.518	19.000	19.164	19.247	19.296	19.329	19.353	19.371	19.385
3	10.1280	9.5521	9.2766	9.1172	9.0135	8.9406	8.8867	8.8452	8.8123
4	7.7086	6.9443	6.5914	6.3882	6.2561	6.1631	6.0942	6.0410	5.9988
5	6.6079	5.7861	5.4095	5.1922	5.0503	4.9503	4.8759	4.8183	4.7725
6	5.9874	5.1433	4.7571	4.5337	4.3874	4.2839	4.2067	4.1468	4.0990
7	5.5914	4.7374	4.3468	4.1203	3.9715	3.8660	3.7870	3.7257	3.6767
8	5.3177	4.4590	4.0662	3.8379	3.6875	3.5806	3.5005	3.4381	3.3881
9	5.1174	4.2565	3.8625	3.6331	3.4817	3.3738	3.2927	3.2296	3.1789
10	4.9646	4.1028	3.7083	3.4780	3.3258	3.2172	3.1355	3.0717	3.0204
11	4.8443	3.9823	3.5874	3.3567	3.2039	3.0946	3.0123	2.9480	2.8962
12	4.7472	3.8853	3.4903	3.2592	3.1059	2.9961	2.9134	2.8486	2.7964
13	4.6672	3.8056	3.4105	3.1791	3.0254	2.9153	2.8321	2.7669	2.7144
14	4.6001	3.7389	3.3439	3.1122	2.9582	2.8477	2.7642	2.6987	2.6458
15	4.5431	3.6823	3.2874	3.0556	2.9013	2.7905	2.7066	2.6408	2.5876
16	4.4940	3.6337	3.2389	3.0069	2.8524	2.7413	2.6572	2.5911	2.5377
17	4.4513	3.5915	3.1968	2.9647	2.8100	2.6987	2.6143	2.5480	2.4943
18	4.4139	3.5546	3.1599	2.9277	2.7729	2.6613	2.5767	2.5102	2.4563
19	4.3807	3.5219	3.1274	2.8951	2.7401	2.6283	2.5435	2.4768	2.4227
20	4.3512	3.4928	3.0984	2.8661	2.7109	2.5990	2.5140	2.4471	2.3928
21	4.3248	3.4668	3.0725	2.8401	2.6848	2.5727	2.4876	2.4205	2.3660
22	4.3009	3.4434	3.0491	2.8167	2.6613	2.5491	2.4638	2.3965	2.3419
23	4.2793	3.4221	3.0280	2.7955	2.6400	2.5277	2.4422	2.3748	2.3201
24	4.2597	3.4028	3.0088	2.7763	2.6207	2.5082	2.4226	2.3551	2.3002
25	4.2417	3.3852	2.9912	2.7587	2.6030	2.4904	2.4047	2.3371	2.2821
26	4.2252	3.3690	2.9752	2.7426	2.5868	2.4741	2.3883	2.3205	2.2655
27	4.2100	3.3541	2.9604	2.7278	2.5719	2.4591	2.3732	2.3053	2.2501
28	4.1960	3.3404	2.9467	2.7141	2.5581	2.4453	2.3593	2.2913	2.2360
29	4.1830	3.3277	2.9340	2.7014	2.5454	2.4324	2.3463	2.2783	2.2229
30	4.1709	3.3158	2.9223	2.6896	2.5336	2.4205	2.3343	2.2662	2.2107
40	4.0847	3.2317	2.8387	2.6060	2.4495	2.3359	2.2490	2.1802	2.1240
60	4.0012	3.1504	2.7581	2.5252	2.3683	2.2541	2.1665	2.0970	2.0401
120	3.9201	3.0718	2.6802	2.4472	2.2899	2.1750	2.0868	2.0164	1.9588
∞	3.8415	2.9957	2.6049	2.3719	2.2141	2.0986	2.0096	1.9384	1.8799

10	12	15	20	24	30	40	60	120	∞
241.88	243.91	245.95	248.01	249.05	250.09	251.14	252.20	253.25	254.31
19.396	19.412	19.429	19.446	19.454	19.462	19.471	19.479	19.487	19.496
8.7855	8.7446	8.7029	8.6602	8.6385	8.6166	8.5944	8.5720	8.5494	8.5264
5.9644	5.9117	5.8578	5.8025	5.7744	5.7459	5.7170	5.6877	5.6581	5.6281
4.7351	4.6777	4.6188	4.5581	4.5272	4.4957	4.4638	4.4314	4.3985	4.3650
4.0600	3.9999	3.9381	3.8742	3.8415	3.8082	3.7743	3.7398	3.7047	3.6689
3.6365	3.5747	3.5107	3.4445	3.4105	3.3758	3.3404	3.3043	3.2674	3.2298
3.3472	3.2839	3.2184	3.1503	3.1152	3.0794	3.0428	3.0053	2.9669	2.9276
3.1373	3.0729	3.0061	2.9365	2.9005	2.8637	2.8259	2.7872	2.7475	2.7067
2.9782	2.9130	2.8450	2.7740	2.7372	2.6996	2.6609	2.6211	2.5801	2.5379
2.8536	2.7876	2.7186	2.6464	2.6090	2.5705	2.5309	2.4901	2.4480	2.4045
2.7534	2.6866	2.6169	2.5436	2.5055	2.4663	2.4259	2.3842	2.3410	2.2962
2.6710	2.6037	2.5331	2.4589	2.4202	2.3803	2.3392	2.2966	2.2524	2.2064
2.6022	2.5342	2.4630	2.3879	2.3487	2.3082	2.2664	2.2229	2.1778	2.1307
2.5437	2.4753	2.4034	2.3275	2.2878	2.2468	2.2043	2.1601	2.1141	2.0658
2.4935	2.4247	2.3522	2.2756	2.2354	2.1938	2.1507	2.1058	2.0589	2.0096
2.4499	2.3807	2.3077	2.2304	2.1898	2.1477	2.1040	2.0584	2.0107	1.9604
2.4117	2.3421	2.2686	2.1906	2.1497	2.1071	2.0629	2.0166	1.9681	1.9168
2.3779	2.3080	2.2341	2.1555	2.1141	2.0712	2.0264	1.9795	1.9302	1.8780
2.3479	2.2776	2.2033	2.1242	2.0825	2.0391	1.9938	1.9464	1.8963	1.8432
2.3210	2.2504	2.1757	2.0960	2.0540	2.0102	1.9645	1.9165	1.8657	1.8117
2.2967	2.2258	2.1508	2.0707	2.0283	1.9842	1.9380	1.8894	1.8380	1.7831
2.2747	2.2036	2.1282	2.0476	2.0050	1.9605	1.9139	1.8648	1.8128	1.7570
2.2547	2.1834	2.1077	2.0267	1.9838	1.9390	1.8920	1.8424	1.7896	1.7330
2.2365	2.1649	2.0889	2.0075	1.9643	1.9192	1.8718	1.8217	1.7684	1.7110
2.2197	2.1479	2.0716	1.9898	1.9464	1.9010	1.8533	1.8027	1.7488	1.6906
2.2043	2.1323	2.0558	1.9736	1.9299	1.8842	1.8361	1.7851	1.7306	1.6717
2.1900	2.1179	2.0411	1.9586	1.9147	1.8687	1.8203	1.7689	1.7138	1.6541
2.1768	2.1045	2.0275	1.9446	1.9005	1.8543	1.8055	1.7537	1.6981	1.6376
2.1646	2.0921	2.0148	1.9317	1.8874	1.8409	1.7918	1.7396	1.6835	1.6223
2.0772	2.0035	1.9245	1.8389	1.7929	1.7444	1.6928	1.6373	1.5766	1.5089
1.9926	1.9174	1.8364	1.7480	1.7001	1.6491	1.5943	1.5343	1.4673	1.3893
1.9105	1.8337	1.7505	1.6587	1.6084	1.5543	1.4952	1.4290	1.3519	1.2539
1.8307	1.7522	1.6664	1.5705	1.5173	1.4591	1.3940	1.3180	1.2214	1.0000

Bibliography

THEORY OF SCIENCE

Kosso, Peter (1992), *Reading the Book of Nature: An Introduction to the Philosophy of Science*, Cambridge University Press, ISBN 9781139172554

National Academy of Sciences (US) (1999), *Science and Creationism: A View from the National Academy of Sciences* (2nd ed.). National Academies Press. p. 2. doi:10.17226/6024. ISBN 978-0-309-06406-4. PMID 25101403.

Säfsten, Kristina, Gustavsson, Maria (2020), *Research Methodology*, Studentlitteratur, ISBN 978-91-44-12230-4

DESIGN OF EXPERIMENTS

Box, George E. P., Hunter, Stuart J., Hunter, William G. (2005), *Statistics for Experimenters*, John Wiley, ISBN 9780471718130

Hicks, Charles R., Turner, Kwenneth V. Jr (1999), *Fundamental Concepts in the Design of Experiments*, 5th ed. Oxford University Press, Oxford, ISBN 0-19-512273-9

McKay, M. D., Beckman, R. J., Conover, W. J. (1979), "A comparison of three methods for selecting values of input variables in the analysis of output from a computer code", *Technometrics*, 21(2): 239–245.

Olsson, A., Sandberg, G., Dahlblom, O. (2003), "On Latin hypercube sampling for structural reliability analysis", *Structural Safety* 25: 47–68.

Shin, Y. J., Xiu, D. B., Shin, Yeonjong, Xiu, Dongbin (2016), "Nonadaptive quasi-optimal points selection for least squares linear regression", *SIAM Journal on Scientific Computing*, 38(1): A385–A411, doi: 10.1137/15M1015868

STATISTICS

Anscombe, F. J. (1948), "The validity of comparative experiments", *Journal of the Royal Statistical Society*, Series A (General). 111 (3): 181–211. doi:10.2307/2984159. JSTOR 2984159. MR 0030181.

Bailey, R. A. (2008), *Design of Comparative Experiments*. Cambridge University Press. ISBN 978-0-521-68357-9

Greenwood, Cindy, Nikulin, M. S. (1996), *A Guide to Chi-Squared Testing*, New York: Wiley, ISBN 0-471-55779-X

International Standard ISO 3534-1 (1993), *Statistics – Vocabulary and Symbols – Part I: Probability and General Statistical Terms*, First Edition International Organization for Standardization, Geneva.

Pearson, Karl (1900), "On the criterion that a given system of deviations from the probable in the case of a correlated system of variables is such that it can be reasonably supposed to have arisen from random sampling", *Philosophical Magazine*, 50(302): 157–175. doi:10.1080/14786440009463897

MEASUREMENT

Baird, D.C. (1995), *Experimentation: An Introduction to Measurement Theory and Experiment Design*, 3rd Edition, Prentice Hall: Englewood Cliffs, 1995.

Bell, S. (1999), *A Beginner's Guide to Uncertainty of Measurement*, Measurement Good Practice Guide No. 11 (Issue 2), ISSN 1368-6550

Czichos, Horst, Smith, Leslie, eds. (2011). *Springer Handbook of Metrology and Testing*, 2nd Edition, Springer. 1.2.2 Categories of Metrology. ISBN 978-3-642-16640-2

PD 6461: Part 1: (1995), *Vocabulary of Metrology, Part 1. Basic and General Terms (International)*, British Standards Institution, London.

UNCERTAINTY

Bevington, Phillip, Robinson, D. (1991), *Data Reduction and Error Analysis for the Physical Sciences*, 2nd Edition, McGraw-Hill: New York.

BIPM, IEC, IFCC, ISO, IUPAC, IUPAP, OIML (1995), *Guide to the Expression of Uncertainty in Measurement*. International Organization for Standardization, Geneva. ISBN 92-67-10188-9, First Edition 1993, corrected and reprinted 1995. (BSI Equivalent: BSI PD 6461: 1995, Vocabulary of Metrology, Part 3. Guide to the Expression of Uncertainty in Measurement. British Standards Institution, London)

BIPM, IEC, IFCC, ISO, IUPAC, IUPAP, OIML (1993), *International Vocabulary of Basic and General Terms in Metrology*, 2nd Edition, International Organization for Standardization, Geneva.

Chatfield, C. (1983), *Statistics for Technology*, 3rd Edition, New York: Chapman and Hall.

Dietrich, C. F. (1991), *Uncertainty, Calibration and Probability*, 2nd Edition, Bristol: Adam Hilger.

EA-4/02 (1999), Expression of the Uncertainty of Measurement in Calibration, European co-operation for Accreditation.

EURACHEM/CITAC (2000), Guide: *Quantifying Uncertainty in Analytical Measurement*, 2nd Edition, Measurement Good Practice Guide No. 11 (Issue 2) 29.

International Standard ISO/IEC 17025 (1999), *General Requirements for the Competence of Testing and Calibration Laboratories*, 1st Edition, International Organization for Standardization: Geneva.

ISO (1993), *Guide to the Expression of Uncertainty in Measurement*. International Organization for Standardization (ISO) and the International Committee on Weights and Measures (CIPM): Switzerland.

Lichten, William (1999), *Data and Error Analysis*, 2nd Edition, Prentice Hall: Upper Saddle River, NJ.

NIST. *Essentials of Expressing Measurement Uncertainty*. http://physics.nist. gov/cuu/Uncertainty/

Taylor, John (1997), *An Introduction to Error Analysis*, 2nd Edition, University Science Books: Sausalito.

United Kingdom Accreditation Service (UKAS) Publication M 3003 (2024), *The Expression of Uncertainty and Confidence in Measurement*, 6th Edition.

Index

VARIORUM COLLECTED STUDIES SERIES

Science in the Public Sphere

Professor Richard Yeo

Richard Yeo

Science in the Public Sphere

Natural Knowledge in
British Culture 1800–1860

Routledge
Taylor & Francis Group
LONDON AND NEW YORK

First published 2001 by Ashgate Publishing

Published 2016 by Routledge
2 Park Square, Milton Park, Abingdon, Oxon OX14 4RN
52 Vanderbilt Avenue, New York, NY 10017

Routledge is an imprint of the Taylor & Francis Group, an informa business

This edition copyright © 2001 by Richard Yeo.

ISBN 13: 978-0-86078-865-2 (hbk)

British Library Cataloguing-in-Publication Data
Yeo, Richard R. 1948–
    Science in the Public Sphere : Natural Knowledge in British
    Culture, 1800–1860. – (Variorum Collected Studies Series: CS726).
    1. Science – Great Britain – History – 19th Century 2. Science – Great Britain – Public
    Opinion – History – 19th century
    I. Title.
    306.4'5'0941'09034

US Library of Congress Cataloging-in-Publication Data
Yeo, Richard R., 1948–
    Science in the Public Sphere: Natural knowledge in British culture, 1800–1860/Richard Yeo.
        p. cm. – (Variorum Collected Studies Series: CS726).
    Includes bibliographical references and index.
    1. Science – Social Aspects – Great Britian – History – 19th Century. 2. Science –
    Philosophy. 3. Natural Theology. I. Title. II. Collected Studies: CS726.
    Q175.52.G7 Y46 2001
    509.41'09'034–dc21   2001046139

VARIORUM COLLECTED STUDIES SERIES CS726

CONTENTS

This volume contains xviii + 302 pages

ACKNOWLEDGEMENTS

These essays were written between 1979 and 1996. Naturally, they reflect some different institutional settings and intellectual influences. The earliest of these publications stem from my doctoral work at the History Department, Sydney University. I take this opportunity to record my debt to the advice and inspiration of my supervisors at that time, Patrick Collinson and Robert (Bob) Dreher. Subsequent publications were completed at the School of Humanities, Griffith University, which also supported periods of study leave in the UK. More recently, I have received funding from the Australian Research Council which greatly assisted the research and writing of chapters IX and X. I would like to thank Rebecca Langlands for help in the preparation of this volume and for compiling the index. Finally, my thanks go to John Smedley of Ashgate Press for his encouragement and advice.

For permission to reprint these essays I am grateful to Science Reviews Ltd (I), Science History Publications (II), Elsevier Science (III), Kluwer Academic Publishers (IV), Taylor and Francis Pty Ltd (www.tandf.co.uk) (V), Cambridge University Press (VI, VII, X), Blackwell Publishers (VIII), The University of Chicago Press (IX), Indiana University Press (XI).

INTRODUCTION

In the early decades of the nineteenth century the activities of some leading men of science could be misperceived and misunderstood by their contemporaries. Consider the experiences of William Buckland, William Conybeare, George Biddell Airy, and William Whewell. When collecting fossils in Ireland in 1813, an innkeeper at first thought Buckland and Conybeare were itinerant workers; and on a subsequent occasion Buckland was not recognised by his own servant, who took him for 'a man with a bag'. In 1826, when Whewell and Airy were measuring the density of the earth, working 1200 feet underground in Dolcoath mine in Cornwall, some of the local villagers regarded them with disapproval as men willingly consorting with demons in the 'bowels of the earth' (Whewell's description of his position). After the basket holding their instruments caught fire, Airy wondered 'whether a superstitious miner had intentionally fired it'.[1]

These instances can stand as parables about the pursuit of science in early nineteenth-century Britain – an indefinite vocation practiced by gentlemen, but one that sometimes required activities that cast doubt on this social status. Charles Babbage played on this theme in his provocative *Reflections on the Decline of Science in England* (1830), drawing the contrast with France where, he said, scientific researchers were both financially supported and honoured by the State.[2] In 1851, Babbage later touched on another aspect of this problem when he remarked that although individuals called themselves astronomers, chemists, geologists, and botanists, 'our language itself contains no *single* term' by which the activities of men of science (the common generic name) could be expressed. Whewell had made the same point in 1834 when offering 'scientist' as the collective noun, a name not widely adopted until the end of the century.[3]

The chapters in this volume consider the assumptions and debates about the nature of science from 1800 to 1860, although some refer back to the eighteenth-century and one or two look forward to late nineteenth-century issues. The focus, however, is on the early Victorian period, interpreted quite generously to include the early decades of the century. This period therefore includes the so called 'second scientific revolution' that signalled the demise of the general category of natural philosophy and the rise of specialist disciplines, perhaps marking the birth of modern 'science'.[4] The closing date reflects the work of Susan Cannon, Robert Young, Frank Turner, and Roy MacLeod who, in different ways, viewed the 1860s as the start of a new era in which the advocates of scientific naturalism rejected the legitimation provided by natural theology, pushed for professional careers, and sought greater financial support from the State.[5]

The three headings used to group the chapters indicate some key areas of debate about the nature of science and its cultural status. The first two – scientific method and natural theology – are major intellectual frameworks; the third refers to the idea of a public sphere in which science was discussed in various genres, such as history and biography, and different forums, such as journals and encyclopaedias.

The stress on scientific method as the defining feature of science, and a guarantor of its conclusions, has been a powerful motif since the Scientific Revolution.[6] It is interesting, therefore, that this theme did not figure prominently in the writings of Cannon and Young, both of whom saw natural theology as the main source of justification (more on this below). Yet it was precisely in the early 1800s that a focus on scientific method crystallised out of more general epistemological discussions conducted by writers in the tradition of British empiricism since John Locke, and the Scottish common sense philosophers such as Thomas Reid and Dugald Stewart. As Larry Laudan pointed out, the 1830s witnessed the publication of John Herschel's *Discourse*, Comte's *Cours de Philosophie Positive* (1830) and Whewell's *History* (1837) and *Philosophy* (1840) of the inductive sciences. These were preceded by Stewart's reflections on Baconian method in his *Elements of the Philosophy of the Human Mind* (vol. 2, 1814) and followed by John Stuart Mill's *System of Logic* (1843), which was indebted in its treatment of scientific reasoning to the concrete examples given by Herschel and Whewell.[7] These works consolidated the issues subsequently treated in philosophical accounts of science, together with the vocabulary in which they were discussed.

In the first half of the nineteenth century these analyses of scientific method were not divorced from other topics, since philosophy of science was not yet a specialised discipline; rather, they were part of discussions in scientific societies that radiated into the public arena constituted by journals, scientific biographies, and histories of science. Indeed, the British Association for the Advancement of Science, founded in 1831, was a major forum in which method was profiled by the scientific community as one basis for the authority of science. One goal of this organisation was the promotion of science to the public; but this aim was complicated by the very diversity of scientific subjects embraced by the Association and displayed in its 'Sections' (seven by 1835).[8] My early essays on this topic argued that appeals to scientific method were a rhetorical device used to affirm the unity of science, in spite of increasing specialisation, and, at another level, its accessibility to non-experts (see chs I, III and IV).[9] Behind much of this discussion was the protean legacy of Baconianism, the quintessential British language for talking about the aims and values of science. Thus the leading quarterly journals carried assessments of Bacon's character by Thomas Babington (Lord Macaulay) which easily carried over into evaluations of his ideas on induction and the use of hypotheses. This emphasis on methodology meant a narrowing of eighteenth-century conceptions of Baconianism as an emblem of Enlightenment, but debates on method were still charged with moral connotations about the proper use of reason (ch II).

Some of the earliest essays and articles in this collection derive from my doctoral work on the role of natural theology as a framework for early nineteenth-century discussions of science. The two dominant historiographic influences here were the writings of Cannon and Young. Despite some significant differences, their work stressed the importance of assumptions from natural theology as the key to an understanding of how scientific debates on the place of man in nature resonated so strongly with wider cultural values, thus ensuring that British science of the early Victorian period remained a matter for public, and not just professional, debate. However, whereas these two authors studied some of the major substantive topics such as the status of miracles, the rise of 'deep' geological time, adjustments to the argument from design, and the manifold issues posed by evolutionary speculation, my emphasis was on the ways in which natural theology supplied a moral and epistemological justification for doing science. William Whewell was my prime example and I approached his writings on the history and philosophy of science from this perspective.[10] Although Whewell was a critic of the conventional design argument, his underlying assumptions about the methodology and epistemology of science were teleological: the undoubted success of science indicated that humanity shared, to some extent, in the Divine mind (chs V and VI). Other commentators on science, such as Herschel, Baden Powell and David Brewster, held similar views, and all believed that the proper use of reason was a matter of moral importance. Their position is highlighted by the contrast with evangelical authors such as Thomas Chalmers, who qualified this confidence in human reason.[11] Some of these themes converged in the debate instigated by Whewell's anonymous *Of the Plurality of Worlds: An Essay* of 1853. As John Brooke has shown, the rejoinder by Brewster, *More Worlds than One* (1854), exposed deep divisions within natural theology, suggesting that this discourse was fragmenting on doctrinal and denominational fault lines at least as much as it was being eroded by scientific naturalism.[12] Moreover, as I have contended, British natural theology flirted with metaphysical notions such as the principle of plenitude, and this allowed images of nature as plastic and self-organising, thus creating problems for the concept of contrivance at the heart of the traditional argument from design (VI).

The title of this collection, *Science in the Public Sphere*, reflects the (belated) impact of Jürgen Habermas' claims about the emergence during the Enlightenment of a distinctive role for the notion of 'the public' as an arbiter of taste and opinion.[13] In principle, participation in the conversation hosted by this public sphere did not depend upon wealth or social rank, but upon a shared commitment to rational debate. Habermas did not consider science in any detail,[14] but it is clear that by the early 1800s men of science were alert to the fact that they needed to address both their peers and the wider audience of educated readers. When telling Herschel about his criticism of the managers of the Royal Society of London, Babbage said that 'with the aid of public opinion I will make them writhe if they do not reform'.[15] Of course, Cannon, Young and other historians have recognised some of the pertinent issues here, but Habermas' work is useful because it draws attention to the political and intellectual conditions necessary to sustain the idea of a public sphere. Cannon and Young spoke of a common intellectual context that made science part of a general conversation among the educated classes. But viewed from a longer perspective, this early Victorian consensus (or 'Truth Complex', as Cannon called it) was already under strain before the corrosive impact of Charles Darwin's *Origin of Species* (1859). Habermas located one manifestation of the early public sphere in eighteenth-century English journals such as the *Tatler* (from 1709) and the *Spectator* (from 1711). These provided opportunities for polite conversation that embraced both aristocracy and bourgeoisie in a civil society.[16] However, we cannot simply transfer Habermas' analysis, without amendment, to early nineteenth-century Britain, a society struggling with the political aftermath of the French Revolution, including more definite working class agitation.[17] As Roger Cooter, Adrian Desmond, Ian Inkster, Alison Winter, and others have insisted, science was not insulated from this political and social turbulence.[18] Indeed, Jack Morrell and Arnold Thackray have shown that this environment was recognised by the managers of the British Association (including many from Cannon's 'Cambridge network') who worked to detach science from political controversy.[19] This strategy contrasts quite clearly with the apologetics of T.H. Huxley and others who later decided to pursue more aggressive campaigns on behalf of science.[20]

The political context of the early Victorian period has implications for the way we view public debates on science. The new 'critical' journals such as the *Edinburgh Review* (from 1802), the *Quarterly Review* (from 1809) and the *Westminster Review* (from 1824) prided themselves on being less 'polite' than the journals of the eighteenth century. Indeed, the explicit Whig, Tory and utilitarian affiliations, respectively, of these three publications made them intellectual clearing houses in which contributors often took definite positions on controversial matters. Reviews of science could.become entangled in these polemics, especially in articles by Brewster and Henry Brougham, two of the most prolific contributors on science, who linked it to campaigns for popular education and State support. This meant that in spite of any consensus on natural theology, science was potentially a point of disagreement in the most influential forum of the day. Moreover, the increasing specialisation of the sciences placed great pressure on the ability of these quarterly reviews to speak to both specialist and general readers.[21]

Another point to note is the variety of public forums in the early Victorian period, such as the presidential addresses of the British Association, biographies and histories of science, and encyclopaedias, as well as the periodical journals. Moreover, each of these had its own conventions. Thus recourse to a bland, undifferentiated public sphere will not register the crucial differences between the anonymous reviews of the quarterly journals and the signed entries in major encyclopaedias (by the early 1800s). In the former, the authority of the review was that of the journal, often cited as a collective 'we'; in the latter, it was the position and expertise of a known member of the scientific community. This difference suggests an intriguing asymmetry: whereas substantial treatises on the sciences in encyclopaedias demanded the authorship of a named specialist, assessments of scientific ideas and discoveries for the educated public and, significantly, more general reflections on the moral, social or religious implications of science, were entrusted to anonymous reviewers.

The chapters in the third section consider some of the ideas about science encountered in this complex public sphere. These include the image of the man of science as moral exemplar and its connection with concepts of method and genius in biographies and histories (VII, VIII, X); the organisation of scientific subjects in encyclopaedias (IX),[22] and the concept of intellectual authority as this was posed in the reception of Robert Chambers' anonymous *Vestiges of the Natural History of Creation* (1844). In fact, this episode nicely illustrates several themes that pervade this volume — the rhetoric of scientific method, the moral and intellectual persona of the man of science, and the need to address the non-specialist interests and concerns of a wider audience in the public domain provided by the major periodical journals (XI).[23]

Collectively, then, these chapters stress that British debates on science between 1800 and 1860 involved ideas and arguments about the nature, history, and ethos of the scientific endeavour; the moral demeanour and personality of the man of science; and the methodological and intellectual status of various forms of knowledge within the hierarchy of specialised scientific disciplines. This is a wider and richer set of issues than usually noticed in work that approaches early Victorian science solely in terms of its relations with religion, or as a drive towards professionalisation. The issues I have addressed in these chapters were discussed both within scientific circles in – personal correspondence and in meetings of societies – and in the wider public sphere offered by quarterly journals, encyclopaedias, histories and biographies. By attending to these debates we can appreciate how the culture of British science in the first half of the nineteenth century began to cast loose some of its earlier theological supports, while still relying on a moral framework to affirm its own norms, method, and value.

RICHARD YEO

Brisbane, June 2001

Endnotes

1   For the best treatment of this phenomenon of mistaken identity, and examples from geology, see Michael Shortland, 'Darkness Visible: Underground Culture in the Golden Age of Geology', *History of Science*, 31 (1994), 1–61.

2   Isaac Todhunter, *William Whewell, D.D. An Account of his Writings with Selections from his Literary and Scientific Correspondence*, 2 vols (London: Macmillan, 1876), vol. 2, 64–69; George Biddell Airy, *Autobiography* (Cambridge University Press, 1896) 68. In a letter penned underground, Whewell reflected on the impression he and Airy had made on the local people: 'There is no end to the number and oddity of their conjectures and stories about us. The most charitable of them take us to be fortune tellers; but for the greater part we are suspected of more mischievous kinds of magic.' Whewell to Lady Malcolm, 10 June 1826, in Janet M. Douglas, *The Life and Selections from the Correspondence of William Whewell, D.D., late Master of Trinity College, Cambridge* (London: C. Kegan Paul, 1881), 104.

3   Charles Babbage, *Reflections on the Decline of Science in England, and on some of its causes* (London: B. Fellowes, 1830), 14–39; see also J.B. Morrell, 'Individualism and the Structure of British Science in 1830', *Historical Studies in the Physical Sciences*, 3 (1971), 183–204; for national comparisons, see Maurice Crosland, *Studies in the Culture of Science in France and Britain since the Enlightenment* (Variorum Collected Studies Series Aldershot: Ashgate, 1995). For an earlier commentary, see John T. Merz, *A History of European Thought in the Nineteenth*

Century, 4 vols (Edinburgh and London: Blackwood, 1896–1914) vol. 1. See also Dorinda Outram, *Georges Cuvier: Vocation, Science, and Authority in Post-Revolutionary France* (Manchester University Press, 1984) for the warning that science still needed public legitimation in France, despite the greater number of scientific posts.

4   Charles Babbage, *The Exposition of 1851: Views of the Industry, the Science, and the Government of England* (London: John Murray, 1851), 189. Whewell suggested this term at the meeting of the British Association for the Advancement of Science in Cambridge in 1833 and made it more public in his review of Mary Somerville's *On the Connexion of the Physical Sciences* (London: John Murray 1834). See [William Whewell,], 'Mrs. Somerville on the Connexion of the Sciences', *Quarterly Review*, 51 (1834), 54–68.

5   Andrew Cunningham and Perry Williams, 'De-centring the "big picture": *The Origins of Modern Science* and the modern origins of science', *British Journal for the History of Science*, 26 (1993),407–32. For earlier comments, see I.B. Cohen, *Revolution in Science* (Cambridge, Mass., 1985); Thomas S. Kuhn, 'Mathematical versus Experimental Traditions in the Development of Physical Science', in *The Essential Tension: Selected Studies in Scientific Tradition* (The University of Chicago Press, 1977), pp. 31–65. For the perceptions of some eighteenth-century observers, see Richard Yeo, 'Classifying the Sciences', in Roy Porter (ed.), *Cambridge History of Science: Volume 4, The Eighteenth Century* (Cambridge University Press, 2001), pp. 272–302.

6   Susan F. Cannon (formerly W.F.), *Science in Culture: the early Victorian Period* (New York: Dawson and Science History Publishing, 1978); Robert M. Young, *Darwin's Metaphor: Nature's Place in Victorian Culture* (Cambridge University Press, 1985) especially ch. 5; Frank M. Turner, *Contesting Cultural Authority: Essays in Victorian Intellectual Life* (Cambridge University Press, 1993); Roy MacLeod, *Public Science and Public Policy in Victorian England* (Variorum Collected Studies Series Aldershot: Ashgate, 1996) and *The 'Creed of Science' in Victorian England* (Variorum Collected Studies Series Aldershot: Ashgate, 2000).

7   See John A. Schuster and Richard R. Yeo (eds), *The Politics and Rhetoric of Scientific Method: Historical Studies* (Boston and Dordrecht: D. Reidel, 1986).

8   Larry Laudan, *Science and Hypothesis: Historical Essays on Scientific Methodology* (Dordrecht: D. Reidel, 1981); Richard Yeo, 'Reviewing Herschel's *Discourse*', *Studies in History and Philosophy of Science*, 20 (1980), 541–52.

9   Jack Morrell and Arnold Thackray, *Gentlemen of Science: early years of the British Association for the Advancement of Science* (Oxford: Clarendon Press, 1981). See also Roy MacLeod and Peter Collins (eds), *The Parliament of Science: the British Association for the Advancement of Science 1831–1981* (London: Science Reviews Ltd, 1981).

10   For a subsequent use of this approach, see Albert Moyer, *A Scientist's Voice in American Culture: Simon Newcomb and the Rhetoric of Scientific Method* (Berkeley: University of California. Press, 1992), esp. 227–37.

11   For a development from this, see Richard Yeo, *Defining Science: William Whewell, Natural Knowledge and Public Debate in Early Victorian Britain* (Cambridge

University Press, 1993). See also Menachem Fisch and Simon Schaffer (eds), *William Whewell. A Composite Portrait* (Oxford University Press, 1991).

12   For a detailed study of this intellectual context, see Boyd Hilton, *The Age of Atonement: The Influence of Evangelicalism on Social and Economic Thought, 1785-1865* (Oxford University Press, 1988).

13   John H. Brooke, 'Natural Theology and the Plurality of Worlds: Observations on the Whewell-Brewster Debate', *Annals of Science*, 34 (1977), 221–86; Pietro Corsi, *Science and Religion: Baden Powell and the Anglican Debate, 1800–1860* (Cambridge University Press, 1988). For the later decades, see W.H. Brock and R.M. MacLeod, 'The Scientists' Declaration: Reflections on Science and Belief in the *Wake of Essays and Reviews*, 1864–5', *British Journal for the History of Science*, 9 (1976), 39–66. On the geological debates that fuelled many of these arguments, see Nicolaas A. Rupke, *The Great Chain of History: William Buckland and the English School of Geology (1814–1849)* (Oxford University Press, 1983) and David Oldroyd, *Sciences of the Earth: Studies in the History of Mineralogy and Geology* (Variorum Collected Studies Aldershot: Ashgate, 1998).

14   Also relevant here is Romanticism. On this, see David M. Knight, *Science in the Romantic Era* (Variorum Collected Studies Aldershot: Ashgate, 1998).

15   Jürgen Habermas, *The Structural Transformation of the Public Sphere: an inquiry into a category of bourgeois society* (Boston: MIT Press, 1989).

16   See Thomas Broman, 'The Habermasian Public Sphere and "Science in the Enlightenment"', *History of Science*, 36 (1998), 123–149. Habermas did discuss the power of modern science and technology in his *Towards a Rational Society* (London: Heinemann, 1970).

17   Babbage to Herschel, 19 March 1830, cited in Morrell and Thackray, *Gentlemen of Science*, 48.

18   Habermas claimed that this development was first apparent in England, which had a constitutional monarchy from the Glorious Revolution of 1688. For a wider case, see Roy Porter, *Enlightenment: Britain and the Creation of the Modern World* (London: Allen Lane, 2000).

19   E.P. Thompson, *The Making of the English Working Class* (Harmondsworth: Penguin, 1968).

20   Roger Cooter, *The Cultural Meaning of Popular Science: Phrenology and the Organization of Consent in Nineteenth-Century Britain* (Cambridge University Press, 1984); Adrian Desmond, *The Politics of Evolution: Morphology, Medicine and Reform in Radical London* (The University of Chicago Press, 1989); Ian Inkster, *Scientific Culture and Urbanisation in Industrialising Britain* (Variorum Collected Studies Series Aldershot: Ashgate, 1998); Alison Winter, *Mesmerized: Powers of Mind in Victorian Britain* (The University of Chicago Press, 1998).

21   Morrell and Thackray, *Gentlemen of Science*; and for insightful studies of the political and institutional contexts of science in the early nineteenth century, see Jack Morrell, *Science, Culture and Politics in Britain, 1750–1870* (Variorum Collected Studies Series Aldershot: Ashgate, 1997).

22   See MacLeod, *The 'Creed of Science' in Victorian England*; Ruth Barton, 'An
      Influential Set of Chaps: The X-Club and Royal Society Politics, 1864–85', *British
      Journal for the History of Science*, 23 (1990), 53–81; Adrian Desmond, *Huxley: The
      Devil's Disciple* (London: Michael Joseph, 1994).
23   For this theme, see Yeo, *Defining Science*, chs 2 and 4; and also chapters IX, X and
      XI in this volume. For other relevant work, see Jonathan R. Topham, 'Beyond the
      "common context": the production and reading of the *Bridgewater Treatises*', *Isis*,
      vol. 89 (1998), 233–62 and the chapters in parts one and two of Bernard Lightman
      (ed), *Victorian Science in Context* (The University of Chicago Press, 1997).
24   For a book-length study, see Richard Yeo, *Encyclopaedic Visions:Scientific
      Dictionaries and Enlightenment Culture* (Cambridge University Press, 2001).
25   For a recent major study, see James A. Secord, *Victorian Sensation: the Extraordinary
      Publication, Reception, and Secret Authorship of Vestiges of the Natural History of
      Creation* (The University of Chicago Press, 2000).

PUBLISHER'S NOTE

The articles in this volume, as in all others in the Collected Studies Series, have not been given a new, continuous pagination. In order to avoid confusion, and to facilitate their use where these same studies have been referred to elsewhere, the original pagination has been maintained wherever possible.

Each article has been given a Roman numeral in order of appearance, as listed in the Contents. This number is repeated on each page and quoted in the index entries.

Scientific Method and the Image of Science
1831-1891

Introduction

The publication of John Herschel's *Preliminary Discourse on the Study of Natural Philosophy* coincided with the formation of the British Association for the Advancement of Science.[1] Both Herschel's *Discourse* and the charter of the British Association were concerned with the need to explain science to the public. In his book, published as a volume in Lardner's *Cabinet Cyclopaedia*, Herschel attempted to give a clear account of scientific method – the hallmark, in his view, of the scientific endeavour. This essay investigates the role of scientific method in the British Association's early attempts to advance the status of science.

Charles Kingsley once declared that the method outlined in Bacon's *Novum Organum* was the one which God had approved for Englishmen.[2] Lord Macaulay, in similar nationalist rhetoric, saw Bacon's precepts as the foundation of Britain's industrial triumphs.[3] Of course, this enthusiasm for Bacon was not universally shared: David Brewster, a Scot, and Justus von Liebig, a German visitor, were outspoken in their dissent from this Fortieth Article of the English faith![4] But the intensity of their disagreement indicates the importance these writers placed upon sound methods of inquiry. This interest in scientific method seems to have been increasing during the early nineteenth century. Lawrence Laudan has remarked that from this period 'entire books rather than prefaces or chapters were devoted exclusively to the subject'.[5] Contemporary writers made similar observations. William Whewell acknowledged that volumes had been written in Britain on general epistemology, but he saluted Herschel's *Discourse* as the first major work to make scientific method its prime concern.[6] In the following decades, the number of publications on this subject expanded enormously, in what Laudan has

I

described as a 'virtual revolution'[7] in the study of the history and philosophy of science. It is perhaps significant that such attention to the methods of science correlated with the increasing organization of the scientific community and the professionalization of its members.[8] The coincidence of Herschel's *Discourse* and the foundation of the British Association therefore symbolizes the importance of scientific method for the relationship between science and its public.

Moral and Social Dimensions of Scientific Method

The Victorians placed great faith in the possibility of intellectual certitude. Although they questioned many traditional values, their doubt 'never involved a denial of the mind as a valid instrument of truth'.[9] They believed in the existence of ultimate truths and the ability of the mind to find them. Accepting the assumptions of natural theology, many men of science viewed man's place in nature as primarily defined by his capacity to understand the world: man was the interpreter of nature.[10] For those who held this theological conception, the philosophy of knowledge carried moral connotations.

In this context, the subjects of method and epistemology were not confined to questions of logic: the proper use of reason was seen as a moral duty and theories about the foundation and limits of reason were related to theological issues concerning the appropriate relationships between man, God and nature. The nineteenth-century debates on the philosophy of knowledge, notably that between empiricism and idealism, were carried on against the background of these assumptions. As two recent writers have contended, this dispute was not simply a conflict in abstract philosophy, but a conflict of values.[11] The proponents of idealism, in defending the concept of efficient causality, were also upholding the dignity of free will and human responsibility; in refusing to derive all knowledge from experience, they were affirming a spiritual reality beyond the material world. But the empiricists also had ideological commitments. For many of them, the battle against metaphysical entities and forces was a holy war in which individual reason struggled against archaic authorities. J. S. Mill, for example, regarded the idealist school, with its claim that intuition could supply knowledge transcending experience, as 'the great intellectual support of false doctrines and bad institutions'.[12] That is, he was adamant that the difference between idealism and empiricism was not 'a mere matter of abstract speculation' but one which was 'full of practical consequences'.[13]

The early nineteenth-century discussion of scientific method,[14] in so far as it can be distinguished from these broad debates over the philosophy of knowledge, also involved moral and social dimensions. Interest in method was sustained by the conviction that truth could be readily attained if proper rules of inquiry were followed; this was reinforced by assumptions of natural theology about the duty to employ the faculty of reason. But concern with method was not restricted to theological discussion.[15] The dispute between

66

Whewell and Mill illustrates this point. Thus, while vehemently disagreeing on epistemology and metaphysics, both Whewell and Mill shared a strong conviction about the need for reliable methods of scientific investigation. And, more generally, they were concerned about the importance of clear thinking in all areas of intellectual discourse. Mill's *System of Logic* was written in the belief that 'no great improvements in the lot of mankind are possible, until a great change takes place in the fundamental constitution of their modes of thought'.[16] Similarly, Whewell spoke of the inductive method as a reforming philosophy and hoped to 'get *the people* into a right way of thinking about induction'.[17] With his friend Richard Jones, the political economist, Whewell had entertained the idea of establishing a periodical as the organ of an inductive school which would embrace the fields of moral philosophy, political economy and science.[18] Indeed, from the 1830s, there is evidence of a concern about rules and principles of inquiry which extended beyond the sphere of natural science. In 1829, for example, Samuel Bailey published his *Essays on the Pursuit of Truth*, defining the general topic as 'the conduct of men in the application of their means and faculties to the investigation of truth'.[19] With Harriet Martineau, who also wrote on this topic, Bailey saw disciplined thinking as a moral and social duty.[20]

These views were founded upon assumptions about the accessibility of scientific method. Indeed, this idea was a significant element in the provincial impetus behind the foundation of the British Association. More specifically, the Baconian conception of science as a collective enterprise provided space for part-time, amateur cultivators of science. Leading men of science saw an important role for such men in the new discipline of geology, where much basic observation and collection of data had to be done.[21] To some extent, this role existed because of the relatively clear distinction between empirical research and theoretical speculation in Baconian philosophy. This assumption supported an egalitarian outlook in which careful methods of observation, accessible to a large number of people, were viewed as a crucial feature of the scientific enterprise.[22]

The following sections of this essay attempt to reveal the ways in which a general concern with method related to the social image of science. It will be argued that an emphasis on method provided a means of resolving two areas of tension implicit in the charter of the British Association: first, the growing tension between the specialization of science and the Association's efforts to facilitate communication among scientists; and second, the tension between the role of the British Association in advancing the progress of science and its educational mission to the layman.[23] In the first case, a focus on method offered a point of consensus in the midst of increasing specialization and allowed the scientific community to promote a unified public image. In the second, attention to scientific *method*, in contrast with superficial knowledge of scientific *facts*, was recommended on moral and educational grounds. In both cases, an emphasis on method as the defining feature of the scientific endeavour suggested a means of attenuating problems inherent in the programme of the British Association. By the late nineteenth century, however, it was apparent that these tensions would remain.

I

Scientific Method and the Unity of Science

One of the major objectives of the British Association was to improve communication among scientists. In his opening address, William Vernon Harcourt argued that contact between scientists from different fields of research was essential to the progress of natural knowledge:

> Nothing, I think, could be a more disastrous event for the sciences, than that one of them should be in any manner dissociated from another; and nothing can conduce more to prevent that dissociation, than the bringing into mutual contact men who have exercised great and equal powers of mind upon different pursuits . . . [24]

By offering a national forum, Harcourt hoped that the British Association would prevent the various branches of science from being insulated from each other. If this separation could be avoided he was confident that science would advance, because the 'chief Interpreters of nature have always been those who grasped the widest field of inquiry . . . '[25]

Harcourt's remarks reflected a concern about the fragmentation of science. This concern was shared by several important men of science as they witnessed the expansion of knowledge and the specialization it entailed. During the first three decades of the century, the major subject boundaries within natural science began to crystallize and could no longer easily be conceived as components of a general natural philosophy. Instead, the idea of individual competency in one field was clearly emerging and, at the institutional level, this was manifested by the appearance of several single-science societies: the Geological, the Astronomical, the Zoological and the Geographical societies were all founded by 1830.[26] With these events in mind, Harcourt warned of the dangers: when a particular science receives special attention 'that science is detached from the central body; first one fragment falls off, and then another; colony after colony dissevers itself from the declining empire, and by degrees the commonwealth of science is dissolved'.[27]

This opinion was strongly supported by William Whewell. As Secretary of the Association at its third meeting at Cambridge in 1833, he repeated his earlier suggestion about the need for Annual Reports as a means of informing members of developments in all branches of science.[28] Whewell was clearly aware of the specialization of recent years and, when reviewing Mary Somerville's book *On the Connexion of the Physical Sciences* in 1834, he had the opportunity to return to this problem. Welcoming the aim of this work, he hoped that it might arrest the process of 'separation and dismemberment'[29] which was affecting the pursuit of science. There was a danger that intensive specialization would produce different mental habits which might vitiate communication within science. And continuing Harcourt's political metaphor, Whewell contemplated the prospect of science disintegrating, like 'a great empire falling to pieces'.[30]

There were several reasons for this anxiety about the unity of science. Firstly, many scientists of the early Victorian period were committed to a metaphysical belief in the unity of truth and thought that each advance in one discipline should be constantly related to other spheres of knowledge. They were opposed to any insulation of research which would weaken these intellectual connections. Furthermore, there was a strong assumption that the progress of science depended upon individuals who could draw upon several fields for their insights. As far as the physical sciences were concerned, Somerville thought that there 'exists such a bond of union, that proficiency cannot be attained in any one without a knowledge of the others'.[31] These convictions continued as philosophic ideals but they were becoming practically impossible even in the early 1830s. Harcourt was conscious of this problem and saw the British Association as providing a partial solution by giving a 'more systematic direction to scientific inquiry'.[32] Finally, the unity of science was crucial to the social image of the scientific community.

In seeking to acheive wider public support for science, the British Association wanted to present its members as a group of men united by a common dedication to the investigation of nature. A consensus of this kind was important, firstly because it maximized the possibility of public recognition and secondly because the critics of science could be best resisted if the British Association appeared to represent a single body of opinion. This point was fully appreciated by Dean William Cockburn, its most virulent opponent, who grudgingly admitted that this united image was successfully advertised. Writing in 1845, he remarked that 'the members of the British Association have always been accustomed to act in strict unison. They discountenance all difference of opinion . . .'[33] But such unity was not easily achieved and, despite the original intentions of the Association, the meetings of its various Sections — seven of them by 1835 — quickly undermined the ideal of a general scientific forum. The Presidential Addresses and Annual Reports continued to celebrate this ideal by stressing the connections between discoveries in various fields, but it was acknowledged that very few individuals could attain an overall perception of the most recent knowledge. In these circumstances, Prince Albert feared that the 'conciousness of its unity which must pervade the whole of Science'[34] would disappear. As Whewell had noted much earlier, the students of nature were losing contact with each other, and when they assembled at York in 1831, the English language contained no generic term which described them.[35] By coining the word 'scientist', he hoped to encourage a sense of common purpose amongst men of science which would enable them to be recognized as a definite group in society.

Nineteenth-century commentators such as Whewell regretted the passing of an age in which men could move freely between the disciplines of science, philosophy and theology. They realized that such breadth of competence was no longer possible, even within the natural sciences. In these circumstances, it was more difficult to find the grounds for a consensus which would retain the feeling of a common identity. And if the members of the scientific community were not impressed with this sense of unity, the chances of conveying such an image to the public would be greatly reduced. As noted previously, Somerville's *On the Connexion of the Physical Sciences*, a work

I

intended to explain science to the educated reader, was seen as an attempt to indicate the unity of science by illustrating the universality of natural laws. But there were doubts as to whether this approach offered sufficient protection against the divisive effects of specialization.

It was this problem which led several writers to look for a strong point of consensus in the *method* of science. In fact, looking back at Somerville's work, Clerk Maxwell suggested that 'the unity shadowed forth in . . . her book is . . . a unity of the method of science not a unity of the processes of nature'.[36] And as early as 1817 Faraday warned that diversity of intellectual interests demanded an agreement upon general rules of inquiry similar to those recommended by Bacon and, more recently, by Isaac Watts.[37] Underlining the significance of method, he remarked that 'in our knowledge *of* knowledge I will venture to say, that it is important to know, rather how to acquire it, than what it is'.[38] This distinction between the process of inquiry and the content of knowledge made it possible to emphasize method as the defining feature of the scientific community.

Such an emphasis upon scientific method was apparent in Herschel's *Preliminary Discourse on the Study of Natural Philosophy*, which aimed to convey some understanding of natural science to the readers of encyclopaedias of science, art and literature.[39] An explanation of the methods of scientific research formed a major part of the book, and it is clear that Herschel considered this treatment as essential to his task. In his view, scientific method was the inductive philosophy promulgated by Bacon 'as the alpha and omega of science, as the grand and only chain for the linking together of physical truths, and the eventual key to every discovery and every application'.[40] For Herschel, science did not really exist before this method was consciously employed and he believed that it was the major factor in the advancement of science.[41] In concluding his lengthy and detailed exposition of scientific procedures, Herschel gave the following summary of his purpose:

> In the foregoing pages we have endeavoured to explain the spirit of the methods to which, since the revival of philosophy, natural science has been indebted for the great and splendid advances it has made. What we have all along most earnestly desired to impress on the student is, that natural philosophy is essentially united in all its departments, through all which one spirit reigns and one method of enquiry applies.[42]

This was a precise formulation of the view that scientific method was the unifying feature of the scientific endeavour and the common bond relating the cultivators of its various branches. In the years following Herschel's *Discourse* this idea came to occupy a prominent place in the view of science which the British Association sought to promote.

The Presidential Addresses of the British Association, which sought to advertise the scientific endeavour before a wider audience, contained considerable references to method and philosophy. In a recent collection of these addresses, the editors stated that 'Victorian science was, in most areas, self-conciously concerned with its own methods. In few other eras before or since, were men of science so given to elaborate analyses of their own practice'.[43] Undoubtedly, there were good intellectual reasons for this interest, but it may also have been supported by social factors. The previous discussion suggests that one of these factors was the need to find a unified image of science at a time of increasing specialization. That is, an emphasis on method offered a point of consensus.

Joseph Agassiz has suggested another reason for the importance of method in this period: namely that it reflected an attempt to bolster confidence in the scientific enterprise at a time of theoretical crisis associated with the attacks upon Newton's optics from Young and Fresnel. Since Newton's work was regarded as the ultimate scientific achievement, a rejection of one of his important theories raised fears about the truth of his mechanics and cast doubt upon the hitherto unquestioned ability of physical science to produce results of lasting certainty. For Agassiz, then, Herschel's *Discourse* was a response to this crisis of faith which threatened to weaken the public prestige of science; it was at least partially a work of propaganda intended to 'reinforce faith in science, in induction, and in Newtonian mechanics'.[44] Herschel aimed to convince the public that men of science did possess a foolproof method, a 'philosophy of success'.[45] In short, method was presented as an element of continuity during theoretical disputes.

Appeals to method could also be used to maintain the conclusions of science against its critics. Geologists, for example, when defending their science against the charge that its theories were sheer conjecture when compared with the 'facts'of the Bible,[46] pointed to the careful procedures by which theories were established. In confronting this issue in his Presidential Address of 1839, Harcourt supported the claims of geology by stating that 'a vast body of the best-informed naturalists have examined, by all the various lights of science and by *undeniable methods of investigation*, the structure of the earth . . .'[47] This was very much an argument from authority, based on the contention that science possessed and used methods of the most reliable kind. And this claim, as suggested by the quoted example, was crucial to science in its struggle with the Church for intellectual authority in the area of natural knowledge.[48]

The focus on method could also serve to demarcate acceptable from unacceptable theories within the sphere of scientific activity. This is apparent in the debate stimulated by the appearance of the *Vestiges of the Natural History of Creation* in 1844. The reviews of this work clearly indicate an attempt by members of the scientific community to ostracize the anonymous author from its ranks, to separate reputable science from irresponsible amateurism. Apart from challenging the substantive doctrine of the book, a significant part of the criticism was directed at its method. Sedgwick and Brewster asserted that the extent of hypothetical speculation in the book was at odds with the requirements of inductive procedure.[49] Herschel and Huxley both referred to the confusion between law and cause in the argument of *Vestiges* and accused the author of philosophical naivety in scientific discourse.[50]

In these several ways then, an emphasis upon method may have acted as a rallying point for the scientific community in the early stages of its profes-soinalization.[51] And it is perhaps significant that when Herschel delivered his Presidential Address at Cambridge in 1845, he could return to the topic of method with some assurance. In reviewing the first fourteen years of the British Association's activities he was pleased to record the 'great deal of attention'[52] which had been recently paid to the method and philosophy of science because this topic was relevant not only to the advancement of science, but to the progress of society:

> If we are ever to hope that science will extend its range into the domain of social conduct, and model the course of human action on that thoughtful and effective adaptation of means to their end, . . . if such be the far hopes and long protracted aspirations of science, its philosophy and its logic assume a paramount importance . . .[53]

Thus, by 1845, in addition to conceiving scientific method as the defining feature of the scientific community, Herschel was recommending it as an approach to social problems.[54] Of course, this view was advanced as early as the seventeenth century, but it gathered strength during the nineteenth century as the scientific community became more organized and as its members sought more public and financial recognition. In these circumstances, it was argued that science possessed reliable procedures and principles which could be extrapolated successfully to other areas. In this sense, the emphasis on method was a major element in the cultural legitimation of science: it was the basis of a claim for intellectual authority beyond the realm of natural knowledge.[55]

The Advancement versus the Promotion of Science

From its inception in 1831, the British Association was keen to disseminate knowledge of science to a wide audience. Harcourt had this in mind when he announced that the Association would aim to 'obtain a greater degree of national attention to the objects of science and a removal of those disadvantages which impede its progress'.[56] David Brewster, one of the other key figures behind the foundation of the Association, believed that lack of public understanding of science was one of the reasons for the lack of patronage and, in his opinion, for the decline of science in Britain. One 'obvious remedy', he suggested, was 'to provide the educated classes with a series of works on popular and practical science . . .'[57] This might cultivate a taste for science which would then be reflected in magazines and journals.

The peripatetic character of the British Association undoubtedly increased the visibility of science and scientists throughout the country. After the relatively small audiences at Oxford and Cambridge the size of the gatherings attracted by the Association grew significantly when it met at the manufacturing towns of the north. Very large crowds were reported at Newcastle in 1838 and the size of such congregations could be swelled by the proclivity of Buckland and Sedgwick for outdoor geological lectures.[58] And Brewster reported that when public lectures were given at Edinburgh in 1834, 'the demand for ladies' tickets could not be met'.[59]

However, the apparent success of these meetings as popular attractions became a source of unease among the leaders of the British Association and a target for the ridicule of its opponents. A. D. Orange has shown how the critics were able to take full advantage of the theatrical aspects of the meetings, focusing upon the lavish dinners, the struggles for accommodation, the dramatic performance of certain lecturers, and the carnival atmosphere which pervaded the annual venue.[60] But, as Orange argues, there were also more serious objections concerning the advisability of bringing scientific debate so openly before uninitiated audiences. This anxiety partly derived from the alleged irreligious connotations of the geological theories favoured by prominent members, but it also had other sources. There was a revulsion at the combination of intellectual discussion and public assemblies. This recent phenomenon was compared unfavourably with the traditional practice of the 'solitary student', emerging from his study to converse only with other scholars.[61] In recommending the continuation of these established modes of intellectual activity, Cockburn condemned the British Association's procedures as an irresponsible departure:

> But to meet a passing crowd and in the confusion of a mixed and multitudinous assembly to proclaim, and to lecture and to dictate, is not a probable mode of elucidating truths or arriving at correct conclusions.[62]

Apart from lowering the quality of scientific discussion, Cockburn was deeply worried about the effects of these public parades of sophisticated theories upon uneducated minds. He doubted whether many listeners could distinguish between fact and hypothesis, and anticipating the next meeting at Birmingham, predicted that 'young men, who know not the difference between a circle and a square, are determined to become philosophers next September, as fast as a horse can carry them'.[63]

Members of the British Association also had reservations about the public display of scientific knowledge. They recognized that it could encourage an emphasis upon those topics which could be dramatically or amusingly illustrated and that it turned attention away from more difficult subjects. This tactic might stimulate interest in science but it could also create an impression of the subject as one of tricks, gadgets and simple experiments.[64] This was the central problem in the social relations of the British Association: in order to gain support and public recognition it had to reach a wide audience, but the popularization involved could produce an image which the scientists did not intend. Indeed, this problem highlighted a tension within the British Association's programme, a tension, or even a contradiction, between the *advancement* of science and its *promotion*.[65] While hoping that these two aims could be pursued simultaneously, some members were keen to ensure that priority was given to the progress of new scientific discoveries rather than to the dissemination of accepted knowledge.

The educational aims of the British Association can be seen in the context of debates about the diffusion of knowledge which occurred during the period of the First Reform Bill. With the establishment of Mechanics' Institutes in the 1820s, the dissemination of scientific knowledge became an important part of educational debate. It was a major subject in Henry Brougham's Society for the Diffusion of Useful Knowledge, founded in 1826.[66] And thus, when Harcourt addressed the first meeting of the British Association he was able to state that 'scientific knowledge has of late years been more largely infused into the education of every class of society'.[67] Indeed, the original aim of utilitarian and Whig reformers was to extend scientific education to the lower classes. It was argued that by attending Mechanics' Institutes artisans would better understand the technological processes of the manufacturing industries in which they worked. Equally important was the prospect of moral improvement and perhaps the possibility of social control which such education offered.[68] Now, this was not the audience which the British Association intended to reach, but the movement towards education of the working classes stimulated discussion about the content of middle and upper class education. A number of writers were worried by the prospect of workers knowing more about science than their social superiors, whose education was dominated by the classics. Some critics of popular education were convinced that such a situation would undermine social stability.[69] On the other hand, men of science such as Lyell and Daubeny viewed the extension of scientific knowledge to the lower classes as an argument for enlarging its place in university curricula.[70] This reasoning also reinforced the views of those in the British Association who believed that it should attempt to improve the scientific understanding of the middle-class audiences at its public meetings. It was hoped that this general diffusion of natural knowledge would remove 'any disadvantages of a public kind' which might impede the progress of science.[71]

There were, however, some early reservations about this programme. At the Edinburgh meeting of 1834 for example, James Forbes urged that the British Association should not be 'confounded with those numerous and flourishing institutions which have sprung up . . . for the simple diffusion of scientific truths'.[72] He went on to stress that 'such *diffusion* does not even properly speaking include any attempt at extension or accumulation: if in many cases it does promote extension, it is indirectly, and beyond a doubt has sometimes had the opposite tendency'.[73] Other members expressed misgivings about the participation of 'laymen', but Forbes' comment was probably the most direct; in effect, it seemed to sabotage any broad educational mission on the part of the organization. This may not have been his intention, but Forbes clearly wished to distinguish between scientific researchers and interested members of the public. In referring to the *Annual Reports,* he said that it should be noted that these reports differ entirely from the short systematic treatises on scientific subjects with which the press teems. They are not primarily intended for the general reader — they are not meant for the purpose of popularizing technical subjects; their main object is to classify existing discoveries as to lead the individual who is prepared to grapple with its difficulties . . . [74]

Forbes, like Whewell and Herschel, seems to have envisaged the British Association as primarily an organization for scientific investigations.[75] He did acknowledge its role towards the wider community but saw this as one of attracting recruits to the scientific fold rather than as one of raising general understanding of science.[76] By underlining this distinction between the advancement and the promotion of science Forbes predicted a major conflict.

The Popularization of Science

The educational objectives of the British Association were confronted by the problems associated with the popularization of knowledge. By the 1830s works of 'popular' science and literature were addressed to a less homogenous audience than they had been in the eighteenth century. The existence of these more varied educational levels made the task of writing popular books and articles a very difficult one. Some popular works were obviously too advanced for a large section of the reading population and, consequently, there was a demand for more simple publications.[77] W.T. Brande, who tried to answer this need by producing *A Dictionary of Science, Literature and Art* in 1842, explained that articles in the existing encyclopedias were too long and too complex. In his publication therefore, he hoped to offer simple, concise accounts. But the danger here was that simplification might go too far and that the result might be the dissemination of superficial, and possibly harmful, knowledge. Brande was aware of this and tried to meet the potential criticism:

> Neither must it be supposed that because these articles are for the most part brief and compendious, they are either flimsy or superficial. On the contrary, they have been compiled throughout with the greatest care. Popularity has not been sought at the expense of science . . . [78]

But it was precisely the case of popular scientific instruction which gave rise to anxieties about the public image.

Early nineteenth-century writers of popular science books found it difficult to satisfy their reviewers. Mary Somerville's works, for example, which were received enthusiastically by the scientific community, were not so well regarded by those concerned with popular education. John Nichol, who was both an academic scientist and a writer of popular textbooks, found

Somerville's treatise on *Celestial Mechanics,* in spite of its mathematical competence, to be totally unsuitable for a popular audience.[79] Similarly, Brewster, while praising her *On the Connexion of the Physical Sciences* as a feat of intelligent exposition, doubted if it would reach a wide audience because it was not sufficiently popular in form.[80] But on the other hand there was criticism of excessive simplification. In reviewing a new *Popular Cyclopaedia of Natural Science* in 1843, the *Athenaeum* was appalled by the profusion of similar publications. The reviewer said that Lardner's books were popular enough, and thought that anything more simple would be the intellectual equivalent of a poor-house diet.[81] Indeed, by the 1840s there were signs of pessimism about the state of popular science education. One writer claimed that the ideals of the Mechanics' Institutes had been too high: the working class did not want difficult science and the effect of exposing them to a 'sudden smattering' of it confirmed the truth that, though a great deal of knowledge steadies the head, a little overturns it.[82]

The supporters of working class education did acknowledge the limits of their programmes. In fact, in his first major pamphlet on this topic, Henry Brougham admitted that lack of leisure time would mean that most students 'must be content with never going beyond a certain point, and with reaching that point by the most expeditious route'.[83] But the response of the critics was that such knowledge was superficial and morally dangerous in that it encouraged false pride and gave the student a mistaken conception of his talents and achievements.[84] However, there was another dimension to the criticism of popular education which was most relevant to the British Association's attempt to foster public support of science.

From the early nineteenth century the correlation of science and utility had been the target of authors who condemned the moral and cultural effects of industrialization.[85] For Coleridge, Carlyle, Julius Hare and H.J. Rose the popularity of science was symptomatic of the material values of the new middle class. These writers were not opposed to natural philosophy as an intellectual endeavour, but they argued that the conception of science which had taken hold of the public mind was mechanistic and materialistic, interested only in the practical benefits which followed from the control of nature. To a significant extent, this Romantic critique was shared by important men of science.

While clearly illustrating the practical social advantages of science, Herschel was careful not to ally himself with an extreme utilitarian attitude. The dignity of science would be degraded, he warned, if it was only considered as a 'mere appendage to and caterer for our pampered appetites'.[86] Whewell was pleased to be able to endorse Herschel's opinion and affirmed that scientific activity was an aspect of man's speculative behaviour which should be intrinsically valued.[87] But this idea of science as a search for truth was largely overwhelmed by the utilitarian passions of the age. Herschel realized that it was the practical, material benefits of science rather than its capacity to enlighten which had thrust it to the centre of public attention.[88] There was therefore a discrepancy between the public image of science and the conception shared by leaders of the scientific community such as

Herschel, Whewell, Forbes, Babbage and Lyon Playfair.

By the 1840s a number of observers felt that the popularization of science was reinforcing this discrepancy by inculcating an empirical and utilitarian image of science which neglected its theoretical dimension. It was argued that popular education and literature emphasized factual information and its practical value to the virtual exclusion of scientific reasoning and principles. There can be little doubt that much knowledge of science was confined to this concrete level, because the aims of popular instruction were seldom higher. John Nichol explained that 'the investigation by which a certain Law of the celestial mechanism was discovered is one thing, but the nature of the Law itself is another, and it is the latter only that the popular teacher undertakes to explain and illustrate by analogy'.[89] This was probably a realistic approach, but some observers were anxious about the consequences of the concentration upon the results, rather than the processes of science.

James Forbes, for example, was alarmed when Macaulay attempted to deny any solid distinction between superficial and profound knowledge.[90] In Forbes' view, this distinction marked the essential difference between acquaintance with scientific conclusions, and a grasp of the methods by which they were achieved and the theoretical framework in which they were incorporated.[91] He therefore urged that the student of science, especially the university undergraduate, should 'look beyond the mere facts with which he becomes acquainted, to the laws and generalizations by which they are connected, and the appropriate ideas or theories by the assistance of which a body of science is constructed and united'.[92] Forbes wanted to stress these distinctions because he believed that popular education should be elementary without being superficial. Other commentators agreed: but they noted that the attractiveness of easily acquired information had taken precedence over the careful study of scientific reasoning, and profound theoretical speculation had been overshadowed by the demand for immediately useful knowledge.[93] Popularization, by catering to these tastes, had encouraged a perception of natural knowledge as a collection of interesting and useful facts, and this view militated against a proper appreciation of science.

In reviewing the state of science in 1847, the *Athenaeum* charged that:

Owing to the want of that cautious habit of examination which is required to elicit truth, we find our patent lists crowded with abortive schemes, and fortunes wrecked in speculations which in the hurry of the age have been stamped as genuine where the smallest effort of logical reasoning would have disclosed their deceptive character ...[94]

Making a similar point A. J. Joyce reprimanded those who 'risk their livelihood in search of inventions when they are not sufficiently educated to know the limits of the scientifically possible'.[95] The plethora of fantastic inventions illustrated the widespread influence of a crudely empirical and self-help approach to technology. Several writers expressed their concern about this situation and tried to temper the empirical account of technological innovation by stressing the importance of basic scientific knowledge. They claimed that elementary science was at work in all inventions, that the best inventions were not made by accident but by prepared minds. And when speaking of preparation they meant sound education in science rather than practical experience.[96] Indeed, given the existing bias towards the practical, it was suggested that there should be a return to first principles; failure to recover this balance would threaten the progress of science. The *Athenaeum* believed that abstract science was already in jeopardy: 'Original research, except in a few rare cases, has been entirely neglected; and in many cases experimental inquiry has been carried on under the pervading influences of the day'.[97] Thus, while science delighted in the discovery of new facts, there had been little reference to 'those great generalizations which advance our knowledge of Nature's ever wonderful workings — and raise man in the scale of intelligence'.[98]

Responses to the Dilemma of Popularization

Leading members of the British Association had similar doubts about the prospects of abstract science in Britain. The British Association had committed itself to the task of gaining public acceptance of science by bringing its researches and discoveries to the notice of a wider section of society. But although acquaintance with science increased in the following decades it did not necessarily lead to a better understanding of the nature and needs of the scientific enterprise. Instead, it could be argued that popularization had created misconceptions by stressing the practical results of science rather than the arduous investigation and theoretical speculation which often preceded major discoveries. Such misconceptions were damaging to the claims of abstract science because they encouraged a simplistic view of the connection between scientific findings and technological applications: practical innovations were regarded either as the product of empirical observation or as the direct and intended result of specific scientific inquiry. This view devalued pure, abstract research which did not appear to have any immediate practical bearing. In attempting to represent the case of abstract science and its financial requirements, Charles Babbage and, later, Lyon Playfair had to challenge these assumptions. Throughout his professional life, Playfair argued that industrial progress depended, in complex ways, upon abstract scientific research. And in order to do this, he tried to replace the popular image of science as useful, factual knowledge with one which stressed the analytical, theoretical and methodological aspects of the scientific approach. In short, he felt compelled to distinguish science from empiricism and scientific method from common sense.[99]

Prominent members of the British Association who sought to promote the educational value of science also stressed the methods of scientific thinking rather than scientific facts. The Duke of Argyll, for example, recommended this distinction as the basis of an improved science education.[100] Addressing the Aberdeen meeting of the Association in 1855, he contended that knowledge of physical laws, by itself, exerted no moral or intellectual influence: 'What we want in the teaching of the young, is, not so much the mere results, as the *methods*, and, above all, the *history* of science'.[101] He believed that a study of the intellectual processes which preceded scientific discovery should be recognized as an 'essential element in every liberal education'.[102] In making this claim he acknowledged that the results of such education depended on how science was taught and he rejected all approaches which merely amounted to a 'cramming of facts from manuals . . .'[103] When lecturing at the Royal Institution in the same year, Whewell had also affirmed the educational value of science. He hoped to promote the 'culture of the inductive habit of mind, or at least an appreciation of the method and its results';[104] and apart from detailed study of some branch of science, he recommended the history of the natural sciences as a means of introducing students to the process of inductive discovery.[105]

In summarizing his case for an extension of education in abstract science, the Duke of Argyll made two points: firstly, that it would contribute to the advancement of science itself, and secondly, that it 'would be an instrument of vital benefit in the culture, and strengthening of the mental powers'.[106] The first point may have satisfied Playfair and the small group of British scientists who warned about the practical dangers of neglecting pure science.[107] The second point had a larger audience, because this emphasis upon scientific method as a self-conscious and rigorous intellectual activity was a crucial element in the debate about the place of science in liberal education. Thus, it was argued that science, like the classics and mathematics, was capable of providing mental discipline, that it developed the rational and observational faculties of the mind. In spite of certain differences in perspective, men of science such as Whewell, Herschel, Sedgwick, Forbes, Babbage, Airy and their younger colleagues such as Playfair, Huxley, Tyndall and Pearson, all affirmed this general moral and educational role of science.[108] These men wanted to spread scientific knowledge throughout the community but they were anxious to ensure that the public did not form a superficial image of science. Addressing a university audience in 1871, Clerk Maxwell described their concern and their intentions:

> the popularization of scientific doctrines is producing as great an alteration in the state of society as the material applications of science are effecting in its outward life. Such indeed is the respect paid to science, that the most absurd opinions may become current, provided they are expressed in language, the sound of which recalls some well-known scientific phrase. If society is thus prepared to receive all kinds of scientific doctrine, it is our part to provide for the diffusion and cultivation, not only of true scientific principles, but of a spirit of sound criticism, founded on an examination of the evidences on which statements apparently scientific depend.[109]

Clerk Maxwell's statement underlined the importance of an emphasis on method for science education and the public image of science. This view rested upon two assumptions: first, that there was agreement on the definition of method within the scientific community, and second, that this method, in its detail, could be grasped without significant scientific experience. But there are grounds for suggesting that these assumptions were no longer valid by the 1870s. Firstly, as the Darwinian debates indicated, the subject of method was no longer a point of consensus but·a point of controversy.[110] Normative judgements about Darwin's methodology were a major element in the attack upon *The Origin of Species.* [111] Similarly, in the argument between Lord Kelvin and the Darwinians over the age of the earth, disagreement focused upon the definition of acceptable proof and evidence.[112]

Secondly, by the middle of the century there was less confidence in the assumption which had informed Herschel's *Discourse*: the notion that scientific method could be explained to any intelligent reader, even one lacking in scientific experience.[113] Several writers believed that this stress upon the accessibility of method created a superficial view of scientific thinking; they doubted whether anything but a misleading idea of method could be achieved in the absence of substantive scientific work. And although Huxley tried to define scientific method as 'trained and organized common sense',[114] others feared that this might reinforce an extreme inductive and empirical view of science. Lyon Playfair claimed that this attitude was a major impediment to proper appreciation of abstract science in England.[115] He referred approvingly to an earlier lecture on political economy in which Richard Whately argued that social science could not be left to common sense, but demanded scientific method.[116] In arguing against Macaulay in 1849, Forbes was anxious to make a similar point: superficial knowledge should not be confused with profound knowledge, common sense should not be identified with trained scientific judgement.[117] And in his *Outlines of Astronomy,* Herschel also felt compelled to make this distinction, warning that science required the student to unlearn many common sense mental habits.[118] Thus, even a work of popular exposition placed considerable demands upon the reader. These strictures severely qualified the assumption that scientific method could be appreciated by those who could not keep pace with scientific discoveries. And, as scientific knowledge expanded, becoming more specialized and more mathematical in character, the accessibility of its method became even more problematic. In these circumstances, it became difficult to separate the methods and expertise of science from the esoteric knowledge it produced – both were beyond the scope of the layman. These examples suggest that an emphasis on method would not resolve the problem of communicating an understanding of science to the public in the late nineteenth century.

Nevertheless, the emphasis on method and its social significance continued to form a major part of the claims made on behalf of science. Writing in 1892, in his *Grammar of Science*, Karl Pearson recorded most of these claims. He distinguished between scientific method and the conclusions or applications of science, arguing that the former offered the greatest educational benefit. Acquaintance with the methods of science would teach the student an 'exact and impartial analysis of facts' and their relations; it would encourage 'a high standard of reasoning, a clear insight into facts and their results . . .'[119] Pearson asserted that science education, by fostering this attitude, provided a most valuable contribution to society:

> I assert that the encouragement of scientific investigation and the spread of scientific knowledge by largely inculcating scientific habits of mind will lead to more efficient citizenship and so to increased social stability. Minds trained to scientific methods are less likely to be led by mere appeal to the passions or by blind emotional excitement to sanction acts which in the end may lead to social disaster.[120]

In short, Pearson contended that scientific method was essential for good citizenship in a democracy.[121]

This was perhaps the highest claim which scientists made on behalf of their activities and it provides a fine example of the importance of method in the cultural legitimation of science. In making this claim, Pearson logically extended the assumptions of earlier scientists and anticipated the views of those who succeeded him. Earlier men of science such as Whewell, Forbes, Herschel and the Duke of Argyll may not have alluded so plainly to democracy, but they strongly recommended the mental discipline of science as part of general education. In the next generation, writers such as Huxley, Tyndall, Galton and Clifford were more strident in their assertions. For them, science clearly offered the surest means to knowledge and the best guide to conduct in politics and ethics.[122] By the 1930s, Pearson's views on the relevance of science to citizenship were being vigorously expressed in the principal weekly of the scientific establishment. In December 1940, *Nature* carried an article on the 'Cultural Significance of Scientific Method' which was specifically related to the contemporary political events:

> The development of a scientific outlook is of fundamental importance for the preservation of democracy and the transformation of educational content and method to encourage that development and not the mere inculcation of scientific facts is one of the most pressing problems in education today. The fate of democracy is bound up with the spread of the scientific attitude. More important even than the contribution of science to the solution of immediate and practical problems, the scientific attitude is the sole guarantee against wholesale misleading by propaganda. It is the only assurance of the possibility of a public opinion intelligent enough to meet present social conditions.[123]

Conclusion

This essay has examined two intellectual problems associated with the aims

I

and objectives of the British Association. Firstly, it has argued that the attempts of the Association to facilitate communication among scientists were complicated by the increasing specialization of science. A concern about the unity of science had been a significant factor in the founding of the British Association, but the meetings of its specialist sections soon began to assume more academic importance than its general forums. Secondly, it has argued that there was a tension between the general educational aims of the Association and its commitment to the advancement of scientific research. It was assumed that the dissemination of scientific knowledge would increase public support for science; however, many observers feared that popularization was creating a superficial image of science which prevented an adequate understanding of scientific research, its intellectual demands and its financial requirements.

Between 1830 and 1870 there was an attempt to resolve these tensions by stressing the importance of scientific method. This emphasis on method as the defining feature of the scientific enterprise made it possible to argue that there was unity in spite of specialization and continuity in spite of theoretical disputes. Such a focus upon the methods rather than the factual results of science allowed a case to be made for the intellectual value of science and its importance for liberal education. More generally, method was emphasized in claims which scientists made about the political significance of their activities, and about the relevance of scientific procedures to all areas of knowledge. As such, the idea of method had a social and rhetorical role in the public relations of science.

The attempts to resolve these tensions were not successful. With increasing specialization, the unity of science was undermined and the gap between scientific advance and public understanding was widened. The idea of a single method as the unifying feature of science could not resolve these problems. Under these circumstances the British Association could not convincingly promote a public appreciation of science, and its difficulty in coming to terms with the tensions within its original programme made the Association a philosophically ambiguous apologist for science during the twentieth century.

Acknowledgements

I am grateful to Roy MacLeod and Peter Collins for their helpful comments on an earlier draft of this essay.

Notes

1. There is some dispute about the publication date of Herschel's book. The title page to the English edition says 1830, but the page with the portrait says 1831. See J.

Agassiz, 'Sir John Herschel's Philosophy of Success', *Historical Studies in the Physical Sciences, 1* (1971), 1, for a claim that 1831 is correct date. For contextual comment on the *Discourse*, see W. F. Cannon, 'John Herschel and the Idea of Science', *Journal of the History of Ideas, 22* (1961), 215-39.

2. C. Kingsley, 'How to Study Natural History', in *Works*, 28 vols. (London: Macmillan,1880-85), Vol. 19, 308.

3. T. Macaulay, 'Lord Bacon', *Edinburgh Review, 65* (1837). 1-104.

4. D. Brewster, *Memoirs of the Life, Writings and Discoveries of Sir Isaac Newton*, 2 vols. (Edinburgh: T. Constable, 1855), Vol. 2, 400-5; J. von Liebig, 'Bacon and Natural Philosophy', *Macmillan's Magazine, 8* (1863), 237-49, 257-67.

5. L. Laudan, 'Theories of Scientific Method from Plato to Mach: a Bibliographical Review', *History of Science, 7* (1968), 29.

6. W. Whewell, 'Modern Science – Inductive Philosophy', *Quartlerly Review, 45* (1831), 377; also 375-6 for his curiosity about the lack of works in English, similar to those by Continental scientists such as Cuvier and Berzelius, which explained science to a wider audience.

7. Laudan, *op. cit.* note 5, 29.

8. See E. Mendelsohn, 'Revolution and Reduction', in Y. Elkana (ed.), *The Interaction between Science and Philosophy* (New Jersey: Atlantic Highlands, 1974), 408.

9. W. E. Houghton, *The Victorian Frame of Mind, 1830-1870* (London and New Haven: Yale University Press, 1957), 13.

10. On this theme, see R. Yeo, 'William Whewell, Natural Theology and the Philosophy of Science in Mid-nineteenth-century Britain', *Annals of Science, 36* (1979), 493-516, esp. 495-8.

11. A. Ellegård, 'The Darwinian Theory and Nineteenth Century Philosophies of Science', *Journal of the History of Ideas, 18* (1957), 391-3; R. Smith, 'The Background of Physiological Psychology in Natural Philosophy', *History of Science, 11* (1973), 97-105 and his 'Essay Review of D. L. Hull, *Darwin and his critics . . .*', *British Journal for the History of Science, 7* (1974), 282-5.

12. J. S. Mill, *Autobiography*, ed. J. Sillinger (London and Oxford: Oxford University Press, 1971), 134.

13. *Ibid.* 162.

14. I have deliberately avoided an attempt to offer a precise definition of this term. In the nineteenth-century debates, the concept of method had broader connotations than it has in contemporary philosophy of science, and I am concerned with the ways these connotations were linked with concerns about the social image of science. On the problem of defining method in the nineteenth-century context, see the discussion following Mendelsohn, *op. cit.* note 8, 427-31.

15. See M. Faraday, 'Observations on the Inertia of Mind', in B. Jones, *The Life and Letters of Michael Faraday*, 2 vols. (London, 1870), Vol. 1, 261-79, for an emphasis on moral, as distinct from religious duty.

16. Cf. Mill, *op. cit.* note 12, 142.

17. I. Todhunter, *William Whewell, D.D. An Account of his Writings*, 2 vols. (London: Macmillan, 1876), Vol. 2, 115. Whewell to R. Jones, February 1831.

18. *Ibid.* 118, Whewell to Jones, 24 April 1831.

19. [S. Bailey], *Essays on the Pursuit of Truth* (London, 1829), iii.

20. *Ibid.* 27-34, 38-9, 67-9; H. Martineau, 'Essays on the Art of Thinking', in *Miscellanies*, 2 vols. (Boston, 1836), Vol. 1, 72, 100, 103.

21. See S. F. Cannon, *Science in Culture: the Early Victorian Period* (New York: Dawson and Science History Publications, 1978), 187-8, for the suggestion that these amateurs wanted the 'professional' scientists to direct them.

22. Cf. Martineau, *op. cit.* note 20, Vol. 1, 72, 100, 103.

23. For another statement of the second of these tensions, see O. J. R. Howarth, *The British Association for the Advancement of Science: a Retrospect 1831-1931*, (London: B.A.A.S. 1922, 2nd edn. 1931), 42-3.

24. W. V. Harcourt, *B.A.R.* York, 1831, 28.

25. *Ibid.*

I

26. See M. J. S. Rudwick, 'The Foundation of the Geological Society of London: its Scheme for Co-operative Research and its Struggle for Independence', *British Journal for the History of Science, 1* (1963), 325-55.

27. Harcourt, *op. cit.* note 24, 28.

28. Whewell, *B.A.R.* Cambridge, 1833, xi-xxvi; Harcourt, *op. cit.* note 24, 36.

29. Whewell, 'On the Connexion of Physical Sciences', *Quarterly Review, 51* (1834), 59.

30. *Ibid.* See [G. Moll], *On the Alleged Decline of Science in England by a Foreigner* (London, 1831), 14, for the view that French science was far more specialized than English science. This point was made in the context of his charge that Babbage's claims about the poor state of English science were rash. For a discussion of the problem of specialization, see A. J. Meadows, *Communication in Science* (London: Butterworth, 1974), 72-9, 210-16, 219-23.

31. M. Somerville, *On the Connexion of the Physical Sciences* (London: John Murray 3rd edn., 1836), Preface.

32. Harcourt, *op. cit.* note 24, 32.

33. W. Cockburn, *The Bible Defended against the British Association* (London: Whittaker, 5th edn., 1845), 34. See D. Brewster, 'British Association for the Advancement of Science', *North British Review, 14* (1850) 278, on the desire of himself and other members to maintain unity.

34. Prince Albert, *B.A.R.* Aberdeen, 1859, lxiii.

35. Whewell, *op. cit.* note 29, 59.

36. J. Clerk Maxwell, 'Grove's Correlation of Physical Forces', *Scientific Papers,* ed. W. D. Niven, 2 vols. (Cambridge, 1890), Vol. 2, 402. Somerville did hint at this, cf. Somerville, *op. cit.* note 31, 412. See also E. Patterson, 'Mary Somerville, FRS', *British Journal for the History of Science, 4* (1969), 311-39 (esp. 331).

37. M. Faraday, *Some Observations on the Means of Obtaining Knowledge* (London: 1817), 8-9. This was a lecture delivered before the City Philosophical Society on 19 February 1817. See L. Pearce Williams, *Michael Faraday* (London: Chapman and Hall, 1965), 12-13, for the influence of Watt's *Improvement of the Mind* on the young Faraday.

38. Faraday, *op. cit.* note 37, 8.

39. On this point, see A. Hughes, 'Science in English Encyclopedias, 1704-1875' *Annals of Science, 7* (1951), 340-70.

40. J. F. W. Herschel, *A Preliminary Discourse on the Study of Natural Philosophy* (London, 1831), 114.

41. *Ibid.* 104-5.

42. *Ibid.* 219.

43. G. Basalla, W. Coleman and R. Kargon (eds.), *Victorian Science: a Self-Portrait from the Presidential Addresses to the B.A.A.S.* (New York: Anchor, 1970), 399. See Prince Albert, *B.A.R.* Aberdeen, 1859, lxii, for reference to the self-consciousness of method as a distinctive feature of science.

44. Agassiz, *op. cit.* note 1, 22.

45. *Ibid.* 2; 32-6.

46. 'Scriptual geologists' attacked geologists such as Buckland and Sedgwick for deserting sound inductive method. See, for example, J. Kidd, *A Geological Essay on the Imperfect Evidences in Support of a Theory of the Earth* (Oxford, 1815), 14, 267-9; G. Penn, *A Comparative Estimate of the Mineral and Mosaic Cosmologies* (London, 1822), 10-11, 26-30, 36-45; W. Cockburn, *A Letter to Professor Buckland concerning the Origin of the World* (London: Hatchard, 1838), 5.

47. Harcourt, *B.A.R.* Birmingham, 1839, 18. (My emphasis).

48. The most obvious battles over this issue occurred later in the century. See W.H. Brock and R. M. MacLeod, 'The Scientists' Declaration: Reflexions on Science and Belief in the Wake of *Essays and Reviews* 1864-5', *British Journal for the History of Science, 9* (1976), 39-76; and F. M. Turner, 'The Victorian Conflict between Science and Religion: A Professional Dimension', *Isis, 69* (1978), 356-76.

49. Sedgwick, 'Vestiges of the Natural History of Creation', *Edinburgh Review, 82*

(1845), 2, 25, 85; J. Clark and T. Hughes (eds.), *Life and Letters of the Reverend Adam Sedgwick*, 2 vols., (Cambridge: Cambridge University Press, 1890), Vol. 2, 85, Sedgwick to Charles Lyell, 9 April 1845; D. Brewster, 'Vestiges of the Natural History of Creation', *North British Review, 3* (1845), 503.

50. Herschel, *B.A.R.* Cambridge, 1845, xxxix, xlii-xliii; T. H. Huxley, 'The Vestiges of Creation', *British and Foreign Medical Review, 13* (1854), 425-39. Cf. also, M. Ruse, *The Darwinian Revolution* (Chicago and London: University of Chicago Press, 1979), 110-11.

51. In using the term 'professionalization' in relation to nineteenth-century British science, I do recognize its problematic character. For recent work on this topic, see J.B. Morrell, 'Science in Culture: Review', *History of Science, 18* (1980), 39-45.

52. Herschel, *op. cit.* note 50, xl.

53. *Ibid.*

54. Herschel had also referred to this possibility in his *Discourse*. Cf. Herschel, *op. cit.* note 40, 72-4.

55. For a recent perspective on this issue, see M. Berman, *Social Change and Scientific Organisation: The Royal Institution 1799-1844* (New York: Cornell University Press, 1978) and J. Habermas, *Towards a Rational Society* (London: Heinemann, 1971) chs. 5 and 6.

56. Harcourt, *op. cit.* note 24, 22.

57. Brewster, 'Herschel's Treatise on Sound', *Quarterly Review, 44* (1831), 476-7. For the early nineteenth-century popular interest in science, see G. Foote, 'Sir Humphry Davy and his Audience at the Royal Institution', *Isis, 43* (1952), 6-12.

58. A. D. Orange, 'Idols of the Theatre: The British Association and its Early Critics', *Annals of Science, 32* (1975), 283. See Brewster, 'The British Association for the Advancement of Science', *North British Review, 14* (1850), 268, for estimates of attendance.

59. Brewster 'The British Scientific Association', *Edinburgh Review, 60* (1834-35), 376.

60. Orange, *op. cit.* note 58, 280-1, 294. Cf. Howarth *op. cit.* note 23, 91-2, 242, for reference to the Red Lions Club.

61. J. H. Bowden, 'The British Association for the Advancement of Science', *British Critic, 25*, (1839), 14.

62. Cockburn, *A Remonstrance, Addressed to His Grace the Duke of Northumberland, upon the Dangers of Peripatetic Philosophy* (London, 1838), 5-6.

63. *Ibid.* 24.

64. Cf. Orange *op. cit.* note 58, 281-3; *The Times*, 28 June 1832, 4; Brewster, *op. cit.* note 59, 389, 393.

65. For the attitude of American scientists to this problem, see S. G. Kohlstedt, *The Formation of the American Scientific Community: The American Association for the Advancement of Science, 1848-1860* (Chicago: University of Illinois Press, 1976), ch. 1.

66. See J. Hays, 'Science and Brougham's Society', *Annals of Science, 20* (1964), 227-41.

67. Harcourt, *op. cit.* note 24, 30. See also J. H. Plumb, *England in the Eighteenth Century* (Harmondsworth: Penguin, 1964), 170.

68. See S. Shapin and B. Barnes, 'Science, Nature and Control: Interpreting Mechanics' Institutes', *Social Studies of Science, 7* (1977), 31-74; I. Inkster, 'The Social Context of an Educational Movement: A Revisionist Approach to the English Mechanics' Institutes, 1820-1850', *Oxford Review of Education. 2* (1976), 277-307.

69. 'Diffusion of knowledge', *Edinburgh Review, 45* (1827), 190-1. This was a review of a work by a 'Country Gentleman' who attacked Brougham's programme for working class education.

70. C. Lyell, 'State of the Universities', *Quarterly Review, 36* (1827), 223; C. Daubeny, 'On the Importance of the Study of Chemistry as a Means of Education', in E. L. Youmans (ed.), *Modern Culture: its True Aims and Requirements* (London, 1867), 57.

71. Cf. note 24, ix.

I

72. J. D. Forbes, *B.A.R.* Edinburgh, 1834, xii.

73. *Ibid.*

74. *Ibid.* xv.

75. For Herschel's attitude to the British Association, see S. F. Cannon, *op. cit.* note 21, 192-6.

76. Forbes, *op. cit.* note 72, xii-xiii. For the view that the best scientists should not be wasted on the task of popularization, see T. Galloway, 'Sir John Herschel's Astronomy', *Edinburgh Review, 58* (1833-34), 165. But the dissemination of scientific knowledge was usually seen as crucial to the aims of the British Association. In his Presidential Address of 1837 Professor Traill asked: 'who can believe that familiarizing large masses of the community with such investigations, and exhibiting how the highest branches of philosophy may be made available to the purposes of life, will fail to promote the vowed purpose of our meetings?'; *B.A.R.* Liverpool, 1837, xxvii.

77. The Mechanics' Institutes were frustrated by the poor reading skills of those who attended. In 1855, the Rev. Richard Dawes, a supporter of popular education, said that 'until recently, our Mechanics' Institutes throughout the country, with the exception of those in some of our large towns, have been entire failures . . .'. The reason, he concluded, was poor elementary education. Cf. R. Dawes, 'Address to the Huddersfield Mechanics' Institute', in *British Eloquence of the Nineteenth Century*, 3rd series (London, 1856), 261-2.

78. W. T. Brande (ed.), *A Dictionary of Science, Literature and Art* (London: Longman, 2nd edn., 1852), vi.

79. J. P. Nichol, Address delivered to the Stirling School of Arts, 10 January 1849, in *The Importance of Literature to Men of Business: A Series of Addresses delivered at Various Popular Institutions* (London, 1852), 217-18, 220. On Nichol, cf. Cannon, *op. cit.* note 1, 232-3.

80. Brewster, 'Mrs. Somerville on the Physical Sciences', *Edinburgh Review, 59* (1834), 156. Cf. also Patterson, *op. cit.* note 36, 329-31, for a description of Somerville as an expositor rather than a popularizer of science.

81. 'Review of Popular Cyclopaedia of Natural Science', *Athenaeum*, 1 July 1843, 604.

82. F. Whitwell, 'Popular Science', *Quarterly Review, 84* (1848-49), 322. In any case, by the end of the 1830s it was apparent that the Mechanics' Institutes were not attracting large numbers of workers, but rather the lower middle classes. See J. F. C Harrison, *Learning and Living 1790-1960; a Study in the History of the English Adult Education Movement* (London: Routledge and Kegan Paul, 1961), 47-74.

83. H. Brougham, *Practical Observations upon the Education of the People, addressed to the Working Classes and their Employers* (London, 1825), 9.

84. The strength of this attack is indicated by the extent to which supporters of popular education felt obliged to rebutt it. See the addresses in *British Eloquence*, 1st series (London, 2nd edn., 1855).

85. T. Carlyle, 'Signs of the times', *Edinburgh Review, 49* (1829); H.J. Rose, *The Tendency of Prevalent Opinion about Knowledge Considered* (Cambridge, 1826); J. and A. Hare, *Guesses at Truth* (London, 1838); 'The State of Society in England', *Athenaeum*, 28 March 1828, 487. For a more recent analysis, see G. A. Foote, 'Mechanism, Materialism and Science in England, 1800-1850, *Annals of Science, 8* (1952), 152-61.

86. Herschel, *op. cit.* note 40, 10; cf. 44-74 for his references to social benefits.

87. Whewell, *op. cit.* note 6, 404-5; and *History of the Inductive Sciences*, 3 vols. (London: J. W. Parker, 3rd edn., 1857), Vol. 1, 13; also R. Hunt, *The Poetry of Science* (London, 1848).

88. Herschel, 'Whewell on the Inductive Sciences', *Quarterly Review, 68* (1841), 177-8, 185. See also G. Foote, 'Science and its Functions in Early Nineteenth Century England', *Osiris, 11* (1954), 438-54 (esp. 445-9).

89. Nichol, *op. cit.* note 79, 220-1.

90. J. D. Forbes, *The Danger of Superficial Knowledge* (London, 1849), 29, 33, 45, 50. Forbes was responding to a speech delivered by Macaulay at the Edinburgh Phil-

osophical Institution on 4 November 1841. I intend to discuss this debate in a future article.

91. *Ibid.* 88.

92. *Ibid.* vii.

93. 'Science in 1847', *Athenaeum*, 12 January 1848, 60. Cf. Orange, *op. cit.* note 58, 294, for a similar remark by John Phillips in a letter to Harcourt.

94. 'Science in 1847', *op. cit.* note 93, 60.

95. A. J. Joyce, 'The Progress of Mechanical Inventions', *Edinburgh Review, 89* (1849), 47.

96. *Ibid.* 51-7, 82; Whitwell, *op. cit.* note 82, 340-4.

97. 'Science in 1847', *op. cit.* note 93, 60.

98. *Ibid.*

99. L. Playfair, 'The Study of Abstract Science Essential to the Progress of Industry', in *British Eloquence of the Nineteenth Century*, 2nd series (London, 1865), 46-86.

100. J. D. Campbell (Duke of Argyll), *B.A.R.* Glasgow, 1855, lxxxi. For his support of abstract science and its practical relevance, cf. lxxx, where he qualifies Whewell's views about practice always preceding theory.

101. *Ibid.* lxxxiii.

102. *Ibid.* lxxxii.

103. *Ibid.* lxxxiv.

104. Whewell, 'On the Influence of the History of Science upon Intellectual Education', in Youmans (ed.), *op. cit.* note 70, 246.

105. *Ibid.* 247-8.

106. Argyll, *op. cit.* note 100, lxxxii. Cf. also C. Daubeny, *B.A.R.* Bristol, 1836, xxxv-xxxvi.

107. See C. Babbage, *On the Economy of Machinery and Manufactures*, (London: Charles Knight, 1832) and *The Exposition of 1851* (London: John Murray, 1851); W. Fairbairn, *B.A.R.* Manchester, 1861, li; Playfair, *B.A.R.* Aberdeen, 1885, 7-16; O. MacDonagh, 'Government, Industry and Science in Nineteenth-century Britain: A Particular Study', *Historical Studies, 16* (1975), 503-18.

108. Cf. Whewell, *op. cit.* note 104, 228-9, 246-9; also the papers by Herschel, Faraday, Daubeny, Prince Albert, Huxley, Spencer and Tyndall in Youmans, *op. cit.* note 70. But the supporters of abstract science were not always in full agreement. There was an important difference between those who campaigned for government assistance (e.g. Playfair, Babbage, Huxley) and those who saw pure science as an academic discipline which should not be associated with technological applications or Government research (Whewell, Forbes, Airy, Clerk Maxwell). For this latter attitude cf. Todhunter, *op. cit.* note 17, Vol. 2, 204-5, Whewell to Forbes, 14 February 1835; also S. F. Cannon, *op. cit.* note 21, 183-9.

109. J. Clerk Maxwell, 'Introductory lecture on experimental physics', in Clerk Maxwell, *op. cit.* note 36, 242. The Duke of Argyll also noted the broad influence of science and warned that if 'her discoveries, and above all, her methods and her history be but partially and superficially understood, the popular mind will be a perpetual prey to the most specious form of error'. Cf. note 100, lxxxvi.

110. I am not suggesting that there was complete agreement on methodological details in the early part of the century; but the absence of any major dispute allowed for the notion of method as the unifying feature of science. Note however, that in 1845, Herschel was anxious to stress that the epistemological dispute between Whewell and Mill did not disturb the general agreement on inductive method. Cf. Herschel, *op. cit.* note 50, xl.

111. For example, it was alleged that Darwin's method, as well as his theory, was morally suspect: by entertaining conjecture and hypothesis he had deserted the safe path of induction and the virtuous qualities it ensured. Cf. Smith, *op. cit.* note 11, 'Essay Review . . .', 281-5; A. Ellegård, *Darwin and the General Reader* (Goteborg Universitets Årskrift, 1958), ch. 9.

112. Cf. Cannon, *op. cit.* note 21, 234-5, 279-80.

113. But even in 1831, Herschel realized that science might become divorced from

I

the public and he stressed the need for simple language to prevail over obscure jargon. Cf. Herschel, *op. cit.* note 40, 70.

114. T.H. Huxley, 'On the Educational Value of the Natural History Sciences', (1854), *Science and Education* (London: Macmillan, 1893), 45.

115. Playfair, *op. cit.* note 99, 58-60 and *op. cit.* note 107, 25-8; C. Daubeny, *B.A.R.* Bristol, 1836, xxxiv-xxxv.

116. Playfair, *op. cit.* note 99, 83; cf. R. Whately, *Introductory Lectures on Political Economy* (London, 3rd edn., 1847), 55-64, 208, 215-17.

117. Cf. Forbes, *op. cit.* note 90.

118. Herschel, *Outlines of Astronomy* (London: Longman, 1849), 1-2; also iv, 5, 8-9.

119. K. Pearson, *The Grammar of Science* (London: A. and C. Black, 3rd edn., 1911), 9. For the idea that 'The unity of all science consists alone in its method, not in its material', cf. 12.

120. *Ibid.* 9.

121. In another passage he contended that 'It is the want of impersonal judgement of scientific method, and of accurate insight into facts, a want largely due to a non-scientific training, which renders clear thinking so rare, and random and irresponsible judgements so common, in the mass of our citizens today', *Ibid.* 25.

122. Cf. T. H. Huxley, 'On the Adviseableness of Improving Natural Knowledge', (1866), in *Method and Results* (London: Macmillan, 1898), 31-2, 38-41.

123. *Nature, 146* (28 December 1940), 817. For a similar support of scientific method, as distinct from scientific facts, cf. G. Orwell, 'What is Science?', in *Collected Essays, Journalism and Letters*, 4 vols. (London: Secker and Warburg 1968), Vol. 4, 27-30.

II

AN IDOL OF THE MARKET-PLACE: BACONIANISM IN NINETEENTH CENTURY BRITAIN

INTRODUCTION

During the seventeenth and eighteenth centuries the name of Francis Bacon was closely associated with the idea of experimental philosophy or natural science. For the leading members of the early Royal Society, he was the guiding moral and intellectual spirit of the scientific endeavour, or, as Abraham Cowley's Ode would have it, a prophetic Moses who led natural philosophers towards the promised land.[1] Similarly, within a European context, one writer commented that Bacon was "the greatest man for the interest of Natural Philosophy that ever was".[2] This association is now almost totally disrupted; practising scientists are less given to public pronouncements about methods and goals than their predecessors and, since the early years of this century, professional philosophers of science have treated his work with nearly complete indifference.[3]

The critical period for this shift in the fortunes of Bacon and Baconianism was the nineteenth century. With the expansion of scientific institutions, the beginning of large scale co-ordination of research, and the application of scientific theory to technology, some of Bacon's most novel and visionary speculations were, in some senses, tangibly realized.[4] On the other hand, in the four decades from 1820, his writings about the proper method of scientific inquiry became, for the first time, the object of serious scholarship. In 1881 it was still possible for one writer to underline Bacon's significance by recording his imprint on language: phrases such as "crucial instance" and "solitary instance" had "become household words in our language, and especially in the vocabulary of scientific men".[5] But this was said against a background of criticism. At first moderate in tone, this revisionism culminated in the 1850s with the aggressive and often virulent denunciations by David Brewster, the Scottish physicist and Justus von Liebig, the German chemist – two 'foreigners' who were not prepared to tolerate the English idolatry which surrounded Bacon's name. Before these indictments came the various assessments of John Playfair, Dugald Stewart, Macvey Napier, John Herschel, William Whewell, T. B. Macaulay and J. S. Mill. As a result of this attention, the earlier image of Bacon as the father of experimental science and the legislator of its methodology was largely abandoned. Although still acknowledged as a significant spokesman for the scientific enterprise, his understanding of scientific method came to be seen as seriously flawed and dangerously misleading.

There was, however, an asymmetry between the declining reputation of Bacon's philosophy of science and the continuing appeal to certain aspects of his thought in a variety of public forums. During the first half of the century in which Bacon's methodology was being critically evaluated by philosophers of science, his precepts on method formed a significant part of the apologetics by which science was promoted and of the language in which scientific debates were conducted. When Charles Kingsley discussed the study of natural history – which for him, included astronomy because it was an observational science – he announced that the Baconian method was the one which God had approved for Englishmen.[6] Most references to Bacon were not so dramatically chauvinistic, but positive statements about his methodology and the merits of observation over hypothesis and theory were nevertheless common elements in the presentation of science, especially in the new organic disciplines, and a feature of the rhetoric by which the aims of these subjects were explained and defended. Even Charles Darwin, who was certainly not opposed to theorizing, recalled in his *Autobiography* that in his early research he had "worked on true Baconian principles, and without any theory collected facts on a whole-sale scale".[7] Statements such as this have been cited as evidence of a prevailing methodological orthodoxy. Charles Gillispie saw it as "a commentary on the excessively Baconian cast of Victorian notions of science",[8] and referred to the "widely additive inductive approach [which] still enjoyed an almost exclusive methodological orthodoxy. Charles Gillispie saw it as "a commentary on the excessively Baconian cast of Victorian notions of science",[8] and referred to the explained the attack on its speculations in terms of the "Baconian ... particularizing spirit of most early nineteenth-century science".[10]

These descriptions of nineteenth century science and scientists as 'Baconian' (or even 'inductivist'), immediately beg difficult, if not intractable, questions of both definition and evidence. These terms have not had a neutral and stable meaning; rather, they have been objects of controversy and multiple interpretation to an extent which renders them practically useless as simple descriptive epithets. One historian has recently advocated a 'no-nonsense' position on the use of Baconianism as an historical category. Susan Cannon suggests that the concept is a fiction of twentieth century historians which they incorrectly impose on nineteenth century science.[11] Clearly, the dangers of anachronism in this area are profourd, and Cannon points to some of the problems involved in the difficult task of assessing the status and role of statements about methodology. However, her recourse to nominalism is arguably too extreme, in that it precludes interesting historical questions. In other words, her various caveats about the confusions surrounding this term, such as the possible gap between nineteenth century interpretations of Bacon and "what modern historians call Baconianism",[12] or the suggestion that the most vociferous 'Baconians' were clergymen, not scientists, can be taken as invitations to study the place of Baconianism in the cultural relations of science.[13]

This essay focuses on certain discourses and debates about Bacon's thoughts from the late eighteenth to the late nineteenth century. In doing so, it attempts to reveal the shifting emphases in attitudes to his work, the various levels at which it was discussed, and the different ways in which it was interpreted and deployed. The concern here is not with the question of whether Bacon's precepts were implemented by practising scientists, but with the extent to which his ideas were involved in debates about the history, philosophy and cultural image of science.[14]

BACONIANISM IN THE SEVENTEENTH AND EIGHTEENTH CENTURIES

Since the nineteenth century, discussions of Bacon and Baconianism have been almost exclusively concerned with the methodology and philosophy of science. The uses of the term 'Baconian' cited by the *Oxford English dictionary* are taken from various deliberations of this period about induction and its contrast with other allegedly less proper methods of investigation.[15] It is important to note, however, that this concentration on Bacon as a theorist of method was preceded by interpretations which also had wider social connotations; indeed, there is a sense in which the nineteenth century focus on Bacon as a methodologist took place alongside attempts to neutralize the implications of these earlier versions of Baconianism. Some brief consideration of these seventeenth and eighteenth century readings of Bacon is therefore necessary.

Bacon's writings on science, its method, organization and social value have been a protean source of political rhetoric. His attack upon the prevailing scholastic philosophy and his appeal to a new means of advancing knowledge became powerful resources for a variety of reformist causes. In the case of the period immediately following Bacon's death, this has been impressively documented by Charles Webster in *The Great Instauration*, which investigates scientific and medical thought in England from 1626 until the Restoration of 1660. Webster delineated the activities of Puritan men of science and medicine who accepted the challenge of the Baconian philosophy and attempted a reform of existing attitudes and institutions. In establishing this connection between Bacon and Puritanism, Webster was confronting the influential seventeenth century interpretation which presented Bacon as the guiding force behind the establishment of the Royal Society in 1662. He suggested that this account was partly designed to conceal the *earlier* association of a "vulgar Baconianism" with the controversial outlook of Puritan reformers.[16] In this public rhetoric, Bacon's methodological precepts were exploited in the service of an image of science as a co-operative endeavour distinct from religious sectarianism and political conflict.[17] These apologetics of the Royal Society successfully removed the earlier reformist associations of Bacon's doctrines in Britain, but these were renewed by the French *philosophes*.

The significant moment in the career of Baconianism during the eighteenth century was its association with the French *Encyclopédie*. The two editors, Diderot and D'Alembert, presented Bacon as the presiding figure behind their project and adopted his classification of knowledge, with some modification, as an organizing device. The Preliminary Discourse to the first volume of 1751 included a detailed account of Bacon's map of knowledge and saluted him as "the greatest, the most universal and the most eloquent of the philosphers".[18] In short, he was the glorious inaugurator of the reformation in the arts and sciences which the *Encyclopédie* set out to describe and disseminate.

Diderot and D'Alembert saw their espousal of Bacon's merits as unusual, and this was not merely a French perspective. The English translator of the Preliminary Discourse, which appeared in London in 1752, hoped that the *Encyclopédie* might "raise a Desire in the *English* Reader to consult those admirable Works; which certainly hitherto have been too much neglected in *England*".[19] This contemporary perception was later endorsed by Dugald Stewart, who believed that the study of Bacon in the early eighteenth century had been very limited; in his judgement, the publication of the *Encyclopédie* marked a new era in Bacon's reputation.[20] But if the French *philosophes* drew attention to Bacon's views on science, they also re-constituted Baconianism as a broad philosophy, founded on a commitment to scientific inquiry, but extending beyond the natural world to the reformation of society.

The character of this new emphasis can be illustrated by comparing the *Encyclopédie* with its main predecessor, Ephraim Chambers's *Cyclopedia*. This *Dictionary of the arts and sciences* of 1728, provided the original stimulus for the French venture which had begun life as a projected translation of the English publication. Chambers's work was informed by a free-thinking suspicion of superstitious and magical beliefs, and in contrast with other encyclopedias of the period, contained no entries on religious or theological subjects.[21] Like its famous successor, the *Cyclopedia* was a compendium of scientific and technical knowledge, an apologia for the contribution of arts and crafts to the reform of society. But there was no appeal to the authority of Bacon as a means of displaying the philosophical foundation of this approach. Chambers did not include an entry on Bacon, and the article on "Induction" was a straightforward account of syllogistic logic with no reference to his views.[22] In the *Encyclopédie*, however, there was an article on "*Logique*" which discussed Bacon's criticism of scholastic philosophy and his proposals for the reform of scientific thinking. Furthermore, it included an entry for "*Baconisme*" which endorsed a polemical philosophy of knowledge and praised Bacon as its founder.[23]

The *philosophes* found much to admire and exploit in Bacon. In particular, they were attracted by the contrast between a widely accessible method and dogmatic authority, between practical, useful knowledge and the sterile debates of the schools. But they also endowed Bacon with an anti-clerical accent. His attack upon scholasticism became a call for the secularization of learning; his emphasis on the importance of empirical knowledge became a sensationalist epistemology; and his programme for a co-operative study of nature became a critique of the monopoly of education and culture by privileged élites. At its most extreme, Baconisme could mean secular thought, utility, religious scepticism and materialism.

As a result of this association with the *Encyclopédie*, Baconianism was conceived in much broader, and more controversial, terms in France than in Britain. French works on Bacon continued to appear after the *Encyclopédie* — writers such as Condillac, Cabanis, Lasalle, de Luc, Joseph de Maistre and Charles de Remusat were involved in debates about the philosophical and theological implications of his thought. At its most polemic, this was a controversy in which Bacon was used by various groups in religious and political arguments: some praised his empiricism and religious scepticism; some attempted to present him as a Christian philosopher; and, perhaps most notoriously, de Maistre condemned him as the intellectual architect of the Enlightenment and the horrors of the French Revolution.[24] German interest developed more slowly, but when Kuno Fischer attempted to bring the thought of the English Chancellor to a German audience, his approach was also wide-ranging. In his *Francis Bacon von Verulam: die Realphilosophie und ihr Zeitalter* of 1856, Fischer presented Bacon as the founder of a philosophical tradition which flourished in the work of Hobbes, Locke and Hume; he regarded these, and all subsequent manifestations of this "realistic philosophy" as varieties of Baconianism.[25] In Britain however, after the debates at the time of the early Royal Society, Baconian ideas were not so explicitly mobilized in any major political or cultural controversy. Nevertheless, there were a number of issues concerning the philosophical and social aspects of science – its relationship with theology, its epistemology, and its utility – in which Baconian themes were involved. No thorough treatment is attempted here, but it is possible to indicate the ways in which British writers had to deal with certain controversial interpretations of Bacon's work.

BACON AND THE CULTURAL RELATIONS OF SCIENCE

In his work of 1878 on *Diderot and the Encyclopaedists*, John Morley cited the high praise bestowed on Bacon by the French *philosophes*. "No more striking panegyric", he wrote, "has ever been passed upon our immortal countryman than is to be found in the Preliminary Discourse. The French Encyclopaedia was the direct fruit of Bacon's magnificent conceptions".[26] As a representative of Victorian secular liberalism, Morley was pleased to acknowledge an English paternity for the *Encyclopédie*, which he saw as a crucial historical influence on subsequent social and intellectual process. But at the time of the Napoleonic wars, ideas with French affiliations were suspect,[27] and for this reason, Baconian concepts which had been put to polemic use by the *philosophes* could still carry controversial connotations. This could influence the way in which some of Bacon's ideas became involved in nineteenth century debates. For example, his dismissal of final causes as barren virgins became relevant to the major dispute about the status of teleological notions in science and natural theology.[28] Critics of traditional natural theology, from the *philosophes* to Comte, exploited these remarks in their attacks upon teleological notions of design in nature. British defenders of natural theology had to counter this use of Bacon. Dugald Stewart, for example, argued that those who saw Bacon as a champion of irreligious causes failed to recognize that his strictures had "a particular reference to the theories of the schoolmen"; he did not question the metaphysical validity of final causes, but rather their place in physical reasoning. With an eye on contemporary debates. Stewart castigated French physiologists, such as Cabanis, who persisted in quoting Bacon as a justification for their materialist doctrines. Secondly, in what became a common ploy, Stewart acknowledged that Bacon had overlooked areas of science in which research had been facilitated by teleological thinking.[29] In this way, readings of Bacon's pronouncements on final causes became part of significant debates on natural theology, positivism and evolutionary theory during the first half of the nineteenth century.[30]

In his *Essay on the origin of human knowledge* of 1740, Condillac nominated Bacon as the first to recognize that all knowledge had its origin in sensory experience.[31] This epistemology was also announced in the Preliminary Discourse to the *Encyclopédie*, and in the following decades, Bacon was linked with Locke as one of the English founders of the empiricist or sensationalist philosophy which French thinkers such as Cabanis, Condorcet and de Tracy had adopted and developed. British attitudes to this genealogy need to be seen in the context of debates about the appropriate epistemology of science, a debate which, like that on the role of teleology, was initially conditioned by efforts to distinguish natural science in Britain from the materialist and irreligious affiliations it was seen to have in France. Opposing assessments of this background informed the conflict between idealist and empiricist, or positivist, philosophies of science; and interpretations of Bacon were often involved.

One of the most striking attempts to dissociate Bacon from the views of his French admirers occurred in Samuel Taylor Coleridge's "Preliminary treatise on method" of 1818.[32] Coleridge saw the "foundations of the modern French school" as deriving from the assumption that all knowledge and science begin and end with the objects of sense".[33] He was concerned that "the monstrous puerilities of Condillac and Condorcet" had been linked with what he considered to be a mistaken reading of Bacon, and his reply was to present Bacon as the "British Plato".[34] Both great philosophers had been misunderstood: Plato did not deny the importance of "plain experience", and Bacon did not reject the contribution of ideas. One reason for such a misreading of Bacon in the nineteenth century, he suggested, was that "those who talk superficially about Bacon's philosophy, that is to say, nineteen-twentieths of those who talk about it at all, know little more than his induction, and the application which he makes of his own method to particular classes of physical facts".[35] But the epistemology which guided Bacon's general approach to knowledge was not what the French *philosophes* assumed. Bacon recognized the need for an "intellectual or mental initiative" as the starting point of scientific knowledge, and agreed with Plato in regarding laws of nature as expressions of underlying metaphysical ideas. For Coleridge, therefore, Bacon had completed the "ideal system" by applying "the same method to external nature, which Plato had before applied to intellectual existence".[36]

In spite of Coleridge's efforts, however, the association between Baconianism and an insensitivity towards metaphysics was reinforced. Writing from the perspective of the *Naturphilosophie* tradition, J. B. Stallo, a disciple of Coleridge, complained that the "materialistic, utilitarian tendencies, which at present pervade every branch of science under color of a misconceived Baconism, have revoked every alliance between philosophical pursuits and the investigation of nature".[37] In his *Biographical history of philosophy*, G. H. Lewes presented Bacon as the intellectual father of Positivism: modern philosophy had begun with the method of Bacon and culminated with that advocated by Comte. For Lewes, Bacon possessed "the *positive* spirit" because he had urged the separation of science and theology and criticized the use of metaphysical reasoning.[38] Lewes therefore reinforced the controversial associations which Baconianism had acquired during the eighteenth century. Some religious critics of scientific naturalism in Britain made use of this, arguing that the sensationalist epistemology of Locke, Condillac and Comte was "in the tradition of Bacon" and that this philosophy was the foundation of the modern scientific movement.[39] Writers such as Whewell, who wanted to defend Bacon's reputation, and at the same time elaborate a non-positivist philosophy of science, had to reject these associations.

One of the most prominent and celebrated features of Bacon's work was his conception of the goal of science as "the relief of man's estate".[40] This emphasis on the practical value of natural knowledge and its connection with craft and technology had been significant in the early Royal Society and in the *Encyclopédie*. Such a theme was also resonant with the values of the commercial and industrial middle classes in nineteenth century England, and a reference to the utilitarian aspects of science was therefore attractive to those seeking to promote it. Thus, in his *Discourse* on natural philosophy in Lardner's *Cyclopaedia*, Herschel gave examples of the practical, and often surprising, applications of scientific theory. But this concept of utility was sensitive because it was often conflated, in Britain, with sensationalism and materialism. In reviewing Herschel's book, Whewell warned that a preoccupation with utility could engender an attitude which failed to look beyond the sensory basis of knowledge and beyond its instrumental uses.[41]

This issue assumed a new dimension in the 1830s when the British Association was seeking to define the public image of science. At the same time, groups such as the Society for the Diffusion of Useful Knowledge, originally sponsored by utilitarians and Whig reformers, were highlighting the social and economic fruits of science. It was suggested, for example, that industry might benefit from artisans who had gained some acquaintance with science in Mechanics' Institutes. Bacon was invoked as the intellectual authority for this outlook and some of his works were published in cheap editions. In 1837, in a famous essay on Bacon, Macaulay promoted this interpretation, claiming that the Baconian philosophy sought to satisfy "vulgar wants", and that its keywords were "Utility and Progress".[42]

This deployment of Bacon posed difficulties for those who wanted to advance an alternative view of the scientific enterprise, one in which the pursuit of pure knowledge, rather than material outcomes, was the ultimate justification. Herschel had been aware of this in 1831, when he said that the dignity of science would be diminished if it was seen as a "mere appendage to and caterer for our pampered appetites".[43] Whewell strongly endorsed this observation and asserted that the search for theoretical knowledge had intrinsic value. In the *Philosophy of the inductive sciences* he explained that he had not imitated those sections of Bacon's writings which contemplated "man's dominion over nature as the main object of Natural Philosophy".[44] While granting that "knowledge is power", Whewell remarked that "its interest for us in the present work is — not that it is power, but that it is knowledge".[45] In reacting against the image of Bacon disseminated by the *philosophes*, and reinforced by Macaulay, writers such as Coleridge, Whewell, Remusat, and Fischer castigated what they called vulgar or pseudo-Baconianism, and claimed that Bacon gave equal attention to the intrinsic worth of knowledge. In these circumstances, the vision of science working for the relief of man's estate was largely repressed in favour of an image of science as apolitical, primarily concerned with pure research, and only incidentally with the practical social consequences of the knowledge it produced. The scholarly debate on Bacon in the nineteenth century focused almost exclusively on his methodology, and the effect of this approach is reflected in the way in which more recent writers have seen themselves as recovering other levels of his work.[46]

While the French *philosophes* applauded Bacon as the prophet of enlightened scientific rationalism and useful technology, the major focus in nineteenth century Britain was on Bacon as the theorist of induction. The development of this concentration has not been adequately chartered, but Laudan has suggested that in the seventeenth century "Bacon was not praised (or condemned) as an inductive philosopher so much as an experimental one".[47] Little detailed work has been done on the ways in which Bacon's philosophy was discussed in the first half of the eighteenth century. There is some evidence that although the moral, political and historical essays were widely read – at least nine editions appeared in the eighteenth century – his writings on science were not extensively studied. When an English edition of *The philosophical works* was published in 1733, the editor referred to the lack of previous translations and suggested that Bacon's work was not receiving the attention it deserved.[48] In *Letters concerning the English nation* of 1733 Voltaire observed that keen interest in his thought was no longer so obvious, and remarked that although the *Novum organum* was the best of his works it was now the most useless and the least read.[49] Henry Hallam later endorsed this view in his *Introduction to the literature of Europe*, published in 1838, claiming that the philosophical works were rarely consulted, and pointing to the fact that no English press had issued a separate edition of the *Novum organum* since the first of 1620.[50] However, these comments need to be treated carefully because it is likely that other Baconian texts on science were studied. Thus, although the *Novum organum* does not appear on Cambridge reading lists during the first half of the century, the *Historia de ventis* and the *Sylva sylvarum* were set.[51] Indeed, it was these works which the contributors to the early English encyclopaedias had in mind when they celebrated Bacon as the founder of experimental philosophy.[52] This may suggest a different hierarchy of concerns: namely, that the natural histories had not yet been displaced by the focus on the logic of induction which dominated nineteenth century debates. However, by the late eighteenth century this interest in Bacon as a methodologist was firmly represented in the writings of the Scottish philosophers.[53]

THE SCOTTISH DEBATE ON BACON

In the mid-eighteenth century, David Hume viewed the contemporary adulation of Bacon as a reflection of the English "national spirit". Writing in his *History of England*, Hume argued that Bacon's role in the history of science was less important than that of Galileo, and he regarded some of the praise bestowed upon him as "partial and excessive".[54] Yet by the end of the eighteenth century Bacon's work had become closely associated with the distinctive features of Scottish moral and natural philosophy. According to Hallam, writers such as Thomas Reid and Dugald Stewart "turned that which had been a blind veneration into a rational worship".[55] More recent historians of science have continued to acknowledge the importance of Bacon for Scottish views on the philosophy of science;[56] but it should not be assumed that there was a naïve or simplistic reception of Bacon in Scotland. Indeed, George Davie has seen some point to the radical, alternative thesis enunciated by Thomas Henry Buckle in his *History of civilization in England*. From the perspective of the positivist historiography which informed his work, Buckle saw eighteenth-century Scottish thought as weakened by its compromise with theology and ecclesiastical institutions; the inductive philosophy of Bacon, the natural ally of the positive or scientific spirit, had been overwhelmed by deductive and metaphysical approaches which represented the very essence of an "anti-Baconian method".[57] Given these different perceptions it is important to discuss some of the issues raised by the Scottish interpretations of Bacon.

Bacon's most influential Scottish champion was Thomas Reid, the leading figure of the Common Sense philosophy. In reacting against Hume's scepticism, Reid found it significant that he had not been "aware of the extraordinary merits of Bacon as a philosopher".[58] So highly did Reid value Bacon's work that he was "very apt to measure a man's understanding by the opinion he entertains of that author".[59] He also set part of the agenda of subsequent debates by stating, or at least renewing, some key theses: the distinction between Bacon's induction and Aristotle's logic; the superiority of his precepts over those of Descartes; the relevance of his method for the moral sciences; and, perhaps of most symbolic importance, the nexus between Bacon and Newton. Reid saw Newton's rules of philosophizing as an expression of Bacon's methodology, and suggested that "in the third book of his *Principia*, and in his *Optics*, [Newton] had the rules of the *Novum Organum* constantly in his eye".[60] Furthermore, he interpreted Newton's "hypotheses non fingo" as an integral part of the Baconian legacy.

Other Scottish writers, especially Stewart and Playfair, continued this affirmation of Bacon's work and reinforced his foundational role in the establishment of modern science. In his Dissertation for the supplement of the fifth edition of the *Encyclopaedia Britannica* in 1817, Stewart remarked that the "merits of Bacon, as the father of Experimental Philosophy, are so universally acknowledged, that it would be superfluous to touch upon them here".[61] Playfair's Dissertation dealt with the progress of mathematical and physical science, and this provided the opportunity to defend Bacon against Hume's assessment. For Playfair, there was no doubt that "more substitutes could be found for Galileo than for Bacon", and that the progress of science since his time had resulted from the application of his method.[62] However, by the second decade of the nineteenth century some of the claims about Bacon's role in the progress of science were coming under critical scrutiny. In 1818, for example, Macvey Napier found it necessary to consider recent qualifications to Bacon's reputation as the "Father of Experimental Philosophy". In a paper read before the Royal Society of Edinburgh, Napier acknowledged that "There are some, however, who seem to think, that there is no good ground for honouring him with this title, either on account of the merits or the effects of his writings".[63]

Napier's essay can be seen as the first major contribution to the nineteenth century reassessment of Bacon's work. But it should be recognized that the terms of reference were not identical to those of later participants in this debate because, for him, the validity of Bacon's method was not the central issue. The attack which Napier detected was directed against the originality and priority of Bacon's methodology, and against its effective influence in the history of science. The sources of the critique which he identified were an article on Galileo by the French scientist, Jean-Baptiste Biot, in the *Biographie universelle*, and a review of Stewart's *Dissertation* in the *Quarterly review*. In both cases, there was dissent from the view that Bacon was "the first who clearly and fully pointed out the legitimate rules and ends of philosophical inquiry" and inaugurated "a new and important era in the history of modern science".[64] One recurring theme was the charge that science had benefited more from the practical examples of inductive method provided by the work of Copernicus, Galileo, Gilbert and Harvey – contemporaries of Bacon who, unlike the Lord Chancellor, made actual contributions to scientific knowledge. In reply, Napier acknowledged that Bacon's fame did not rest on his substantive knowledge of the physical world but on his systematic formulation of the rules for attaining it. At this level, although he granted that a reformation of scientific thinking was already in progress, and that "the Inductive Method had been happily *exemplified* in the discoveries of some of his contemporaries", he insisted that no previous writer had attempted to "systematize the true method of discovery; or to prove, that the *Inductive*, is the *only* method by which the genuine office of philosophy can be exercised".[65]

In this respect, his affirmation of Bacon's significance was stronger than other Scottish commentators. Stewart, for example, while noting that before the *Novum organum*, the method which it recommended had been only "followed accidentally, and without any regular, or preconceived design", nevertheless stressed the ways in which Bacon's writings were "powerfully influenced by the circumstances and character of his age", and concluded that science would eventually have progressed without his work.[66] Napier was also disappointed with James Macintosh's failure to distinguish between the methodology of Aristotle and Bacon. Writing in the *Edinburgh review*, Macintosh said that "the method of induction, which is the art of discovery, was so far from being unknown to Aristotle that it was often faithfully pursued by that great observer".[67] Napier complained that this overlooked the crucial ways in which Bacon improved upon simple and spurious notions of induction, and stressed that Baconian philosophy was not merely the revival of an earlier method but the announcement of a novel set of procedures.[68]

More severe than these qualifications concerning the originality of Bacon's work was the attack which cast doubts upon his claim to have grasped the limits of human understanding and the proper goals of science. In criticizing what it saw as Stewart's excessive praise of Bacon, and his neglect of Locke's merits, the *Quarterly review* argued that Bacon's favourable attitude towards alchemy and all manner of natural magic showed that he did not rise above the age in which he lived.[69] However, this criticism focused more on the *subjects* which Bacon chose to investigate rather than the methods which he prescribed for science. This allowed Napier to reply that, firstly, Bacon was not alone in expressing an interest in alchemy – Boyle and Newton were notable examples – and secondly, that Bacon wished to reform the arcane practices of the alchemists by subjecting all assertions about the powers and essences of nature to a rigorous and public method.[70]

Napier's major objective was not a detailed defence of the validity of Bacon's methodology but the demonstration of its influence on major scientific figures. Accordingly he cited the favourable views of Sprat, Glanville, Oldenburg, Boyle and other seventeenth century natural philosophers and, conversely, used the criticism of opponents such as Henry Stubbe and his complaint about "a Bacon-faced generation" as further evidence of the contemporary currency and influence of Bacon's ideas.[71] Like Reid, he thought it undeniable that Newton "was guided by the principles which Bacon taught; and that his philosophy derives its imperishable character from his rigid adherence to them".[72] Whereas Stewart had said that Bacon was rarely studied in France until the publication of D'Alembert's Preliminary Discourse, Napier argued that his influence in Europe was extensive, and referred to the testimony of Descartes, Gassendi and others. In underlining the substantive effects of Bacon's teachings Napier was explicitly rejecting the appraisal offered by the *Quarterly review*, which suggested that Bacon's contribution to the progress of physical science could be compared to the role of the husbandmen in Aesop's fable: that is, although his specific directions were not efficacious, his general call for experimental inquiry did produce results.[73] Again, Napier was displeased to see that Macintosh, who defended Bacon's high status, implied that his merit consisted not in the enunciation of a new method but in the stimulation and popularization of a "new spirit" of inquiry.[74] Napier wanted to be more specific; and while he admitted that "the adoption of the Inductive Method would have taken place in time", maintained that Bacon provided the clearest expression of the correct method and "did *more* to forward its general adoption than any other person".[75]

Napier's essay can be taken as a guide to the early nineteenth century Scottish debates about Baconianism, but only after its particular context is specified. At a general level it was well-received: while there was a reluctance to extend what Thomas Young called the "apotheosis" of Bacon,[76] there was strong support, especially in Whig circles, for Napier's defence of his reputation. James Mill wrote from London thanking him for writing on Bacon:

His is a battle which I often have to fight in conversation, at least; for Englishly-educated people are all hostile to him, as they ... are hostile to everybody who seeks to advance the boundaries of human knowledge, which they have sworn to keep where they are. Your learned and valuable collection of facts will make me triumphant.[77]

But it should be noted that Napier did not deal with the full range of issues covered by Scottish discussions on the philosophy of science. Thomas Brown, who was then Stewart's assistant in moral philosophy at Edinburgh, indicated this when he told Napier that Bacon's belief that "*essences* of all material substances were capable of being discovered by the inductive process" suggested, as the *Quarterly review* contended, that his philosophy did not recognize the true limits of human knowledge.[78] There was also no treatment of the status of hypothesis, analogy and theory in Baconian methodology, a topic which had acquired new currency since Stewart and Brown began to revise Reid's antagonism to the danger of hypothetical conjecture. On all of this Napier was silent; and in this respect his essay did not address the issues which came to occupy the centre of debates over the relevance of Baconianism to natural science. Indeed, it is worth proposing that Napier's contribution had at least as much to do with contemporary disputes about the nature of *moral* science as it did with the philosophy and history of the physical sciences.

Some of the major Scottish commentators on Bacon were moral philosophers, not men of science, and their interest in Baconian method was closely linked with their efforts to establish the philosophical foundations and procedures of the moral sciences. The project undertaken by the Common Sense philosophers was that of applying the successful method of the physical sciences to the investigation of human nature, while avoiding the twin dangers of scepticism and materialism. Bacon's method was seen as the key to the secure establishment of moral science, and Stewart elaborated the rationale in his *Elements of the philosophy of the human mind*, first published in 1792. Noting that this subject lacked a "proper method", he remarked that this was not surprising because physical science had only progressed since it had been guided by Bacon's philosophy. The reformation which his work inaugurated should now be extended to the study of the mind by applying "the method of induction".[79]

Crucial to this scheme, of course, was the conviction that Bacon's method had in fact been instrumental in producing major advances in natural knowledge. In the context of this concern for the foundation of the moral sciences, Napier's exhaustive documentation of Bacon's influence in the history of science can be seen as an attempt to vindicate the success of a method which could be extrapolated to the study of human nature. Napier was involved in this debate over the status of moral science and had earlier defended Stewart against the critique of Francis Jeffrey, who had argued that the study of mind could not be an experimental subject.[80] This controversy also surfaced in the article on Bacon because the *Quarterly review*, while accepting the merits of Bacon's work with respect to "physics", had questioned its relevance to the understanding of mental phenomena. In strongly rejecting this distinction, Napier insisted that "the principles of philosophizing are the same, whether the investigation relates to the laws of matter or the laws of mind; and thus the logic of the *Novum Organum* cannot be useful with reference to the one, without having the same character with reference to the other".[81] It is possible that this interest in such a transfer of method dictated the absence of any detailed consideration of the various interpretations of Bacon's pronouncements on hypothesis, analogy and the role of mathematics. Napier either believed, or wanted to imply, that there was a consensus on the nature of Baconian method and its efficacy in the advancement of the physical sciences. It is therefore necessary to go beyond his essay in order to analyse Scottish attitudes to Baconian methodology.

In his work on *Scottish philosophy and British physics*, Richard Olson regards Baconianism as the initial framework for the development of the philosophy of science presented by the Common Sense school.[82] But within this tradition, as Olson shows, the attitude to Bacon's work was not static: whereas Oswald, Beattie and Reid emphasized his warnings about the various idols, especially the danger of conjectural systems, later writers such as Stewart and Brown pursued a more sophisticated interpretation of the role of analogical and hypothetical thinking. In referring to Reid's position, Stewart remarked that "the indiscriminate zeal against hypotheses, so generally avowed at present by the professed followers of Bacon, has been much encouraged by the strong and decided terms in which, on various occasions, they are reprobated by Newton".[83] With appropriate diplomacy, he suggested that Newton's references need not be taken literally, and proceeded to defend the controlled use of hypothesis in science, indicating that such a method was not inconsistent with that of Bacon. For example, he made good use of Hooke's defence of conjecture and theory against the strictures of the early Royal Society, praising his sagacity in contrast to those who "have been more disposed to follow the letter of some detached sentences, than to imbibe the general spirit of Bacon's logic".[84] With this more liberal approach to the Baconian texts, Stewart was able to endorse the view of Reid's great enemy, David Hartley, who regarded hypotheses as essential to the progress of science. He also justified the use of hypotheses by situating them within the province of analogical reasoning, which was itself vindicated by the unifying relationships which permeated nature and were central, in Stewart's opinion, to Bacon's conception of the unity of truth.[85] This position did not amount to a total renunciation of Reid's teaching because Stewart distinguished between arbitrary fictions and hypotheses sanctioned by analogy; the link with empirical data was retained because admissable analogies and hypotheses had to be based on sound knowledge of one area before being applied to another. He also distinguished between the heuristic status of hypotheses and the ontological status which might be ascribed to theories based on inductive generalizations from observed phenomena.[86] Thus, in Stewart's work, the method of hypothesis was asserted without any direct confrontation with the authority of Bacon. "In thus apologizing for the use of hypotheses", he explained, "I only repeat in a different form the precepts of Bacon, and the comments of some of his most enlightened followers".[87]

Scottish scientists were also able to interpret the Baconian heritage in a manner which allowed a legitimate role for hypotheses. Indeed, even before Stewart's contribution, John Robison, Professor of Natural Philosophy at Edinburgh, allowed some exceptions to his otherwise unqualified attack on hypothesis and conjecture. In the entry on "Philosophy" in the *Encyclopaedia Britannica*, Robison devoted most of the space to an exposition of Baconian thought, laying stress on the danger of conjectural systems. But he did acknowledge a role for hypotheses in science provided that they were capable of being tested by experiment, and concluded that "in this way conjectures have their great use, and are the ordinary means by which experimental philosophy is improved".[88] Thus, even in the late eighteenth century it was possible for Scottish writers to qualify Reid's more extreme pronouncements. It is therefore not surprising that in 1827 the young James Forbes could write a student essay on Bacon in which he drew upon the writings of Stewart, Playfair and Gregory to indicate the role of hypothesis and analogy in science, while still affirming the continuing relevance of Bacon's system.[89]

By the 1820s, therefore, leading Scottish philosophers and men of science continued to award high prestige to Bacon's work and his impact on science. Although Hume's earlier complaint about excessive praise had been reactivated, largely by foreign commentators such as Biot, and by the *Quarterly review*, this had been met by Napier, whose defence seems to have been generally accepted. Some of the leading writers associated with the *Edinburgh review* such as James Mill, James Macintosh and Henry Brougham were receptive to Bacon's cause and saw contemporary political and social relevance in his work. A decade after the publication of Napier's essay Brougham approached him to write commentaries on the *De augmentis scientiarum* and the *Novum organum* for the Library of Useful Knowledge.[90] But although this suggests a continuing interest in Bacon, it is important to recognize that there were different levels in the debates about his work.

The early Scottish interest in Bacon was closely associated with debates about the establishment of moral science. Thus, Napier's attempt to vindicate the efficacy of Baconian method in the advancement of physical science was linked with his belief in the feasibility of extending it to the science of mind. In the case of his essay, the strategic demands of this debate may explain the absence of any detailed treatment of methodology. Such a discussion did however occur in the writings of Stewart, Brown and Hamilton, where there was a sophisticated treatment of the role of hypothesis, analogy and theory in the physical sciences.[91] Stewart argued that while Reid had correctly warned against the danger of transferring hypothesis based on the material world to the sphere of moral phenomena, he had been too restrictive concerning the use of hypotheses within natural science.[92] Stewart's position, together with that of Brown, Gregory and Forbes, also indicates that a positive attitude to the role of conjecture and imagination could be advanced without an attack on Bacon's reputation as a methodologist. However, this liberal reading could also suggest indifference, and in his biographical memoir of Sir John Leslie, Napier regretted what he saw as a complete neglect of Bacon's ideas:

> Notwithstanding the contrary testimony, explicitly recorded by the founders of the English Experimental School, he denied all merit and influence to the immortal delineator of the Inductive Logic.[93]

In 1830 this heterodoxy was repeated by another of Scotland's famous experimental scientists — David Brewster.

Brewster's denigration of Bacon's significance occurred in the context of his admiration for Newton. In his view, the worship of Bacon had become idolatry, and the claims about the importance of his work to the advancement of science threatened to "depose Newton from the high priesthood of nature".[94] Brewster severed the sacred link between Baconian method and Newton's scientific achievements, arguing that neither the great natural philosopher nor any of the major early contributors to science owed any clear debt to Bacon's writings. This was a direct denial of Napier's position, and Brewster was conscious of repudiating a contemporary consensus. Several years before the appearance of his short biography of Newton, he confessed his heresy in a letter to Maria Edgeworth.

> The opinion so prevalent during the last thirty years, that Lord Bacon introduced the art of experimental inquiry on physical subjects, and that he devised and published a method of discovering scientific truth, called the method of induction, appears to me to be without foundation, and perfectly inconsistent with the history of science.[95]

In fact, Brewster extended the parameters of Napier's argument by confronting not only the question of the influence of Bacon's ideas but the substance of his method, claiming that it was useless as a means of successful scientific inquiry. Writing to Forbes in 1830, he said, "Forget all you have heard of Lord Bacon's Philosophy. Give full reins [sic] to your imagination. Form hypotheses without number".[96] When he repeated these views in his larger *Memoirs of the life, writings and discoveries of Sir Isaac Newton* of 1855, they were consonant with a revisionism endorsed by other commentators; in 1831 they were seen as eccentric. Referring to this criticism of Bacon, the *Edinburgh review* remarked that "with one or two exceptions, he is the only man of science of any considerable name, who has laboured to detract from the glory of the great reformer of philosophy".[97] By the mid-nineteenth century, positive appraisals of the conjectural and imaginative element in scientific thinking were contrasted with Bacon's position and, together with other issues, formed the basis of an explicit rejection of his methodology. Thus, although Brewster's views on the use of hypotheses were not radically different from those of other Scottish writers, his open attack on the substance of Bacon's method anticipated a new phase in the British reassessment of the Baconian legacy.

THE MID-NINETEENTH CENTURY DEBATE ON BACON

In 1866 the Oxford logician, Thomas Fowler, described the 1830s as a "rich decade" for writings on Bacon.[98] This epithet was appropriate because Brewster's *Life of Newton* and Herschel's *Preliminary discourse*, which appeared in 1831, both referred to Bacon's place in the history of science and offered opposing estimates of his significance. In 1837, Macaulay reviewed Basil Montagu's edition of Bacon's works and attacked both his character and his philosophical reputation. In 1840 Whewell published the *Philosophy of the inductive sciences*, a work which in some senses sought to modernize Bacon's programme by drawing upon the examp. 's of scientific discovery which had occurred since the early seventeenth century. By examining these works, the relationships between them, and the responses they provoked, it is possible to analyse the issues at stake in a debate which led to the demise of Bacon's status as the founder of the method by which science advanced.

John Herschel's *Preliminary discourse on the study of natural philosophy* confirmed the centrality of Bacon's ideas to inquiries about the methodology of science. The frontispiece carried his portrait and Herschel actually stated that before "the publication of the Novum Organum of Bacon, natural philosophy, in any legitimate and extensive sense of the word, could hardly be said to exist".[99] It was only after Bacon's clear exposition of inductive philosophy that the method of ascending from particular observations to general laws became the means by which science progressed. In reviewing the *Discourse* the *Athenaeum* accepted this as uncontroversial, saying that "it need scarcely be mentioned here, that to Lord Bacon we are indebted for the great general principle of the method now universally pursued in the investigation of nature, namely, that of induction".[100] The value of Herschel's treatise was seen to consist in its attempt to provide a new statement of Bacon's philosophy in the light of examples from the progress of knowledge. Whewell regarded the *Discourse* as a modern, and much needed, addition to the project which Bacon had initiated: namely, the attempt to formulate a satisfying account of the "mental process exemplified in modern science".[101] He complained that neither the general "philosophy of mind" pursued by the Scots, nor the traditional syllogistic logic expounded by Richard Whately, had provided an adequate analysis of the inductive mode of discovery manifested in the progress of the physical sciences. In spite of all the paeans to Bacon, Whewell believed that there had been little substantial explanation of inductive philosophy suitable for a wide audience. Herschel's book promised to supply this need.[102]

The Baconian aspect of Herschel's *Discourse* has been emphasized by Joseph Agassi who regards it as a major reassertion of inductivism at a time when its status as an account of scientific progress was under threat. Agassi contends that Herschel was responding to a crisis produced by the optical revolution – the displacement of Newton's theory of light by those of Young and Fresnel – which raised doubts about the ability of physical science to produce results of lasting certainty. In his view, the *Discourse* responded by reinforcing "faith in science, in induction, and in Newtonian mechanics", and reaffirming the notion of "a foolproof scientific method".[103] Furthermore, he suggests that Herschel's praise of Bacon should be seen in the context of contemporary doubts about his significance. Agassi says that "both Bacon's alleged greatness and his methodology were equally challenged".[104] However, this statement requires qualification. As argued above, the earlier critiques, such as those to which Napier responded, were directed against the originality of Bacon's theory of induction and against the influence of his ideas on the progress of science; the validity of the inductive method he presented was never attacked.

This distinction between Bacon's general reputation and the efficacy of his method is important for a balanced assessment of Herschel's text. On the one hand, Agassi is certainly correct in underlining the extent to which Herschel stressed Bacon's foundational role in the history of science; but his interpretation of Herschel's exposition of Baconianism as the reassertion, at a time of scientific crisis, of a method previously under attack, is problematic because it is not clear that such an assault existed before 1830. Although the *Discourse* can legitimately be read as a new statement of Bacon's ideas, there is a sense in which Herschel allowed scope for their reformulation or qualification in a way that defenders such as Napier had not countenanced. For example, Herschel did not repeat the earlier standard references to the nexus between Bacon and Newton; in fact, like Brewster, he awarded a distinct and singular place to Newton's achievement: "Whichever way we turn our view, we find ourselves compelled to bow before his genius and to assign to the name of Newton a place in our veneration which belongs to no other in the annals of science."[105] Secondly, Herschel moved away from Bacon's focus on the method of discovery, diminishing the importance of the means by which facts were discovered in comparison with the method by which they were verified. "In the study of nature", he explained, "we must not, therefore, be scrupulous as to *how* we reach to a knowledge of such general facts: provided only we verify them carefully when once detected".[106] In this sense he foreshadowed the central concern of subsequent philosophy of science.

In contrast with Susan Cannon's portrayal of Herschel as the representative of the early Victorian scientific élite for whom Baconianism was *passé*, Agassi presents him as a "conservative *par excellence*", one of the last exponents of the orthodox Baconian doctrine of prejudice.[107] He claims that Herschel followed the eighteenth century writer, Isaac Watts, who warned against the danger of preconceived notions, and celebrated the concept of the pure or empty mind as the guarantee of true and certain knowledge. Because of this approach, he argues, a refuted theory for Herschel was always a prejudice and never a true part of science.[108] Although Cannon does not explicitly consider formal methodological statements by major scientists, she strongly resists any account of their scientific style as Baconian: the interest in gathering new data, she explains, should not be taken to indicate the priority of factual observation over theory in their philosophy of science. Cannon suggests that the Baconianism of the *Discourse* advocated "a method of analysis, generalization and deduction" which did not exclude a role for hypothesis and theory. Having said this, however, she goes on to claim that the philosophy of early nineteenth century British scientists, including Herschel, was not Baconian but Humboldtian.[109] Agassi cites Herschel's refusal to accept the use of hypotheses as "essential to the method of science" as testimony to his conservative Baconianism;[110] Cannon finds a "sophisticated view of hypotheses" which typifies the more liberal Humboldtian tradition.[111]

Some of this confusion can be removed by recognizing that Herschel presented two accounts of how the transition from observation to laws and theories can occur. John Losee suggests that "Herschel was aware that many important scientific discoveries do not fit the Baconian pattern";[112] that some laws and theories were not derived from observations by the inductive methods described by Bacon; and that there was scope for bold hypotheses. However, his attitude on this subject was ambivalent, ranging from an enthusiastic reference to the "exercise of pure reason" and the success of a "well imagined hypothesis", to warnings about the need to restrain speculation within limits and abandon hypotheses "when they have served their turn".[113] Agassi has emphasized the conservative pole of Herschel's ambivalence and, by judging his attitude to hypotheses against that of Popperian philosophy, has overlooked the ways in which Herschel, like Stewart and Brown, was advancing a more liberal conception of the role of hypotheses while maintaining a commitment to the Baconian framework.[114] Cannon, on the other hand, fails to note that this positive attitude was a feature of early nineteenth century readings of Bacon,[115] a fact which lowers the explanatory value of her appeal to Humboldtianism.

The impact of Herschel's *Discourse* on practising scientists such as Faraday and Darwin has been recognized, and its significance for the philosophy of science in Britain was acknowledged by both Whewell and Mill.[116] It would, however, be misleading to imply that Herschel's presentation of Baconian methodology was accepted unproblematically by other writers. It is important to distinguish between Bacon's work as a framework for the promotion and analysis of inductive method, and the appraisal of specific Baconian precepts.

These two levels were present in Whewell's response. During the early 1830s, together with Herschel and Jones, he considered the possibility of a journal embracing "moral philosophy, political economy and science" which would reform current thinking by explaining the inductive philosophy.[117] Herschel's *Discourse* and Jones's long-awaited anti-Ricardian treatise on rent were seen as crucial documents in a campaign for the propagation of "the true faith". In discussing his review of Herschel's book with Jones, Whewell remarked: "I do not know whether you looked at it in that view, but I intended it to be as good an attempt as I could make to get *the people* into a right way of thinking about induction."[118] In this context, he welcomed Herschel's praise of Bacon as the nearest comprehensive expounder of the principles of the inductive method and its profound importance for the progress of knowledge. Bacon's distinction between the Anticipation and Interpretation of Nature, his emphasis on observation and experiment, his appreciation of the complementary relation between induction and deduction – all of these, in Whewell's opinion, constituted important contributions to the philosophy of science which still had to be absorbed. However, his enthusiasm was more restrained than Herschel's and, in commenting on Bacon's inductive table, with its positive and negative instances, he doubted the relevance of "the formalities" it described, and said that "these directions of the master are noticeable principally as a curiosity".[119] Whewell was also diffident about Herschel's suggestion of using the *Novum organum* as the vehicle for the promotion of inductive philosophy in general education. While endorsing the goal, Whewell told Herschel that

> I have always doubted, however, whether Bacon's book could do much more than inspirit its readers with an admiration of the dignity and vigour which the prosecution of this object infuses into the mind. I do not think it complete and methodical enough to teach much to students; and it is clear, I think, that it requires ... to be accommodated to the present state of thought and knowledge....[120]

Thus, the praise of Bacon in Herschel's *Discourse* should not be seen as evidence of an uncritical acceptance of his thought amongst English men of science; it does, nevertheless, indicate the way in which Bacon's work was taken as the intitial framework for the consideration of the philosophy and methodology of science. The strength of this view can be judged by the response to Macaulay's attack on Bacon.

Perhaps even more than Herschel's *Discourse*, Macaulay's essay in the *Edinburgh review* of 1837 – a review of Basil Montagu's edition of the collected works – had a lasting, although confusing, influence on subsequent debates. Macaulay's extensive discussion was structured around a distinction between Bacon the philosopher and Bacon the politician, or, "Bacon seeking for Truth and Bacon seeking for the Seals".[121] In rejecting Montagu's favourable assessment of Bacon's life and character, Macaulay suggested that this distinction was like that "between the soaring angel, and the creeping snake". This condemnation of Bacon's public, political life, although usually in a less aggressive form, became a standard theme in most subsequent treatments, including the entries in the *Encyclopaedia Britannica* and the *Dictionary of national biography*.

In dealing with Bacon's thought, Macaulay was emphatic in recording the great significance of the inductive philosophy, which he regarded as the vehicle of material and intellectual progress.[122] However, he rejected the image of Bacon as the sole architect of the new science, noting that the challenge to scholasticism was already under way: Bacon was a Bonaparte, not a Robespierre. More controversially, he argued that the praise of Bacon as the discoverer of the inductive method was mistaken: "Not only is it not true that Bacon invented the inductive method; but it is not true that he was the first person who correctly analyzed that method and explained its uses".[123] Macaulay awarded some credit to Aristotle here, but his main point was that induction was tantamount to the common sense which had been practised "ever since the beginning of the world by every human being".[124] He did not doubt the ingenuity of Bacon's analysis of this method in the second book of the *Novum organum* but questioned its relevance and utility by suggesting that "the inductive process like many other processes, is not likely to be better performed merely because men know how they perform it".[125] Macaulay was sceptical about the value of systematizing the rules of this method, claiming that such an exercise could not reduce all minds to the one level. In his view, the crucial difference between the average person and the gifted scientist was not the knowledge of inductive method but the judgements involved in its practice. Arguing that Baconian rules and precepts were too general to allow mechanical application, he asked, for example, "What is slight evidence?", "What collection of facts is scanty?".[126] Questions such as these, he suggested, were integral to scientific research but could not be accommodated by simple rules. There was, therefore, a sophisticated edge to Macaulay's critique which anticipated later rejections of any simple generalizations about the mode of scientific discovery.

In 1837, however, the immediate response from major writers on the philosophy of science was decidedly negative. Brougham wrote to Napier, who had become editor of the *Edinburgh review*, saying that Macaulay had committed a great blunder, and knew nothing of the inductive philosophy.[127] Whewell regarded the essay as "scanty and superficial" and as indicative of the need to educate the public about the inductive sciences.[128] There were two related issues here. Firstly, Macaulay's reduction of induction to common sense suggested that he had failed to appreciate the distinction between enumeration and exclusion as inductive procedures. Indeed, it is likely that his article compounded the misconception about Bacon's position. In his *System of logic*, J. S. Mill felt it necessary to clarify the issue, explaining that Bacon had specifically rejected the kind of induction "natural to the mind when unaccustomed to scientific methods".[129] Without accepting the method which Bacon *did* endorse, Mill acknowledged that "it was, above all, by pointing out the insufficiency of this rude and loose conception of Induction that Bacon merited the title as generally awarded to him, of Founder of the Inductive Philosophy".[130]

Secondly, there were fears that Macaulay's rhetoric would reinforce popular prejudices about the superiority of common sense over theory and method. Brougham ridiculed Macaulay's view as tantamount to the suggestion that "all men balance themselves in order to walk and, therefore, there is no science of mechanics...".[131] Mill argued that it was precisely this lack of systematic methodology which handicapped many areas of thought, and he endorsed Whately's efforts to distinguish between method and common sense in the field of political economy. In this way, major writers such as Herschel, Whewell, Powell and Mill all defended the value of Bacon's attempt to systematize the procedures of scientific practice.[132] But this support did not prevent a critical assessment of Bacon's methodology, one which culminated in a reformulation of his historical significance. Macaulay foreshadowed the direction of this shift when he contended that Bacon's great contribution was not his analysis of the inductive method but his success in convincing men that it could be employed to acquire *new* knowledge and control of nature.[133]

WHEWELL'S ASSESSMENT

Whewell gave a balanced assessment of Bacon's significance in his *Philosophy*. Indirectly endorsing some of Napier's claims about the influence of Bacon's works on European men of science, Whewell noted the favourable responses of Gassendi, Hooke, Oldenburg, Boyle and Barrow, and referred to the connection between the Baconian programme and that of the Royal Society.[134] He acknowledged that there were other thinkers and scientists who either called for a reform in attitudes to knowledge or assisted this process by the example of their work, but he accepted the conventional claims for Bacon as "not only one of the Founders, but the supreme Legislator of the modern Republic of Science". To this extent he agreed with the historiography of Herschel's *Discourse*, and stated that:

> Bacon stands far above the herd of loose and visionary speculators who, before and about his time, spoke of the establishment of new philosophies. If we must select some one philosopher as the Hero of the scientific revolution in scientific method, beyond all doubt Francis Bacon must occupy the place of honour.[135]

However, Whewell's appraisal distinguished between the status of Bacon's writings as the most comprehensive promotion of a reforming philosophy and their specific methodological prescriptions. This distinction became standard in subsequent commentaries and Whewell's was the first detailed, critical treatment of Baconian methodology.

Bacon's greatest achievement, in Whewell's estimate, was the new direction he imparted to physical investigation, away from anticipations and premature systems towards "graduated and successive induction" as the method for the interpretation of nature.[136] Although Galileo had seen the importance of this procedure he had not produced the general and systematic expression which distinguished Bacon's work. It was this stress on successive stages of generalization which Whewell identified when he said that "the general Baconian notion of the method of philosophizing" was exemplified in Newton's discovery of universal gravitation.[137] He also dismissed Macaulay's suggestion that Bacon merely described the inductive process which "all men instinctively practise", pointing to the complex and technical procedures which Bacon formulated. Furthermore, Whewell qualified the eighteenth century image, repeated by Macaulay, which presented Bacon as answering the scholastic fixation on ideas with an emphasis on facts, gathered through observation and experiment. He granted that this was the most obvious message of Bacon's work, but contended that on closer examination his philosophy did not neglect the importance of ideas. For example, the *Novum organum* urged the need for a reform of the conceptions underlying current terminology, thereby recognizing that clear ideas were necessary for scientific progress.[138] Whewell thought that the contrast between Plato and Bacon as respective champions of the intellectual and the empirical was misplaced, because Bacon "had the merit of showing that Facts and Ideas must be combined".[139] In this way, Bacon almost appeared as the point of synthesis of the two motifs in Whewell's historiography of the philosophy of scientific knowledge:

> He held the balance, with no partial or feeble hand, between phenomena and ideas. He urged the Colligation of Facts, but he was not the less aware of the value of the Explication of Conceptions.[140]

Indeed, there is a sense in which Whewell judged Bacon's philosophy to be premature rather than mistaken; it was inevitably incomplete because it could not draw on the historical examples of actual discovery. If these had been available, Whewell believed that Bacon would have been led to cultivate more carefully "the ideal side" of his philosophy.[141]

Against the background of this general appraisal Whewell made specific criticisms of several Baconian concepts. Bacon's example of the study of the nature of heat was flawed by the priority it gave to the search for the *causes* of phenomena over the quantitative expression of the *laws* which governed them. This was a case in which the history of science since Bacon's time showed how gradual the inductive process must be.[142] Whewell also agreed with Herschel in doubting the practical value of Bacon's instances as guides to inductive research. However, in contrast with the *Discourse*, he questioned the relevance of Bacon's well-known theory of idols and the doctrine of prejudice to the study of natural philosophy, suggesting that these warnings were more applicable in the case of moral and social subjects.[143]

More generally, it is important to recognize that Whewell, unlike Herschel, made a specific criticism of Bacon's failure to appreciate "the sagacity, the inventive genius, which all discovery requires". He rejected the notion of a mechanical method which ignored this factor and the antagonism towards all conjecture which it supported. For Whewell, Newton's achievements, in spite of his own methodological pronouncements, demonstrated the crucial role of hypotheses in discovery. But Whewell was not simply placing greater stress on the heuristic value of hypotheses – he was advancing a different theory of induction, one which asserted the "necessity of a *conception* which must be furnished by the mind in order to bind together the facts".[144] This difference was highlighted in his dispute with Mill, when he insisted that Kepler superinduced the conception of an ellipse upon the observations; that such a conception was not merely a sum of the observations but something additional supplied by the mind which allowed the same facts to be *"seen under a new point of view"*.[145] Whewell saw this process as essential to all inductive discoveries. This point is relevant to any adjudication of Whewell's position in the debate about Bacon's methodology. For example, although Agassi recognizes that Whewell stressed the role of imagination, he nevertheless posits an affinity between Herschel and Whewell in terms of a commitment to the existence of a foolproof method capable of guaranteeing scientific certainty.[146] However, Whewell's attitude to this assumption was more complex than Herschel's or Mill's, precisely because he emphasized the unpredictable role of imagination and genius in providing the idea through which inductive discoveries were accomplished. This central feature of Whewell's philosophy of science was therefore in fundamental conflict with the Baconian theory of induction, but it was not until the 1850s that he explicitly acknowledged it.

The authoritative edition of Bacon's collected works, produced by James Spedding, Robert Ellis and Douglas Heath, appeared in 1857. This enterprise had its roots in earlier nineteenth century debates, insofar as Spedding had commenced Baconian studies in order to answer Macaulay's attack, and Ellis, who wrote most of the philosophical introduction, was clearly influenced by Whewell's writings on science.[147]

In his preface to the philosophical works, Ellis highlighted the connection between Bacon's methodology and his conception of scientific inquiry as a search for the 'natures' or essential forms of things. He followed Whewell and Herschel in asserting that science attempted to discern the *laws* of phenomena and not the *forms* of nature, but he developed this criticism by showing how Bacon's theory of eliminative induction presupposed a world composed of a finite number of elementary forms or natures. This exposure, he argued, made Bacon's claims for the certainty and the mechanical and accessible character of his method, even more transparent. Ellis then deployed Whewell's philosophy of science, claiming that the process by which scientific truths have been established could not be presented in the manner which Bacon described.

> In all cases this process involves an element to which nothing corresponds in the tables of comparence and exclusion; namely the application to the facts of observation of a principle of arrangement, an idea, existing in the mind of the discoverer antecedently to the act of induction.[148]

Whewell endorsed Ellis's critique and stated, more explicitly than before, that Bacon's approach was, in important respects, mistaken.

> Bacon's method, as well as his object, is vitiated by a pervading error: the error of supposing that to be done by method which must be done by mind; ... that to be done by rule which must be done by a flight beyond rule; ... that to be a work of mere labour which must also be a work of genius.[149]

Although he continued to prefer Bacon's philosophy to that of Comte, Whewell finally accepted the assessment, suggested by Ellis, which carefully specified the character of Bacon's significance:

> It is neither to the technical part of his method nor to the details of his view of the nature or progress of science that his great fame is justly owing. His merits are of another kind. They belong to the spirit rather than to the positive precepts of his philosophy.[150]

In spite of these criticisms, however, Whewell's own work betrayed a debt to the Baconian vision. His *Philosophy of the inductive sciences* contained a new version of Bacon's aphorisms, and when the third edition appeared as three separate volumes, one of them was entitled *Novum organon renovatum*. In the preface to this work he asked whether it was possible that "discoveries are made by an *Organ* which has something uniform in its structure and workings?"[151] This question proceeded from the original rationale of his project, from the expectation that the history of science should "afford us some indication of the most promising mode of directing our future efforts".[152] Whewell claimed that historical inquiry demonstrated the importance of appropriate ideas which brought facts together in new relationships. By charting these ideas and showing their crucial role in particular stages of scientific development, he believed that it was possible to establish some general principles which could guide future thinking. The fact that injunctions derived from these principles would not automatically improve the scientific powers of "such persons as Kepler, or Fresnel, or Brewster", did not reduce the value of "attempts to analyze and methodize the process of discovery".[153]

However, we have seen that Whewell vigorously rejected the Baconian notion of an accessible, mechanical set of rules which produced reliable knowledge. There was therefore a tension in Whewell's work between his acknowledgement that there was no such simple *art* of discovery and his aim of constructing a *philosophy* of discovery, founded upon historical evidence, which would reveal the process by which physical science had advanced.[154] Whewell granted that the aspirations of this latter project were less grandiose than those of Bacon's seventeenth century followers, but he insisted that the maxims it suggested, even if they did not guarantee the discovery of new truths, would prevent the repetition of error.[155] And when justifying his hopes for a philosophy of discovery he consciously aligned himself with Bacon against those who wished to restrict the term 'induction' to the sphere of logical proof. Writing to Augustus de Morgan, he defended what he called "*Discovers' Induction*" and, referring to his Inductive Tables, remarked that: "I do not wonder at your denying these devices a place in Logic; and you will think me heretical and profane, if I say, *so much the worse for Logic.*"[156]

Whewell's position illustrates the possibility of different responses to different aspects of Baconian philosophy. Another attempt to broaden the scope of the debate was made by James Spedding, who called for a distinction between Bacon's theory of induction and his project for a Natural History. In Spedding's opinion, the discussion had become narrowly focused on the second book of the *Novum organum* to the exclusion of the co-operative collection of observations which was the foundation of the Baconian programme. Spedding debated the practicality of such a collaborative scheme with Ellis, who argued that the division of labour on which it depended assumed an impossible separation of observation and theory. In reply, Spedding claimed that observations made without conscious reference to a theory could still be useful, and cited the Admiralty *Manual of scientific inquiry* which instructed naval officers to collect astronomical, meteorological, geological and other forms of scientific data.[157] Whewell was very critical of Spedding's position, arguing that "subordinate observers" must always work under the guidance of theory;[158] and yet, when explaining the notion of an "art of Discovery" to de Morgan he appeared to have second thoughts.

The projects of Solomon's House, and Cowley's College, and the like, are not quite visionary, as the British Association has shown. And though such machinery can only collect facts at first, collected facts will suggest discoveries; especially now, that we know in a good degree the way of extracting laws from facts.[159]

Thus although Whewell was the most comprehensive and incisive of Bacon's critics, his final assessment was ambivalent. In spite of its defects, Whewell saw great merit in the central features of Bacon's work.

EXPERIMENTAL SCIENTISTS VERSUS BACON

One of the most aggressive attacks upon the value of Bacon's methodology was made by David Brewster. As mentioned earlier, he rejected any substantive link between Newton's work and Bacon's method, asserting that the latter had never been "tried by any philosopher but himself". In addition to this historical claim, Brewster rejected the logic of induction which Bacon proposed. While not denying that it was a novel system, he viewed it as an example of the absurdity of attempting to advance discovery by any artificial rules.[160]

Brewster's position raises some problems which are worth discussion. At first glance, his anti-Baconianism is puzzling because his own scientific work and his interventions in scientific controversies appear to bear the hallmarks of the Baconian ideal. Brewster's research in the field of optics was pervaded by an experimentalism which, in the judgement of one recent commentator, was not guided by any particular theory, but produced "phenomena for which any theory must, in the end, account".[161] An emphasis upon the need to relate all scientific theory to empirical data also informed his warnings about the danger of hypotheses in the debates about the undulatory theory of light.[162] Secondly, the Baconian interest in applied science or technology, which, according to George Davie, was an important facet of the Scottish tradition, also informed Brewster's critique of Whewell's *History*. In his view, Whewell's perspective betrayed a systematic neglect of the achievements of Scottish scientists, such as Watt and Black, which had resulted in practical and socially useful applications.[163] Thirdly, Brewster's confessed distrust of the role of mathematics in scientific theory aligned him with Bacon. He complained that the wave theorists, especially those trained at Cambridge, placed abstract mathematical analysis above experimental data; and he accused its supporters of "scarcely ever [having] made an experiment on the subject".[164] From the other side, George Airy argued that Brewster's experimental work lacked theoretical sophistication, and remarked that "there are many persons at Cambridge who understand the subject much better than he does, though they have made few experiments".[165] Yet Brewster did not avail himself of the Baconian rhetoric which this opposition between mathematics and experiment appeared to invite. Instead, he attacked Baconianism more strongly than Whewell, who was one of the leading supporters of the wave theory of light.

Having noted this, it is necessary to be clear about the way in which Brewster's position was problematic. For example, his opposition to the wave theory of light should not be interpreted as a wholesale rejection of hypotheses – a position which would then be difficult to reconcile with his anti-Baconianism. Brewster did not totally oppose the use of hypotheses; in fact, in his review of the *Cours de philosophie positive* in 1838, he defended the value of the undulatory hypothesis against Comte's condemnation.[166] In his view, hypotheses were useful as heuristic, organizing devices, and as sources of new experiments; but he distinguished between hypotheses and true causes, and castigated Whewell for presenting the ether, upon which the undulatory theory depended, as a reality. To this extent, Brewster's methodological position was consonant with the Scottish tradition as espoused by Stewart and Brown.[167] Unlike these writers, however, Brewster made an explicit assault on Bacon.

Brewster's fundamental criticism of Baconian method was that its theory of induction did not adequately represent scientific practice or the process of scientific discovery. He claimed that Bacon's directives for the collection and classification of facts were useless because "it is impossible to ascertain the relative importance of any facts, or even to determine if the facts have any value at all, till the master fact which constitutes the discovery has crowned the zealous efforts of the aspiring philosopher". This master fact, or key relationship, then organized previously insulated phenomena. In this respect, Brewster's critique of Bacon resembled Whewell's emphasis on the role of imagination:

> The impatience of genius spurns the restraints of mechanical rules, and never will submit to the plodding drudgery of inductive discipline.... Conscious of having added to science what had escaped the sagacity of former ages, the ambitious discoverer ... forms innumerable theories to explain it, and he exhausts his fancy in trying all its possible relations to recognized difficulties and unexplained facts.[168]

Brewster warned that this inventive faculty should be restrained by experimental testing, but he underlined the gap between the actual processes leading to great discoveries and the logic of induction expounded by Bacon. Like Whewell, he referred to the case of Kepler for illustration of the essentially unpredictable features of scientific thought, and remarked that if these were not so obvious in the work of Newton, "it is because he kept back his discoveries till they were nearly perfected and therefore withheld the successive steps of his inquiries".[169]

However, Brewster believed that Whewell had not completely exorcized the Baconian dream of a mechanical art of discovery. In his reviews of Whewell's major works, Brewster criticized what he interpreted as an improper attempt to codify the process of creative scientific research. He took particular exception to Whewell's assertion that all fundamental discoveries depended upon a "distinct and well pondered idea" which ordered observed phenomena into new relationships.[170] On this view, Whewell explained, "*No scientific discovery* can, with any justice, be considered *due to accident*". Apparently chance discoveries, resting upon very few observations, or even upon a single fact, nevertheless depended upon previous "clear and appropriate ideas" in the mind of the observer.[171] Apart from viewing the epistemology associated with this perspective as a new form of scholasticism, Brewster charged Whewell with ignoring the historical examples which showed that "many of the finest discoveries in science have been the result of pure accident".[172] Moreover, he argued that the notion of "appropriate ideas" was useless as a guide to further research because the appropriateness of a conception could not be known until after a discovery was made. In his opinion, discoveries took place when new phenomena gave rise to new conjectures which then suggested further experiments and observations. All attempts to reduce this process to rules capable of simple application were misguided, and instead of Whewell's effort to derive methodological precepts from the historical record, Brewster offered moral precepts – patience, single-mindedness and perseverance.[173] In a later review of Whewell's *History*, James Forbes supported this critique, remarking that Bacon's quest for an Art of Discovery was a failure.[174] It is significant that Forbes who, as a young Scottish graduate had been impressed by Whewell's views on science and education, should join with Brewster, the scourge of Cambridge, in questioning the value of Bacon's philosophy.

The most unqualified rejection of Bacon's reputation came from the German chemist, Justus von Liebig, whose *Ueber Francis Bacon von Verulam* appeared both in German and English in 1863. The editor of *MacMillan's magazine*, the journal which published his critique, frankly admitted its extreme nature and, before printing, consulted Whewell about the desirability of a reply.[175] There were two levels to Liebig's attack: Bacon's conception of the aims of science, and his understanding of its methods.

Unlike Brewster, Liebig made the Baconian emphasis on utility a major point of criticism alleging that Bacon confused "usefulness" with "truth", thereby promulgating an impoverished account of the impulses behind scientific research.[176] In his estimate, this was the source of the atheoretical and empirical frame of mind which he regarded as the greatest obstacle to the progress of English science, or more pertinently, to the acceptance of his own chemical theories by English agriculturalists. Having confessed this personal interest, Liebig went on to castigate Macaulay as a typical purveyor of this Baconian error, and to affirm, like Whewell, an idealist image of the scientific enterprise as the pursuit of truth.[177] It is worth noting the contexts which may have influenced these reactions to Bacon. Brewster was engaged in a long-running campaign against the ascendency of Cambridge scientists and their passion for abstract mathematical science; given this, it is not surprising that he did not select Bacon's ideas on the utility of Knowledge as his central target. Liebig's response, on the other hand, can be seen in the context of his efforts to defend the priority of abstract research – a task which he believed had been successfully accomplished in Germany, where the position of science in universities had been secured, but not in Britain, where it was often judged in narrow utilitarian terms.[178]

Secondly, Liebig rejected all positive evaluations of Bacon's methodology. He suggested that Bacon had been popular with progressive sections of seventeenth century English society because of his attack upon traditional authority, but that he had never influenced scientists. For Liebig, the Baconian method of studying nature was legalistic: it searched for positive and negative qualities of phenomena as if for witnesses; it did not deal with quantities and was therefore at odds with the central feature of science.[179] In spite of Bacon's call for experimentation, Liebig alleged that his understanding of its role in science was simplistic because he failed to recognize the deductive dimension in scientific inquiry: "An experiment not preceded by a theory – that is, by an idea – stands in the same relation to physical investigation as a child's rattle to music."[180] Like Brewster, he underlined the role of imagination in inductive research and suggested that science involved both reason and fantasy.[181] Thus, by the 1850s, leading experimental scientists such as Brewster and Liebig were the most outspoken debunkers of the man who had been saluted as the father of experimental philosophy.

THE VICTORIAN VERDICT

Victorian scholars such as Spedding, Ellis and Fowler considered these attacks extreme, and in some respects misplaced; but there was general agreement about the central weakness of Bacon's methodology. The *Encyclopaedia Britannica* provided a summary of this consensus in 1875:

> The inductive formation of axioms by a gradually ascending scale is a route which no science has ever followed, and by which no science could ever make progress. The true scientific procedure is by hypothesis followed up and tested by verification; the most powerful instrument is the deductive method, which Bacon can hardly be said to have recognised.[182]

In the eighteenth century, Bacon's negative attitude towards hypotheses had been seen as part of his laudable attack on metaphysical and scholastic systems and the various prejudices which prevented the progress of knowledge. Even his neglect of mathematics as a tool of scientific explanation had not been singled out as a major deficiency. Indeed, Diderot regarded it as virtue because, in contrast with D'Alembert, he believed that the dominance of mathematics in science was excessive; he predicted that the new organic sciences would condemn this as a form of metaphysics.[183] By the mid-nineteenth century, however, an account of scientific inquiry which ignored the role of hypotheses and mathematical technique was not sustainable. Herschel, Whewell and Mill agreed on this and it became an orthodoxy within the philosophy of science. By 1872, in his *Principles of science*, W. S. Jevons felt able to state, without much argument, that the great scientific discoveries had been achieved "by the opposite method to that advocated by Bacon".[184] Significantly, as we have seen, this criticism also appeared in statements by practising scientists, where it was often associated with a scepticism about the value of all systematic methodology. T. H. Huxley, for example, wrote that:

> No delusion is greater than the notion that method and industry can make up for lack of motherwit either in science or in practical life.... Bacon's 'via' has proved hopelessly impracticable; while the 'anticipation of nature' by the invention of hypotheses based on incomplete inductions, which he specially condemns, has proved itself to be a most efficient, indeed an indispensable, instrument of scientific progress....[185]

By the second half of the century, this demise of Baconianism, both as a publicly acknowledged guide to the practice of science, and as a formal methodology, was largely complete. This decline cannot adequately be understood without an appreciation of the kind of shifts which occurred in the philosophy of science and the ways in which these might have related to developments in scientific thought. During the eighteenth century, in debates about theories involving imperceptible fluids, the use of hypothesis was castigated as a departure from the teachings of both Bacon and Newton. However, during the nineteenth century the "method of hypothesis" achieved a certain orthodoxy.[186] Larry Laudan has argued that Herschel and Whewell, following the earlier work of Stewart, effected a radical shift in methodology which justified the use of hypotheses at the same time as they were defending the undulatory theory of light. He cites this as an example of the way in which changes in epistemology and methodology reflect the need to legitimize prior substantive developments in scientific theory.[187] From this perspective, the nineteenth century critique of Bacon, insofar as this mainly focused on his inadequate treatment of hypotheses and mathematics in science, could be ascribed to the emergence of new styles of scientific practice such as mathematical physics and optics. That is, Baconian methodology had to be severely qualified because it conflicted with the character of successful science.

This explanation is feasible, but some caveats are in order. While it is true that some of the major protagonists in the debate over the wave theory of light were also involved in the appraisal of Bacon, the match between those who supported the new theories and those who attacked Bacon is not perfect. Brewster was an aggressive debunker of Bacon, yet he also criticized the undulatory hypothesis and its supporters. Conversely, Herschel and Whewell, who promoted the wave theory, did not totally reject Bacon's methodology, but tried to adjust it to contemporary science. Nevertheless, it is arguable that the direction of this reinterpretation was partly determined by the increasing use of hypothetical and mathematical constructs in physical science.

There was also another sense in which these developments diminished one of the defining features of Baconianism. Once conjecture and hypothesis rather than inductive generalization from observation were admitted as intrinsic to scientific inquiry, the Baconian conception of methodology as the systematic presentation of an art of discovery became problematic. The focus of attention in the philosophy of science moved away from the process of discovery, which was seen as a region of unpredictable factors, to the elaboration of methods of verification and evaluation. Whewell attempted to salvage something of Bacon's programme, arguing that a philosophy of discovery was desirable, but his own theory of induction, which stressed the role of imaginative leaps, was radically at odds with Bacon's, and with the notion of a simple art or engine of discovery. By the end of the century Bacon no longer had a central place within the philosophy of science.

As we have seen, the major issue in these discussions of Bacon was his theory of induction and the key text his *Novum organum*. In 1818 Coleridge said that this was "venturing his reputation on a tottering basis", since the project of obtaining universals from particulars was impractical and would expend "the life of an antediluvian patriarch".[188] Coleridge thought there was more to Bacon than induction; but the Victorian debate largely took this as its focus and rejected Bacon's methodology as an account of scientific inquiry. By 1875 the *Encyclopaedia Britannica* had to ask a question about the source of Bacon's reputation:

> If, then, Bacon himself made no contributions to science, if no discovery can be shown to be due to the use of his rules, if his method be logically defective, and the problem to which it was applied one from its nature incapable of adequate solution, it may not unreasonably be asked, how has he come to be looked upon as the great leader in the reformation of modern science?[189]

The answer took the form of a distinction between Bacon's method and his wider intellectual influence. Napier allowed the possibility that Bacon was not the sole discoverer of inductive method, but did not question the validity of this method, and defended his importance as its greatest publicist. With various emphases, the Victorians took the view that Bacon's methodology, although an original formulation, was an inadequate account of scientific thinking. However, they continued to affirm his historical significance, arguing that this derived from the framework he had provided for philosophical discussion of science and from his unrivalled ability as a popularizer of the scientific enterprise. Most of the major assessments of Bacon in the late nineteenth century took this form. At the same time, the appraisal of his political behaviour and moral character underwent some redemptive process. Huxley continued to refer to "that sneak Bacon", but other commentators softened their tone. Thus, writing to de Morgan about recent attitudes to Bacon, Whewell was able to speculate that from being "the greatest, brightest, meanest of mankind" in the eighteenth century, his modern epitaph might be "ambitious theorist, but honest, man".[190]

BACONIANISM AND THE RHETORIC OF SCIENCE

The last section has provided an account of the nineteenth century debate on the philosophy of science in which Bacon's methodology was considered, and largely rejected, as an account of scientific practice. During the course of this debate, however, Baconian notions about proper scientific procedure continued to be influential in scientific controversies and in the public apologetics of science. Appeals to Baconian method remained common into the second half of the century – for example, much of the criticism of Darwin's *Origin of species* included a condemnation of its departure from the precepts of Baconian induction. As mentioned earlier, Susan Cannon has criticized historians who have employed Baconianism as a descriptive or explanatory tool with reference to nineteenth century British conceptions of science, arguing that no major men of science accepted Bacon's account of scientific method. The previous section suggests that this generalization requires qualification; but furthermore, it is necessary to distinguish between the different levels of discourse on method in which Baconianism was involved.

Three such levels can be delineated. Firstly, there is the comparatively formal discussion within works on the philosophy of science dealing with induction, deduction, hypotheses, verification and other issues. In the nineteenth century, the systematic treatises of Herschel, Whewell, Mill and Jevons would fall within this category. However, this level, although apparently abstract, should not be divorced from substantive developments in scientific practice. In the assessment of Baconian method this is most obviously exemplified by the interventions of Brewster and Liebig, but it is also arguably a factor in the more systematic presentations of Herschel and Whewell.

Secondly, there is the level at which statements about method function as argumentative devices within scientific controversies, such as those concerning the wave theory of light or Darwin's theory of evolution. Apart from substantive attacks upon a particular doctrine, participants in such disputes often advance methodological critiques designed to show infringements of 'normal' or 'orthodox' scientific procedure. This appeal to method is often most noticeable in cases which involve conflicts between disciplines, such as the argument between geologists and physicists about the age of the earth. In these situations, the recourse to methodological statements is often connected with institutional struggles within the scientific community and can therefore include social as well as intellectual factors. Thirdly, there is the level at which method enters into the public apologetics of science, serving to advertise its political and cultural value.[191] The following section attempts to reveal the ways in which Baconianism acted as a rhetorical resource at this last level of discourse.

As noted earlier, Baconian philosophy has been a resource not only for discussions of the philosophy of science but for statements about its social and political significance. As such, it was exploited by Puritan reformers, the apologists of the Royal Society, and the French *philosophes*. Bacon's contrast between a widely accessible method and the dogma of authority provided the terms for an egalitarian conception of knowledge and a critical attitude to established institutions. This dichotomy informed some of the enlightenment attacks against the Church in Europe, which was presented as a bastion of scholastic prejudice. The idea of a democratic method assumed a more radical aspect in the Jacobin assaults on the learned academies when, combined with a celebration of practical craftsmen – another Baconian preoccupation – it was opposed to the abstruse terminology of the physical sciences. On the other hand, Bacon's stress on empirical observation and collaborative endeavour could be used to present scientific method as the model for political stability. This strategy was adopted by the early Royal Society, and it had a similar appeal in Britain after the Napoleonic wars as a means of separating science from theological controversy, and from the sensitive political issues of an industrializing society.[192]

The political uses of Bacon in nineteenth century Britain were not extensive and were never radical. However, Baconian doctrines were linked with the more moderate English version of reform: the assault upon intellectual arrogance and indolence, and the stress upon careful methods of inquiry as the key to the redemption of man's estate, were compatible with the reforming zeal of both evangelicals and utilitarians. From the 1820s, both these groups were able to draw upon Baconian philosophy as a justification for a new form of practical, empirical instruction for the lower classes. The Scottish evangelical preacher, Thomas Chalmers, saw Bacon's method as encouraging the virtues of patience and humility, and he urged the careful collection of social data as the basis for a solution of social problems.[193] Henry Brougham, the Whig chancellor, believed that the writings of Bacon, if properly studied, would inculcate sound attitudes to knowledge and encourage wide participation in science.[194] Even Whewell, later a Tory appointment as Master of Trinity, caught some of this spirit in 1831 when he wrote to Richard Jones, the political economist, describing Baconianism as a reforming philosophy, and hoping to "get *the people* into a right way of thinking about induction".[195] But these various uses of Bacon never involved the controversial interpretations advanced by earlier commentators. It was probably this contrast which led the radical publisher, Richard Carlile, to suggest that Bacon, Newton and Locke, had been captured by conservative forces and were now "claimed as the patrons of superstition".[196] It is perhaps symptomatic of this appropriation that Sir James Macintosh urged Basil Montagu to give up his study of William Godwin, the anarchist thinker, to edit a new edition of Bacon's works.[197]

During the eighteenth century, the Baconian attack on the systems and dogmas of traditional institutions was an appealing polemical weapon; but by the nineteenth century it was put to more conservative uses. The idea of an accessible method was not usually contrasted with the dogmas of established authority, but with the dangers of intellectual speculation in science. Appeals to Bacon were most commonly associated with attacks upon hypotheses and theoretical conjecture, and with a celebration of factual observation. This is the outlook which Kingsley had in mind when he described the Baconian method as God's gift to Englishmen. Alternatively, Liebig saw the same disposition as a national affliction preventing an adequate space for theory in English science.

There was a strong emphasis on this priority of observation over hypothesis in the new organic sciences, in which the need for large collections of data could be presented as a feasible starting point. Several historians have noted the prominence of Baconianism, as the intellectual authority for such a programme, in the early years of the Geological Society of London. The public statements of the Society violently rejected the eighteenth century theories of the earth as fantastic speculation; the work of Burnet, Woodward and Whiston was condemned for its failure to honour strict inductivism. This unfavourable portrait of early theories of the earth informed a historiography which, with its story of the gradual emancipation of a sober, inductive science from the chaos of myth and superstition, became a set piece in the accounts of geology in textbooks and encyclopaedias.[198] It was largely structured around themes popularized by Bacon's attack upon scholasticism and the habitual fallacies of human reasoning. The influence of these formulations is discernible in Charles Lyell's historical introduction to his *Principles of geology*, in which the errors of past theories about the earth were ascribed to Baconian idols such as abstract philosophical systems defended by entrenched élites, and to anthropomorphic projections.[199] On the evidence of this historical background, Lyell agreed that the emphasis of the Geological Society on careful collection of data was salutary, although he remarked that "the reluctance to theorise was carried somewhat to excess".[200]

In this Baconian rhetoric of the Geological Society, there was a dichotomy between method and theory. Apart from allowing the activities of its members to be safely distanced from the controversial speculations of the past, this contrast made it possible to present geology as a subject open to many people. The method of Bacon was conceived as one in which observations could be made and collected independent of theory, thus making it a procedure accessible to those without training or prior knowledge of the subject. One spokesman suggested that the absence of these could be an advantage since the minds of such novices would be unbiased.[201] Geology, more so than other branches of science, was presented as one which invited large-scale involvement of people from various backgrounds, and since the principal requirements were seen to be those of careful observation and recording of data, "the miner, the quarrier, the surveyor ... and even the Traveller in search of general information" were all seen as potential contributors.[202]

This egalitarian image of science was often expressed in a further contrast between method and genius. On this view, the great scientific achievements did not result from the creative efforts of gifted individuals, but from the proper application of Bacon's method. This rhetoric became part of the justification of the programme for popular science education sponsored by Brougham's Society for the Diffusion of Useful Knowledge. Thus, while recognizing the greatness of Newton, accounts of his work which appealed to the notion of genius were displaced by references to his common qualities. Thus, one writer reported that the famous natural philosopher had said that "if there was any mental habit or endowment in which he excelled the generality of men, it was that of patience in the examination of the facts and phenomena of his subject". This approach was clearly compatible with the philosophy of 'self help' proselytized by Samuel Smiles, because this re-definition of genius as "common sense intensified" reduced all achievement, both intellectual and economic, to concepts such as hardwork, humility and perseverance.[203]

But it was also closely related to Baconian themes: Newton's discoveries, it was agreed, had resulted from his careful use of the inductive method, a method which did not require exceptional mental ability but rather rigorous application of rules. The case of Kepler was more difficult to fit with this perspective since, as the biography in the Library of Useful Knowledge explained, his discoveries seem to have resulted from his good fortune in "seizing truths across the wildest and most absurd theories". This example was not interpreted as a challenge to the stress on gradual induction as the method of science, but as a warning about "the danger of attempting to follow his [Kepler's] method in the pursuit of truth".[204] In these ways, a denigration of the role of individual genius in the history of science reinforced the contention that progress in science resulted from the co-operative and essentially egalitarian process of observation and induction.

Thus, while Baconian theory may not have been accepted as a philosophical account of science or a guide to its practice, it was still important as a rhetorical resource in controversies about substantive theories and in statements about the cultural value of science. But by the last quarter of the nineteenth century, this use of Baconianism also began to decline, partly because its demise as a philosophy of science had become more complete, but also because of changes in the social conditions under which science was pursued. The reference to Baconianism in public statements about science related, in part, to the need to demarcate scientific inquiry from conjectures and speculation which had controversial theological connotations. But with increasing secularization of education – for example, the abolition of the religious tests at Oxbridge – some of the earlier rhetorical uses of Baconianism became redundant. These changes removed the institutional conditions which had demanded accommodation between religion and science: clerical scientists disappeared and natural theology became less necessary. In this situation, scientists did not have to be so defensive about the cultural implications of their activities. An early indication of this shift is apparent in the debates in the British Association about the role of amateur observers. Whewell, Herschel and Brewster, who did not agree about the general merits of Bacon's work, were united in criticizing the more egalitarian conceptions of research which were being justified in Baconian terms.[205] It is arguable that the greater emphasis on hypothesis and imagination in the philosophy of science of the late nineteenth century reflects not only the demise of Baconian methodology, but the growing confidence and autonomy of a professional scientific community.[206]

CONCLUSION

During the nineteenth century, the most widely quoted passage of Bacon's works was that which contained his doctrine of idols. Those who could not accept the positive side of his programme could find various uses for his classification of the obstacles which stand in the way of truth. In the first book of the *Novum organum*, which assessed the current state of knowledge and its weaknesses, Bacon claimed that there were four perennial sources of error. The idols of the Tribe and the Cave derived, respectively, from the nature of the human mind in general, and from the peculiar prejudices of each man's mind. The idols of the Theatre were external in origin, stemming from the perpetuation of entrenched philosophical systems. The fourth class, the idols of the Market Place, were, in Bacon's view, "the most troublesome of all". These were idols "which have crept into the understanding through the alliance of words and names. For men believe that their reason governs words; but it is also true that words react on the understanding...".[207]

There is a sense in which the term Baconianism is one of the greatest idols of the Market place. As Macaulay remarked, even Bacon's most famous works were "much talked of, but little read".[208] This asymmetry is one of the factors which makes the status and significance of Bacon's ideas so difficult to assess. But furthermore, when his writings *were* read, they sponsored various interpretations and, in turn, different intellectual and political uses. Charles Webster was certainly correct in observing that during the seventeenth century Bacon's works were "as flexible and appealing as the Bible";[209] and this also applies to their career since that time. Like similar historiographic categories, such as Newtonianism and Darwinism, Baconianism is an unavoidable but problematic concept; and its degree of complexity is compounded by the fact that it does not refer to substantive data or theory, but to the methodology of science, to the institutional organization of research, and to the social and moral dimensions of the scientific enterprise. In order to understand the discourse on Baconianism as a historical phenomenon, historians need to heed Bacon's warning, to look beyond names and terms so as to recognize the various levels at which Baconianism has been discussed and deployed.

This article has attempted to analyse some of the interpretations, images and uses of Bacon's ideas in different historical situations. It deals mainly with the nineteenth century, but sets this against the background of Baconianism in earlier periods. This perspective reveals the way in which the discussion of Bacon's work narrowed in compass from an interest in the full range of his ideas on natural histories, collaborative research, technology, and the social and moral relations of science, to his views on methodology. Undeniably, the latter was a major part of the Baconian programme, but it is important to recognize that a concentration on Bacon as the theorist of induction correlated with a suppression or avoidance of eighteenth century readings of Bacon as the author of a radical philosophy capable of transforming not only natural knowledge, but established social institutions and values.

Apart from this negative process of selection, it is possible to suggest other reasons for the interest in Bacon's methodology. By the 1830s the philosophy of science began to emerge as a distinct and formal area of discourse, and Bacon's writings provided a framework, distinct from epistemology, for the discussion of scientific method. Although Baconian methodology was qualified, it is important to recognize that most early commentators tried to read Bacon in a way which avoided outright rejection of his philosophy. The early Scottish debate was concerned more with his originality and historical influence than with the validity of his method, and when this came to be the focus in the work of Herschel, Whewell, and Mill, only certain aspects were explicitly attacked. All of these writers defended Bacon against Macaulay in what became an affirmation of the need for systematic analysis of methodological procedures in all areas of thought; and some found it possible to make the role of hypotheses compatible with Bacon's work. There were tensions in the evaluations of Bacon, partly because of the protean nature of his writings, but more especially because his concept of an art or method of discovery, although judged to be flawed, still had attraction. The positions of Whewell and Spedding illustrate this and Whewell's work, in particular, reflects a deep reluctance to abandon Bacon; his criticisms existed alongside the suggestion that Bacon's philosophy was premature rather than mistaken. Ultimately, however, the damaging attacks on Bacon came from practising, experimental scientists such as Brewster and Liebig. Their rejection of his account of scientific discovery coincided with a shift of interest in philosophy of science to methods of evaluation, a development which needs to be related to the increasing use of hypothetical constructs in science. Bacon's reputation then came to rest not on the merits of his methodology, but on his role as a promoter of the scientific spirit.

It is also possible that the early nineteenth century interest in methodology was related to the contemporary debates about the cultural and institutional status of science.[210] Certainly, statements about scientific method were prominent in the public pronouncements of scientific societies and associations. Here again, Baconian themes were circulated, especially in relation to issues such as the accessibility of science, the division of labour, and the role of observation and theory. Spokesmen for new scientific disciplines such as geology were particularly concerned to legitimate their activities, and this often resulted in expressions of Baconian doctrines which were simultaneously undergoing qualification in the more formal works on philosophy of science. The prevalence of these references has led some historians to speak of a Baconian style or commitment as characteristic of certain areas of nineteenth century British science.[211] In reaction to this, at least one author has questioned the value of Baconianism as a term for the description or explanation of attitudes to science in this period.[212] This article has sought to open territory between these two positions by clarifying the different levels of discourse about Baconian ideas, and by revealing their various roles in the philosophy and the social relations of science.

ACKNOWLEDGEMENTS

The author wishes to thank the Editor and the anonymous referees for their valuable comments on earlier drafts of this article. The research for this topic was supported by the School of Humanities, Griffith University.

REFERENCES

1. A. Cowley, in T. Sprat, *The history of the Royal Society of London* (London, 1667), Stanza V, no page.
2. Quoted in M. Hunter, *Science and society in Restoration England* (Cambridge, 1981), 14.
3. For a generally critical account, see K. Popper, "On the sources of knowledge and ignorance", in *Conjectures and refutations* (London, 1974), 3-32. Although neglected by most recent

philosophers of science, there have been some sympathetic treatments of Bacon. See M. Hesse, "Bacon", in C. C. Gillispie (ed.), *Dictionary of scientific biography*, i (New York, 1970), 372-7; M. Horton, "In defence of Francis Bacon: A criticism of the critics of the inductive method", *Studies in the history and philosophy of science*, iv (1973), 241-78. Alexandre Koyré suggested that historians of science should also be sceptical about Bacon's link with the scientific revolution: "Bacon, the founder of modern science, is a joke.... In fact Bacon understood nothing about science. He was credulous and completely uncritical". A. Koyré, *Galileo studies,* trans. by J. Mepham (Brighton, 1978), 39. See also T. Kuhn, *The essential tension* (Chicago, 1977), 116-17. For more general accounts, see P. Rossi, *Francis Bacon: From magic to science* (London, 1968); L. Jardine, *Francis Bacon: Discovery and the art of discourse* (London, 1974); A. Quinton, *Francis Bacon* (Oxford, 1980).

4. Appeals to Bacon were apparent at the time of the foundation of the British Association for the Advancement of Science, which was presented as a realization of the vision of collaborative scientific research. See J. Morrell and A. Thackray, *Gentlemen of science: Early years of the British Association for the Advancement of Science* (Oxford, 1981), 159, 264-7.

5. T. Fowler, *Bacon* (London, 1881), 200.

6. C. Kingsley, "How to study natural history", in *Works* (28 vols, London, 1880-85), xix, 308.

7. C. Darwin, *Autobiography,* ed. by G. de Beer (Oxford, 1974), 71. See also N. C. Gillespie, *Charles Darwin and the problem of creation* (Chicago and London, 1979), 43, 65.

8. C. Gillispie, *The edge of objectivity* (Princeton, N.J., 1960), 314.

9. C. Gillispie, *Genesis and geology* (New York, 1951), 128.

10. M. Millhauser, *Just before Darwin: Robert Chambers and "Vestiges"* (Middletown, Conn., 1959), 144.

11. S. F. Cannon, *Science in culture: The early Victorian period* (New York, 1978), 73.

12. *Ibid.*, 74.

13. *Ibid.*, 229.

14. For a general account see P. Rossi, "Francis Bacon", in P. Wiener (ed.), *Dictionary of the history of ideas* (4 vols, New York, 1973), i, 172.

15. J. Murray *et al.* (eds), *The Oxford English dictionary* (12 vols, Oxford, 1933), i, 617.

16. C. Webster, *The Great Instauration: Science, medicine and reform 1626-60* (London, 1975), 499. See also pp. 492-500.

17. See C. Webster, "The origins of the Royal Society", *History of science*, vi (1967), 106-28; J. Ben David, *The scientist's role in society: A comparative study* (New York, 1977), 72-74; P. Wood, "Methodology and apologetics: Thomas Sprat's *History of the Royal Society*", *The British journal for the history of science*, xiii (1981), 1-26. In various ways, these authors question the historiography which accepted seventeenth century statements about Baconianism as unproblematic evidence of prevailing methodological commitments.

18. J. D'Alembert, *Preliminary discourse to The encyclopedia,* trans. by R. N. Schwab (New York, 1963), 74. For the observations on Bacon's "division of the sciences", *ibid.*, 143-64. See also L. K. Luxembourg, *Francis Bacon and Denis Diderot* (New York, 1967).

19. *The Plan of the French Encyclopaedia: or universal dictionary of arts, sciences, trades and manufactures* (London, 1752), 178-9.

20. D. Stewart, *Preliminary dissertation exhibiting a general view of progress of metaphysical, ethical and political philosophy since the revival of letters in Europe,* in *Encyclopaedia Britannica* (7th edn, 21 vols, Edinburgh, 1830-42), i, 56.

21. P. Shorr, *Science and superstition in the eighteenth century: A study of the treatment of science in two encyclopaedias of 1725-1750* (New York, 1932).

22. E. Chambers, *Cyclopaedia* (2nd edn, 5 vols, London, 1738), i, no page.

23. D. Diderot and J. D'Alembert, *Encyclopédie, ou dictionnaire raisonné des sciences, des arts et des métiers* (27 vols, Paris, 1757-80), ii, 8-10. For identification of author, see J. Lough, *The*

II

Encyclopédie (London, 1971), 141.

24. See J. de Maistre, *Examen de la philosophie de Bacon* (2 vols, Paris, 1836); also J. Lively (ed.), *The works of Joseph de Maistre* (London, 1965); for the defence of Bacon, see J. A. de Luc, *Précis de la philosophie de Bacon* (Paris, 1802).

25. K. Fischer, *Francis Bacon of Verulam: Realistic philosophy and its age*, trans. by J. Oxenford (London, 1857), pp. xii-xvii. For his account of the nineteenth century French debate, see pp. 326-40.

26. J. Morley, *Diderot and the Encyclopaedists* (2 vols, London, 1878), i, 116.

27. See for example R. A. Soloway, *Prelates and people: Ecclesiastical social thought in England, 1783-1852* (London, 1969), 26-29, 34, 45; E. R. Norman, *Church and society in England 1770-1970* (Oxford, 1976), 16, 20-23; N. Garfinkle, "Science and religion in England, 1790-1800: The critical response to the work of Erasmus Darwin", *Journal of the history of ideas*, xvi (1955), 376-88.

28. F. Bacon, *The advancement of learning*, ed. by A. Johnston (Oxford, 1974), 94-95 and *Novum organum*, ed. by F. H. Anderson (New York, 1960), 121-2.

29. D. Stewart, *Elements of the philosophy of the human mind* (2 vols, Edinburgh, 1814), ii, 443-61.

30. See W. Whewell, *Indications of the creator* (2nd edn, London, 1846), 64-66, 84-87, 91-95. Even Baden Powell, who was critical of traditional teleology, felt it necessary to distinguish Bacon's views on final causes from the more extreme attack of Comte; see B. Powell, "Comte's system of positive philosophy", *Monthly chronicle*, iii (1839), 227-38, pp. 229, 235. Authorship of anonymous reviews has been identified in W. H. Houghton (ed.), *Wellesley index to Victorian periodicals* (3 vols, Toronto, 1966-76).

31. See T. Fowler (ed.), *Bacon's Novum organum* (2nd edn, Oxford, 1889), 106.

32. This appeared as the general preface to the *Encyclopaedia metropolitana*, which was based on Coleridge's classification of knowledge. The editor, H. J. Rose, was a High Church theologian who saw Coleridge's philosophy as an answer to the materialist and empiricist heritage of the Enlightenment. See D. Forbes, *The liberal Anglican idea of history* (Cambridge, 1952), for this perspective.

33. S. T. Coleridge, "General introduction; or preliminary treatise on method", in *Encyclopaedia metropolitana* (29 vols, London, 1845), i, 32.

34. *Ibid.*, 27.

35. *Ibid.*, 24-25.

36. *Ibid.*, 32.

37. J. B. Stallo, *General principles of the philosophy of nature* (Boston, 1848), vii.

38. G. H. Lewes, *A biographical history of philosophy* (4 vols, London, 1845-46), iii, 52, 263. See also A. Finch, *On the inductive philosophy, including a parallel between Lord Bacon and A. Comte as philosophers* (London, 1872).

39. J. Doherty, "Flaws in the philosophy of Bacon", in H. Edward (ed.), *Essays on religion and literature* (London, 1874), 309. For a similar view, see F. H. Laing, *Lord Bacon's philosophy* (London, 1877).

40. Bacon, *Novum organum*, ed. by Anderson (ref. 28), 115-16, 118-19. For the ambivalence attached to the idea of utility in the Restoration, see Hunter, *op. cit.* (ref. 2), ch. 4.

41. J. Herschel, *A preliminary discourse on the study of natural philosophy* (London, 1831), 44-74; W. Whewell, "Modern science — inductive philosophy", *Quarterly review*, xiv (1831), 374-407, pp. 404-5.

42. T. B. Macaulay, "Francis Bacon", *Edinburgh review*, lxv (1837), 1-104, pp. 65, 101. On the utilitarian conception of science and its uses, see W. Bowring, "Education of the people", *Westminster review*, vii (1827), 269-317, and *idem*, "Scientific education of the upper classes", *Westminster review*, ix (1828), 328-73.

43. Herschel, *op. cit.* (ref. 41), 10.
44. W. Whewell, *The philosophy of the inductive sciences* (2 vols, London, 1840), i, p. xiii.
45. *Ibid.*, ii, 576, 464. See also J. Herschel, "Whewell on the inductive sciences", *Quarterly review*, lxviii (1841), 177-238, pp. 177-8, 185.
46. See W. Whewell, "Francis Bacon", *Edinburgh review*, cvi (1857), 287-322, pp. 320-2 for praise of Remusat's non-utilitarian reading. For Whewell's philosophical response to utilitarianism see R. Yeo, "William Whewell, natural theology and the philosophy of science in mid-nineteenth-century Britain", *Annals of science*, xxxvi (1979), 493-516. See Fischer, *op. cit.* (ref. 25), 382-405 for criticisms of Macaulay. For an emphasis on the practical facets of Baconianism, see Webster, *op. cit.* (ref. 16), 494-6, 519-20; Rossi, *op. cit.* (ref. 3), ch. l; and B. Farrington, *Francis Bacon: Philosopher of industrial science* (London, 1949).
47. L. Laudan, *Science and hypothesis: Historical essays on scientific methodology* (Dordrecht, Boston and London, 1981), 34.
48. P. Shaw (ed.), *The philosophical works of Francis Bacon, methodized and made English* (3 vols, London, 1733), i, pp. iii-viii.
49. Voltaire, *Letters concerning the English nation* (new edn, London, 1767), 64-65.
50. H. Hallam, *Introduction to the literature of Europe* (4 vols, London, 1838-39), iii, 225-7; also B. Montagu, "Bacon's *Novum organum*", *Retrospective review*, iii (1821), 141-67, p. 141, on the greater popularity of the *Essays*.
51. See C. Wordsworth, *Scholae academicae: Some account of the studies at the English universities in the eighteenth century* (Cambridge, 1877), 78. For the number of editions see R. W. Gibson, *Bacon: A bibliography of his works and of Baconiana to the year 1750* (Oxford, 1950).
52. *A new royal and universal dictionary* ... (London, 1769), entry on Bacon; T. H. Crocker (ed.), *The complete dictionary of the arts and sciences* (3 vols, London, 1764).
53. In the eighteenth century, references to Bacon were often associated with praise of Newton's anti-conjectural method. See the works of Pemberton, Turnbull (who was Reid's teacher) and Goldsmith cited in Laudan, *op. cit.* (ref. 47), 103-8.
54. D. Hume, *The history of England* (new edn, 6 vols, Boston, 1851), iv, 525.
55. Hallam, *op. cit.* (ref. 50), iii, 227.
56. L. Laudan, "Thomas Reid and the Newtonian turn of British methodological thought", in Laudan, *op. cit.* (ref. 47), 86-110; R. Olson, *Scottish philosophy and British physics 1750-1880* (Princeton, 1975).
57. G. Davie, *The democratic intellect: Scotland and her universities in the nineteenth century* (Edinburgh, 1961), 189-90, 321; T. H. Buckle, *Introduction to the history of civilization in England* (new and revised edn, London, n.d.), 834.
58. D. Stewart, *Account of the life and writings of Thomas Reid* (Edinburgh, 1802), 27.
59. Reid to R. Gregory, quoted *ibid.*, 41-42.
60. T. Reid, *Works* (Edinburgh, 1863), 711-13.
61. Stewart, *op. cit.* (ref. 20), 32.
62. J. Playfair, *Dissertation third: Exhibiting a general view of the progress of mathematical and physical science since the revival of letters in Europe*, in *Encyclopedia Britannica* (7th edn), i, 468-74.
63. M. Napier, "Remarks illustrative of the scope and influence of the philosophical writings of Lord Bacon", *Transactions of the Royal Society of Edinburgh*, viii (1818), 373-425, p. 373.
64. *Ibid.*, 374-6.
65. *Ibid.*, 383, 386, 391.
66. Stewart, *op. cit.* (ref. 58), 44, and *idem*, *Elements of the philosophy of the human mind* (new edn, Edinburgh, 1843), part ii, 545.

67. J. Macintosh, "Stewart's introduction to the Encyclopaedia", *Edinburgh review*, xxvii (1816), 180-244, p. 186.
68. Napier, *op. cit.* (ref. 63), 385, 387.
69. "Stewart's Dissertation", *Quarterly review*, xvii (1817), 50-53.
70. Napier, *op. cit.* (ref. 63), 378-80.
71. *Ibid.*, 396-409.
72. *Ibid.*, 404.
73. "Stewart's Dissertation" (ref. 69), 52.
74. J. Macintosh to M. Napier, 8 January 1832, in M. Napier (ed.), *Selections from the correspondence of the late Macvey Napier* (London, 1879), 33-34; Napier, *op. cit.* (ref. 63), 384-5.
75. Napier, *op. cit.* (ref. 63), 391.
76. T. Young to M. Napier, 23 April 1818, Napier Papers, British Library, Add Ms. 34,612, ii, f. 188. I am grateful to Geoffrey Cantor for this reference.
77. J. Mill to M. Napier, 30 April 1818, in Napier (ed.), *op. cit.* (ref. 74), 18-19. For the general Scottish interest in Bacon, and its political dimension, see J. C. Robertson, "A Bacon-facing generation: Scottish philosophy in the early nineteenth century", *Journal of the history of philosophy*, xiv (1976), 35-45, p. 41, and S. Collini, D. Winch and J. Burrow, *That noble science of politics: A study in nineteenth-century intellectual history* (Cambridge, 1983), 48-49, 52-53.
78. T. Brown to M. Napier, 14 April 1818, in Napier (ed.), *op. cit.* (ref. 74), 17.
79. D. Stewart, *Elements of the philosophy of the human mind* (London, 1792), 8-9.
80. Robertson, *op. cit.* (ref. 77), 43-48; Davie, *op. cit.* (ref. 57), 176.
81. Napier, *op. cit.* (ref. 63), 382.
82. Olson, *op. cit.* (ref. 56), 6, 15.
83. D. Stewart, *Collected works*, ed. by W. Hamilton (10 vols, Edinburgh, 1854-58), iii, 299.
84. *Ibid.*, 303.
85. *Ibid.*, 289, 298, 301.
86. *Ibid.*, 305, 309.
87. *Ibid.*, 302.
88. J. Robison, "Philosophy", *Encyclopaedia Britannica* (7th edn), xvii, 426-46, p. 443.
89. J. Forbes, "On the inductive philosophy of Bacon, his genius and achievements" (Moral philosophy essays, 1827-28), 24-25, 27, 39-40, Forbes Papers, St Andrews University. See also the reference to this in Olson, *op. cit.* (ref. 56), 226.
90. H. Brougham to M. Napier, 18 March 1827, in Napier (ed.), *op. cit.* (ref. 74), 47-48.
91. Olson, *op. cit.* (ref. 56), chs 4-5.
92. Stewart, *op. cit.* (ref. 83), iii, 314-16.
93. M. Napier, *Treatises on various subjects of natural and chemical philosophy by Sir John Leslie, with a biographical memoir* (Edinburgh, 1838), 49. On Leslie's criticism of "dull experiment", see G. N. Cantor, "The academy of physics at Edinburgh 1797-1800", *Social studies of science*, v (1975), 109-34, pp. 133-4.
94. D. Brewster, *The life of Sir Isaac Newton* (London, 1831), 330. See also D. Brewster (ed.), *Edinburgh encyclopaedia* (18 vols, Edinburgh, 1830), iii, 181-6, which contained some critical comments. Brewster told Napier that "Macaulay's article is splendid. It would have killed Playfair, who took me to task for inserting a similar view of Bacon's character (written by Dr Lee) in the Edinburgh Encyclopaedia". Brewster to Napier, 27 July 1837, in Napier, *op. cit.* (ref. 74), 194.
95. D. Brewster to M. Edgeworth, 26 April 1824 in M. Gordon, *The home life of Sir David Brewster* (2nd edn, Edinburgh, 1870), 24.
96. Quoted in Olson, *op. cit.* (ref. 56), 238; see also pp. 178-86 for the view that Brewster's

methodological views did not radically depart from those of Stewart and Brown.
97. B. H. Malkin, "Brewster's life of Newton", *Edinburgh review*, lvi (1832), 1-37, p. 37.
98. Fowler (ed.), *op. cit.* (ref. 31), 152.
99. Herschel, *op. cit.* (ref. 41), 104-5; also p. 188.
100. *Athenaeum*, 15 January 1831, 38-39, p. 38.
101. Whewell, *op. cit.* (ref. 41), 377.
102. *Ibid.*, 379, 398.
103. J. Agassi, "Sir John Herschel's philosophy of success", *Historical studies in the physical sciences*, i (1971), 1-36, pp. 1-2.
104. *Ibid.*, 20-21.
105. Herschel, *op. cit.* (ref. 41), 272; also p. 188.
106. *Ibid.*, 164.
107. Cannon, *op. cit.* (ref. 11), 232; W. F. Cannon, "John Herschel and the idea of science", *Journal of the history of ideas*, xxii (1961), 215-39, esp. pp. 221-2; Agassi, *op. cit.* (ref. 103), 8.
108. Agassi, *op. cit.* (ref. 103), 11-12.
109. Cannon, *op. cit.* (ref. 11), 73-74, 228-9.
110. Agassi, *op. cit.* (ref. 103), 27.
111. Cannon, *op. cit.* (ref. 11), 93.
112. J. Losee, *A historical introduction to the philosophy of science* (London, 1972), 116.
113. Herschel, *op. cit.* (ref. 41), 190, 196.
114. Olson, *op. cit.* (ref. 56), 264-70.
115. In claiming that Herschel adopted the "sophisticated view of hypotheses" which she identifies with Humboldtianism, Cannon notes its Scottish roots. Cannon, *op. cit.* (ref. 11). However, the previous discussion of the Scottish debate shows how this view of hypotheses was presented as compatible with Baconianism. It is interesting that William Gladstone, a non-scientific reader of Bacon, compiled detailed notes on the *Novum organum* in 1834 and concluded that the Baconian account of induction need not exclude "that elasticity of spirit" which stimulated hypotheses. Gladstone Papers, 44, 723. F196-231, British Library.
116. M. Faraday to J. Herschel, 10 November 1832, in *The selected correspondence of Michael Faraday*, ed. by L. Pearce Williams (2 vols, Cambridge, 1971), i, 235; Darwin, *op. cit.* (ref. 7), 38; J. S. Mill, "Herschel's Discourse", *Examiner*, 20 March 1831, 179-80, p. 180; Whewell, *op. cit.* (ref. 41), 377, 398.
117. W. Whewell to R. Jones, 24 April 1831, in I. Todhunter, *William Whewell D.D.: An account of his writings* (2 vols, London, 1876), ii, 118.
118. W. Whewell to R. Jones, February 1831, in *ibid.*, 115.
119. Whewell, *op. cit.* (ref. 41), 381, 399-400.
120. W. Whewell to J. Herschel, 9 April 1836, in Todhunter, *op. cit.* (ref. 117), ii, 234; J. Herschel, "Views on public education", *Philosophical magazine*, viii (1836), 432-8, p. 434.
121. Macaulay, *op. cit.* (ref. 42), 39.
122. *Ibid.*, 65-68. See also his "Mill's Essay on government", *Edinburgh review*, xlix (1829), 159-89, pp. 101-2, for his criticism of utilitarianism as a pre-Baconian philosophy.
123. Macaulay, *op. cit.* (ref. 42), 88; also p. 73.
124. *Ibid.*, 81.
125. *Ibid.*, 89-92.
126. *Ibid.*, 91.
127. H. Brougham to M. Napier, 28 July 1837, in Napier (ed.), *op. cit.* (ref. 74), 197.
128. W. Whewell to R. Jones, 6 September 1837, in Todhunter, *op. cit.* (ref. 117), ii, 258.
129. J. S. Mill, *A system of logic* (London, 1843), 376.
130. *Ibid.*, 378. See also Bacon, *Novum organum*, ed. by Anderson (ref. 28), 98-99.
131. H. Brougham, in Napier (ed.), *op. cit.* (ref. 74), 196. But there is evidence of a continuing

support for Macaulay's attack on methodology. In a review of Baconian scholarship, one writer praised Macaulay and remarked that "No logic of any kind ever taught a man to reason". See W. H. Smith, "Spedding's life of Bacon", in *Blackwood's magazine*, xciii (1863), 480-99, p. 497.

132. See Whewell, *op. cit.* (ref. 41), 379, for reference to the "plebeian notions, which represent the Baconian method ... as something obvious"; Herschel, *op. cit.* (ref. 41), 113-14; B. Powell, *An historical view of the progress of the physical and mathematical sciences from the earliest ages to the present time* (London, 1834), 196-212; R. Whately, *Bacon's Essays* (2nd edn, London, 1857), pp. xiii-xiv. J. S. Mill, "Whately's Logic", in *Westminster review*, ix (1828), 137-72, pp. 141-8.

133. Macaulay, *op. cit.* (ref. 42), 92-94.

134. W. Whewell, *On the philosophy of discovery* (London, 1860), 165-6, 170-80. I have used this later work which contains the chapter on Bacon from the *Philosophy of the inductive sciences* and additional material.

135. *Ibid.*, 126-8.

136. *Ibid.*, 130-2.

137. *Ibid.*, 182.

138. *Ibid.*, 135.

139. *Ibid.*, 145.

140. *Ibid.*, 135.

141. *Ibid.*, 135-6.

142. *Ibid.*, 140-1.

143. Whewell, *op. cit.* (ref. 46), 316-17.

144. Whewell, *op. cit.* (ref. 134), 138.

145. *Ibid.*, 253. For Whewell's remarks about the currency of a more narrow view of induction and its association with interpretations of Bacon and Newton, see his *The philosophy of the inductive sciences* (2nd edn, 2 vols, London, 1847), ii, 56-58.

146. Agassi, *op. cit.* (ref. 103), 2, 12.

147. See J. Spedding, *Evenings with a reviewer* (2 vols, London, 1848). Ellis became seriously ill in 1853 and could not continue his work. He died in 1859.

148. R. Ellis, "Preface to the philosophical works", in J. Spedding, R. Ellis and D. Heath (eds), *The works of Francis Bacon* (14 vols, London, 1857-74), i, 25-38. For more recent and less dismissive treatments of Bacon on 'forms', see C. D. Broad, *The philosophy of Francis Bacon* (Cambridge, 1926), 30-46; M. Hesse, "F. Bacon", in D. J. O'Connor (ed.), *A critical history of western philosophy* (London, 1964), 145-9; Quinton, *op. cit.* (ref. 3), 45-46, 63-65.

149. Whewell, *op. cit.* (ref. 134), 151.

150. Ellis, *op. cit.* (ref. 147), 64.

151. W. Whewell, *Novum organon renovatum* (London, 1858), pp. iv-v.

152. W. Whewell, *History of the inductive sciences* (3rd edn, 3 vols, London, 1857), i, 4; also p. viii.

153. Whewell, *op. cit.* (ref. 44), ii, 186.

154. *Ibid.*, i, 9.

155. W. Whewell, "Remarks on the review of the 'Philosophy of the inductive sciences' in the *Athenaeum*" 22 September 1840, Trinity College, Cambridge, Whewell Papers, 266.C.80), 7.

156. W. Whewell to A. de Morgan, 18 January 1859, in Todhunter, *op. cit.* (ref. 117), ii, 417. See also G. Gore, *The art of discovery* (London, 1878).

157. Spedding *et al.* (eds), *op. cit.* (ref. 148), i, 370-90. This is an account of a dialogue between Ellis and Spedding on the question of Bacon's major claim to fame. For a discussion of seventeenth century programmes, see M. B. Hall, "Solomon's house emergent: The early Royal Society and co-operative research", in H. Woolf (ed.), *The analytic spirit: Essays in*

the history of science (Ithaca, 1981), 177-94.
158. Whewell, op. cit. (ref. 46), 313-16 and op. cit. (ref. 95), 153-6.
159. W. Whewell to A. de Morgan, 14 February 1859, in Todhunter, op. cit. (ref. 117), ii, 418.
160. D. Brewster, Memoirs of the life, writings and discoveries of Sir Isaac Newton (2 vols, Edinburgh, 1855), ii, 404. Augustus de Morgan also rejected any link between Bacon and Newton, remarking that "If Newton had taken Bacon for his master, not he, but somebody else, would have been Newton". See his "The works of Francis Bacon", Athenaeum, 18 September 1858, 332-4, 367-8, p. 368. Another critical work is A. Lasson, Über Bacon von Verulam's wissenschaftliche Principien (Berlin, 1860).
161. I. Hacking, Representing and intervening (Cambridge, 1983), 157.
162. G. N. Cantor, Optics after Newton: Theories of light in Britain and Ireland, 1704-1840 (Manchester, 1983), 173-86.
163. D. Brewster, "Whewell's History of the inductive sciences", Edinburgh review, lxvi (1837), 110-51, pp. 146-51; Davie, op. cit. (ref. 57), 172, 175.
164. Quoted in Cantor, op. cit. (ref. 162), 178-9.
165. G. Airy to J. Herschel, 16 July 1831; Herschel Papers, i, no. 19, Royal Society of London.
166. D. Brewster, "M. Comte's Course of positive philosophy", Edinburgh review, lxvii (1838), 217-308, pp. 306-7.
167. Cantor, op. cit. (ref. 162), 179-81.
168. Brewster, op. cit. (ref. 160), 404.
169. Ibid., 405. George Davie suggests that Brewster distinguished between "a superficial and a profound version of empiricism", and that he and Forbes believed Whewell did not sufficiently acknowledge this. See op. cit. (ref. 57), 185. But it is not clear from the evidence supplied that Whewell would not have accepted this distinction.
170. Brewster, op. cit. (ref. 163), 121.
171. Quoted in D. Brewster, "Whewell's Philosophy of the inductive sciences", Edinburgh review, lxxiv (1842), 265-306, p. 291.
172. Brewster, op. cit. (ref. 163), 120-1.
173. Brewster, op. cit. (ref. 171), 302.
174. J. D. Forbes, "The history of science and some of its lessons", Fraser's magazine, lvii (1858), 283-94, p. 292.
175. D. Masson to W. Whewell, 15 May 1863; Trinity College, Cambridge, Whewell Papers, Add. Ms. 89^{145}.
176. J. von Liebig, Ueber Francis Bacon von Verulam und die Methode der Naturforschung (Munich, 1863), 45-46.
177. Ibid., pp. iii-iv; J. von Liebig, "Lord Bacon as natural philosopher", Macmillan's magazine, vii (1863), 237-49, 257-67, pp. 264-5.
178. See O. Sonntag, "Liebig on Francis Bacon and the utility of science", Annals of science, xxxi (1974), 373-86.
179. Von Liebig, op. cit. (ref. 177), 246-9, 262-3.
180. Ibid., 263.
181. J. von Liebig, "Induction and deduction", Cornhill magazine, xii (1865), 296-305, p. 299, 304-5.
182. R. Adamson, "Bacon", Encyclopaedia Britannica (9th edn, 25 vols, Edinburgh, 1875-89), iii, 200-18, p. 216. For further late nineteenth century recognitions of the weaknesses in Bacon's philosophy, see Fowler (ed.), op. cit. (ref. 31), 133; R. W. Church, Bacon (London, 1884), 187-98. It should be noted however that the earlier confidence about his method could still be found in popular expositions. See J. James, The philosophy of Lord Bacon and the systems which preceded it (Bradford, 1860), 11-12; E. Goulburn, Bacon: The first principles of his philosophy stated in a popular form (London, 1860), 21.
183. T. L. Hankins, Jean d'Alembert: Science and enlightenment (Oxford, 1970), 89-92.

184. W. S. Jevons, *The principles of science: A treatise on logic and scientific method* (2nd edn, London, 1877), 507. See also de Morgan, *op. cit.* (ref. 160), 367-8. Bacon's neglect of mathematics was noted even by his most gentle critics. See Powell, *op. cit.* (ref. 132), 211.

185. T. H. Huxley, "The progress of science" (1887), in C. Bibby (ed.), *The essence of T. H. Huxley* (London, 1967), 42. See also C. Bernard, *An introduction to experimental medicine* (New York, 1927), 50-51: "I do not think ... it is very profitable for men of science to discuss definitions of induction and of deduction, nor, for that matter, the question whether we advance by one or other of these so-called processes of mind."

186. Laudan, *op. cit.* (ref. 47), 4-18.

187. *Ibid.*, 127-36.

188. Coleridge, *op. cit.* (ref. 33), 25.

189. Adamson, *op. cit.* (ref. 182), 217. Compare the more favourable assessments in earlier editions of the *Encyclopaedia Britannica* (3rd edn, Edinburgh, 1788-97), ii; (8th edn, Edinburgh, 1853-60), iv, 357.

190. T. H. Huxley to C. Darwin, in L. Huxley (ed.), *The life and letters of T. H. Huxley* (2 vols, London, 1900), ii, 14. W. Whewell to A. de Morgan, 18 January 1859, in Todhunter, *op. cit.* (ref. 117), ii, 417. For typical late nineteenth century accounts of Bacon's character, method and historical significance, see Adamson, *op. cit.* (ref. 182); T. Fowler, "Bacon", *Dictionary of national biography* (London, 1885), ii, 349-60. Apart from Brewster, it was foreign commentators who made the most extreme attacks. The American writer, William Draper, said that Bacon was "a pretender in science, a time-serving politician, an insidious lawyer, a corrupt judge, a treacherous friend, a bad man". W. Draper, *The history of the intellectual development of Europe* (2 vols, New York, 1860), ii, 259-60.

191. For various approaches to this question, see J. A. Schuster and R. R. Yeo (eds), *The politics and rhetoric of scientific method: Historical studies* (Dordrecht, forthcoming). For some references to the way in which Baconian ideas of method were deployed in the Darwinian debates, see A. Sedgwick, "Darwin's origin of species", *Spectator*, 7 April 1860, 335; R. Owen, "Darwin on the Origin of species", *Edinburgh review*, cxi (1860), 487-532, pp. 495-6; A. Ellegard, *Darwin and the general reader* (Goteburg, 1958), ch. 9; D. Hull (ed.), *Darwin and his critics* (Cambridge, Mass., 1973). This issue is relevant to the present article but would require more extensive treatment.

192. See C. C. Gillispie, "The *Encyclopédie* and the Jacobin philosophy of science", in M. Clagett (ed.), *Critical problems in the history of science* (Madison, 1959), 255-89; Wood, *op. cit.* (ref. 17); Ben-David, *op. cit.* (ref. 17); Morrell and Thackray (eds), *op. cit.* (ref. 4), 19-34.

193. A. Chalmers, *Evidence and authority of the Christian revelation* (2nd edn, Edinburgh, 1815), 189-229.

194. H. Brougham to M. Napier, 18 March 1827, in Napier, *op. cit.* (ref. 74), 47-48.

195. W. Whewell to R. Jones, February 1831, in Todhunter, *op. cit.* (ref. 117), ii, 115.

196. R. Carlile, *An address to men of science* (London, 1821), 23.

197. I owe this information to Andor Gomme, writing in the *Times literary supplement*, 11 February 1983, 139-40. See B. Montagu, *The works of Francis Bacon, Lord Chancellor of England: A new edition* (16 vols, London, 1825-34).

198. R. Laudan, "Ideas and organisation in British geology", *Isis*, lxviii (1977), 527-38, pp. 530-1; R. Porter, *The making of geology: Earth science in Britain 1660-1815* (Cambridge, 1977), 138, 146. J. Phillips, "Geology", *Encyclopaedia metropolitana* (ref. 33), vi, 530-4.

199. C. Lyell, *Principles of geology* (2nd edn, 3 vols, London, 1832), i, 1-42. For a contemporary reference to Lyell's account of past errors as indicative of Bacon's relevance, see T. Martin, *Character of Lord Bacon: His life and works* (London, 1835), 154-5. See also R. Porter, "Charles Lyell and the principles of the history of geology", *The British journal for the history of science*, ix (1976), 91-103, pp. 91-98.

200. Lyell, *op. cit.* (ref. 199), i, 81.

201. [W. H. Fitton], "Transactions of the Geological Society", *Edinburgh review*, xxix (1817), 70-94, p. 70.

202. "Geological inquiries", *Philosophical magazine*, xlix (1817), 421-9, pp. 421-2. But it should not be assumed that these public statements about an accessible method represented the methodology of all members of the Geological Society. In any case, Baconian ideas of a division of labour could be deployed to stress the need for a specific role for theorists. For a reference to the task of generalization which could be "performed only by philosophers", see *ibid.*, 421.

203. G. L. Craik, *The pursuit of knowledge under difficulties* (2 vols, London, 1846), i, 2; S. Smiles, *Self help* (new edn, London, 1860), 46-47. For a similar contrast between method and genius see H. Martineau, *Miscellanies* (2 vols, Boston, 1836), i, 72-73, 103-4.

204. J. Drinkwater, *Lives of eminent persons* (London, 1833), 15, 54. This work, and that cited by Craik above, were published for the Society for the Diffusion of Useful Knowledge. The society sponsored the publication of cheap editions of Bacon's works and commentaries on them. See J. Hoppus, *An account of Bacon's Novum organon scientiarium* (2 parts, London, 1840); G. L. Craik, *Bacon: His writing and his philosophy* (3 vols, London, 1846-47). The method/genius dichotomy was sometimes associated with a contrast between Bacon and Descartes as alternative models for scientific thinking. See Chalmers, *op. cit.* (ref. 193), 200; W. Whewell, *Astronomy and general physics considered with reference to natural theology* (Cambridge, 1834), 303-41, for a discussion of the contrasting moral effects of inductive and deductive habits.

205. See Morrell and Thackray, *op. cit.* (ref. 4), 269-72. See F. M. Turner, "The Victorian conflict between science and religion: A professional dimension", *Isis*, lxix (1978), 356-76; J. H. Brooke, "The natural theology of the geologists: Some theological strata", in L. J. Jordanova and R. S. Porter (eds), *Images of the earth: Essays in the history of the environmental sciences* (Chalfont St Giles, Bucks, 1979), 39-64.

206. For an example of this confidence, see K. Pearson, *The grammar of science* (London, 1892), ch. 1.

207. Bacon, *Novum organum*, ed. by Anderson (ref. 40), 56. See J. M. Robertson (ed.), *Works of Francis Bacon* (London, 1905), pp. xiv-xv, for the continuing appeal of Bacon's analysis of error.

208. Macaulay, *op. cit.* (ref. 42), 102.

209. Webster, *op. cit.* (ref. 17), 117.

210. See R. Yeo, "Scientific method and the image of science 1831-1891", in R. M. MacLeod and P. Collins (eds), *The parliament of science: The British Association for the Advancement of Science, 1831-1981* (Northwood, 1981), 65-88.

211. Laudan, *op. cit.* (ref. 198), 527, 530, 536; Gillispie, *op. cit.* (refs 8 and 9). Gillispie suggested that the "critical instinct for mathematical clarity ... saved French science ... from submersion in the Baconianism which vulgarised the English tradition...". C. Gillispie, "Science in the French Revolution", in B. Barber and W. Hirsch (eds), *The sociology of science* (Middletown, Conn., 1962), 91.

212. Cannon, *op. cit.* (ref. 11).

III

REVIEWING HERSCHEL'S *DISCOURSE*

HERSCHEL'S *Discourse* has carried a large symbolic load as the epitome of the nature and ethos of early Victorian science. It was often cited by contemporaries as the crucial text in the inspiration and direction of their scientific vocations, and more recently it has been identified as the treatise which first clearly inaugurated the study of the methodology of science as a separate subject.[1] In a sense, these two characterizations are related, because Herschel's readers in the scientific community responded to his text as the first to encapsulate both the excitement of scientific inquiry and the rigour of its methods. Michael Faraday told Herschel that "I read it as all others did with delight. I took it as a schoolbook for philosophers and I feel that it has made me a better reasoner and even experimenter and has altogether heightened my character and made me if I may be permitted to say so a better philosopher".[2] A similar fervour is discernible in Charles Darwin's instruction to W. D. Fox: 'If you have not read Herschel in Lardner's Cyclo — read it directly'.[3] In 1843 J. S. Mill acknowledged the active influence of the *Discourse* in his "attempt to methodize the process of investigating truth".[4]

The effect of this collective testimony on later generations has arguably meant that the work has been accepted as a convincing statement of an early-nineteenth-century consensus. While there are some grounds for this repu-

---

[1]C. J. Ducasse, "John F. W. Herschel's methods of experimental inquiry", in: R. M. Blake, C. J. Ducasse and E. Madden (eds), *Theories of Scientific Method: The Renaissance through the Nineteenth Century* (Seattle: University of Washington Press, 1960), p. 153; L. Laudan, "Theories of Scientific Method from Plato to Mach", *History of Science* **7** (1968), 29.

[2]Faraday to Herschel, 10 November 1932, in: L. Pearce Williams (ed.), *The Selected Correspondence of Michael Faraday*, 2 vols (Cambridge: Cambridge University Press, 1971), vol. 1, p. 235.

[3]Darwin to Fox, 15 February 1831, in: F. Burkhardt and S. Smith (eds), *The Correspondence of Charles Darwin* (Cambridge: Cambridge University Press, 1985), vol. 1, p. 118. See also M. Ruse, "Darwin's Debt to Philosophy", *Studies in History and Philosophy of Science* **6** (1975), 159–181 for Darwin's use of Herschel.

[4]Mill to Herschel, 1 May 1843, Herschel Papers, Royal Society.

Reprinted from *Studies in History and Philosphy of Science*, 20, Richard Yeo, 'Reviewing Hersschel's Discourse', 1989, with permission from Elsevier Science

542

tation of the *Discourse*, it is also possible to show that it has sponsored a range of images about the leading features of early Victorian science. Thus for S. F. Cannon, to be "scientific" in the second quarter of the century was to be like John Herschel. In so far as Cannon spelt out the methodological implications of this position, it was one that involved a "sophisticated view of hypotheses" as opposed to so-called Baconian method. Cannon's Herschel was a leading member of the scientific élite who had embraced the Humboldtian programme of international collection of comparative data in fields from physical geography to physical astronomy, combined with the use of modern mathematical techniques.[5] On the other hand, for Joseph Agassi, Herschel was a "conservative *par excellence*"; the *Discourse* expressed the orthodox Baconian doctrine of prejudice which made failed speculation tantamount to error and rejected its place in true science.[6] More recently, Larry Laudan has argued that Herschel was one of the leading figures in the revival of "the method of hypothesis" and the formulation of a new set of methodological criteria for good science.[7] Whereas Agassi saw Herschel as defending the traditional inductive method, Laudan presents him as a forerunner of a new methodological regime. The tensions between these various interpretations illustrate the danger of taking the *Discourse* as a text that unproblematically records a contemporary consensus on the nature of scientific inquiry.

This paperback reprint of the first edition carries a useful Foreword in which Arthur Fine sketches the structure of the *Discourse* and some of the tensions within it. For example, he notes that Herschel himself was aware of the problem of balancing theoretical language and observational reports, and of the necessary circularity involved in the task of classifying phenomena and establishing "lawlike" connections (xvi). Fine recognizes that the framework of the book is Baconian, both in its affirmation of science as grounded in empirical investigation and in its concern with the social impact of natural knowledge (xi–xii). But as he also notes, Herschel modified Bacon's account in several ways. This should suggest the need for special care in using it as evidence of early-nineteenth-century opinion, but this is more fully under-lined when the *Discourse* is situated in its historical context. It is then possible to see that Herschel's text, in representing a novel attempt to promote and explain the scientific enterprise, also raised issues on which full agreement

---

[5] S. F. Cannon, *Science in Culture: the Early Victorian Period* (New York: Science History Publications, 1978), pp. 73–74, 232, 228–229; W. F. Cannon, "John F. W. Herschel and the Idea of Science", *Journal of the History of Ideas* **22** (1961), 215–239.

[6] J. Agassi, "Sir John Herschel's Philosophy of Success", *Historical Studies in the Physical Sciences* **1** (1971), 1–36 (pp. 1–2, 20–21).

[7] L. Laudan, *Science and Hypothesis: Historical Essays on Scientific Methodology* (Dordrecht: D. Reidel, 1981), pp. 129–131, 189.

was most unlikely.[8] The following discussion will attempt to provide some perspective on the *Discourse* by recalling the manner of its production, approaching its contents through William Whewell's review and, finally, considering its social and rhetorical dimensions.

In his Presidential Address to the British Association in 1845 Herschel referred to the attention which had recently been devoted to "the philosophy of science". He mentioned the work of his two major admirers, Whewell and Mill, noting the degree to which they had extended the subject since "some other works of prior date" — presumably including his own. Taken together, he said, these two writers had left "the philosophy of science, and indeed *the principles of all general reasoning* in a very different state from that in which they found them." But it is evident that, for Herschel, this picture of a subject steadily progressing was complicated by the epistemological conflict between Whewell's "peculiar and *a priori* point of view" and Mill's attempt to present "experience as the ultimate foundation of all knowledge". Consequently, he sought to underline what he took to be their agreement on "the most essential features" of "the inductive philosophy".[9] Although he did not say what these were, it is worth suggesting that Herschel was trying to advertise a consensus among leading commentators about the *methodology* of science, even if its *epistemology* was a site of contention. This distinction was not an always clear one in contemporary writing, but it is useful in reminding us that while both Whewell and Mill acknowledged the significance of Herschel's work in the development of their own views, their subsequent epistemological conflict altered the parameters of the debate on the philosophy of science. Since their dispute it has become difficult to avoid reading the *Discourse* as a part of this famous controversy. However, it is worth noting that although he responded to the contrast between idealism (or rationalism) and empiricism in his review of Whewell's works in 1840, this was not the dominant polarity in Herschel's book, and no reviewers responded to it in these terms.[10] Both Whewell and Mill, together with other reviewers, received the book as a modern version of Bacon's attempt to delineate the methods of successful scientific inquiry,

[8]Other analyses of the text such as those by Ducasse, *op. cit.* note 1 , pp. 153–182 and the substantial introduction by Michael Partridge to an earlier reprint (New York: Johnson, 1966) also scrutinize inconsistencies in Herschel's work. It is worth commenting that both this Chicago reprint and the one by Johnson lack bibliographical references to studies on Herschel and his intellectual context. In addition to those mentioned in this review, see S. S. Schweber (ed.), *Aspects of the Life and Thought of Sir John Frederick Herschel*, 2 vols (New York: Arno Press, 1981).

[9]J. Herschel, *British Association Reports* (London: Murray, 1846), p. xl.

[10]J. Herschel, "Whewell on the Inductive Sciences", *Quarterly Review* 68 (1841), 177–238. Some views advocated by Herschel in this review differed from those in the *Discourse*: for example, in 1830 he contrasted abstract mathematical concepts and empirical truths; in 1841 he sought to derive both from experience. Both positions differed from the doctrine of the "fundamental antithesis" developed by Whewell. For other reviews of the book, see *Monthly Review* 1 (1831), 286–293; *The Athenaeum*, 15 January 1831, 38–39.

enhanced by a grasp of detailed examples. Mill said that it demonstrated the "advantage of *systematic* investigation" and the weakness of "empiricism under the name of *common sense*".[11] To some extent the response did suggest a consensus on the general principles of scientific method — the kind of agreement that Herschel recommended in 1845. But the one detailed review by another member of the scientific élite — Whewell's in the *Quarterly Review* — indicated, sometimes by omission, that any consensus on method, beyond fairly general endorsements of observation, experiment and induction, was difficult to sustain.

It is worth noting that such a "classic" text as this owed its production to a particular set of historical conditions — political, educational and commercial — which underlay the enormous "diffusion of knowledge" during the 1820s and 1830s. Dionysius Lardner, the editor and proprietor of the *Cabinet Cyclopaedia* in which the *Discourse* appeared, related the plan of the venture and its problems, telling Herschel that the scientific articles were to be of a more "general and popular character" than those in the *Encyclopaedia Metropolitana* — such as those on light, sound and physical astronomy which Herschel had written or was engaged to write. "They will stand related to strict mathematical treatises", he said, "in the same manner as the Système du Monde does to the the Mécanique Céleste . . ." Each of the major subject divisions was to be introduced by a "preliminary discourse in the composition of which I shall endeavour to avail myself of the highest talents of the age". However, the enticement of authors was a delicate matter, as Lardner explained. "My chief difficulty in Science is to find profound men who like yourself are able and willing to write a popular work."[12] Given all this, Herschel clearly deserved the 250 pounds Lardner offered him to tackle what was probably the hardest task — an introductory treatise on the province and methods of "natural philosophy". The choice of this term allowed the contrast with another introductory discourse in the series, William Swainson's *A Preliminary Discourse on the Study of Natural History* of 1834. Herschel's volume was intended to cover the physical sciences, and his third section used the word "physics" in this new sense.[13] But Herschel also discussed subjects usually included under natural history, and his methodological prescriptions were meant to apply to all sciences.

Whewell's review of the *Discourse* was also difficult to write and he

[11]J. S. Mill, "Herschel's Discourse", *Examiner* 20 March 1831, 179–180.

[12]D. Lardner to Herschel, 18 July 1828; 19 January 1830; 28 July 1828, Herschel Papers, Royal Society. It is worth noting that the original edition had *two* title pages: the one reprinted in this Chicago edition, bearing a portrait of Bacon, and a second, headed by the title of the series — "The Cabinet Cyclopaedia". This was placed before the Contents page. The first title page gave as the date of publication 1830; the second 1831.

[13]Herschel used the terms "physics" and "natural philosophy" almost interchangeably throughout the book.

admitted his dissatisfaction with the result. One reason for this was that it represented a novel genre for the English reader — a supposedly popular text on "the history of philosophy of physical science". In contrast with the Continent where men such as Cuvier and Berzelius surveyed the progress of science, no individuals of similar authority had sought to extend the results of science to "a wider circle" or to connect them with "the general body of knowledge". But Whewell noted that Herschel's book contained not only a survey of the state of natural knowledge but an account of the principles on which it rests and the "maxims by which its researches have been and must be successfully conducted". Furthermore, it attempted this in a new way. While "volumes upon volumes" had been written on "the nature of human knowledge and the laws of human thought", there had been inadequate attention to the mental processes exhibited in the progress of science. Herschel's *Discourse* was thus "one of the first considerable attempts to expound in any detail" the rules and doctrines of successful scientific method.[14] As such, it was a welcome alternative to the treatments of induction in general treatises on logic, such as those of Richard Whately. There was a polemical aspect to this contrast, since Whewell, together with Richard Jones, was annoyed by Whately's views on political economy and his insistence on embracing induction within syllogistic logic.[15] But more generally, Whewell was referring to the fact that issues of scientific method had not been given separate and detailed treatment; rather, they were incorporated in more general treatises on metaphysics and epistemology, such as those of the Scottish philosophers. Herschel's *Discourse* did not entirely avoid epistemological questions, but it certainly highlighted methodology; as recent commentators have noticed, Herschel was more secure, and his examples most apposite, when he discussed observation, experiment, classification, hypothesis and verification.[16] Whewell was therefore confronted with the task of composing a popular essay for a general audience on a subject which he saw as a new field, rather than one in which a consensus had been established.

It is possible to classify Whewell's review into points of agreement, difference, and silence. Not unexpectedly, the overall tone of the review was positive. Whewell praised the *Discourse* as the remedy for what he regarded as an appalling confusion about the meaning of Baconian method and some of its crucial terms, such as "induction". Writers such as Thomas Reid had contributed to this by speaking of "the *inductive principle*", by which they

[14]W. Whewell, "Modern Science — Inductive Philosophy", *Quarterly Review* **45** (1831), 374–377.
[15]See R. Jones to Whewell, 24 February 1831, Whewell Papers, Trinity College.
[16]Ducasse, *op. cit.* note 1, p. 181. Prior to Herschel, Dugald Stewart gave considerable attention to certain methodological issues in Part 2 of his *Elements of the Philosophy of Mind*. See Stewart, *Collected Works*, W. Hamilton (ed.), 11 vols (Edinburgh, 1854–1860), vol. 3, chap. 4.

546

mean the "instinctive belief in the permanent uniformity" of the laws of nature. Others envisaged induction as simple collection of data. Herschel, however, offered an extended account of induction as "the process of considering a class, or two associated classes of phenomena, as represented by a general *law*, or single conception of the mind".[17] Secondly, by providing concrete examples of this method of investigation, Herschel was able to show the "mutual dependence and contrast of induction and deduction" in the process by which the "vast structure of science" was built up. Through these careful examples Whewell hoped that Herschel had made it less likely that the public would confuse purely deductive sciences — or those "consisting in the consequences of a few axioms" and governed by syllogistic logic — with those that rightly claimed the "name of inductive sciences". Thirdly, Herschel's discussion gave due emphasis to the importance of mathematics in allowing the expression of precise quantitative laws. Fourthly, it *partially* recognized the way in which a "distinct terminology" formed the language of science and enabled the transmission of the labours of one generation to the next.[18]

Throughout the first two sections of the *Discourse*, Herschel sought to analyse and codify the method by which natural science had advanced: "What we have all along most earnestly desired to impress on the student is, that natural philosophy is essentially united in all its departments, through all which one spirit reigns and one method of inquiry applies" (219). This affirmation of an identifiable method as the clue to the success of science arguably avoided contentious epistemological questions, and, judging from Whewell's response, Herschel's account of the salient features of inductive method provided at least a general point of agreement. Beyond this however, there were tensions between Herschel's views and those of his friendly reviewer.

In the third section of his book Herschel acknowledged that while "one method of inquiry" pervaded all science, it was necessary to look at "the subdivision of physics into distinct branches, and their mutual relations" (221). The chapters in this section of the *Discourse* are primarily organized in terms of objects of study in nature, moving from forces and the structure of objects to the motion of sound and light, then to cosmical phenomena, back to the constituents of the earth, and concluding with imponderable forms of matter. This produced a discussion of the sciences concerned with these natural phenomena, beginning with pneumatics and hydrostatics, moving through optics, astronomy and geology and ending with heat, magnetism and electricity.[19] But while it is clear that Herschel regarded Newtonian

[17]Whewell, *op. cit.* note 14, pp. 378–379.
[18]*Ibid.*, pp. 380–391. For the wider debate, see R. Yeo, "An Idol of the Market-Place: Baconianism in Nineteenth Century Britain", *History of Science* 23 (1985), 251–298.
[19]Here Herschel essentially follows the arrangement of subjects in: Thomas Young's, *A Course of Lectures on Natural Philosophy and the Mechanical Arts* (London, 1807).

astronomy, with its incorporation of the demonstrable laws of dynamics, as the most perfect of the sciences (272), the concept of a hierarchy was not obvious in his survey of the branches of science. When this image did appear it came in an unexpected form. Thus Herschel remarked that "geology, in the magnitude and sublimity· of the objects it treats, undoubtedly ranks, in the scale of the sciences, next to astronomy" (287). Rather than drawing strong distinctions between subjects, Herschel seemed to be more concerned with showing how inquiries in different fields could overlap to produce more general laws.[20] He hailed the convergence of magnetism and electricity, "which had long maintained a distinct existence" (324), as a great event, and suggested future useful interactions between botany and chemistry and botany and geology (345). Herschel stressed that "no natural phenomenon can be adequately studied *in itself alone*, but, to be understood, must be considered *as it stands connected with all nature*" (174, 259); and he expected that this idea of unity would be realized in "higher orders" of generalization embracing several sciences (360).

Whewell's account presented an image of science which stressed difference rather than unity or interaction. He thought readers would be interested to see the sciences arranged in their order of growth from those just beginning to form inductive generalizations to those which had reached what Herschel called "fundamental laws" (200); or more figuratively, "from the tottering girl to the full-formed matron".[21] Although Whewell explained that limits of space prevented a detailed comparison with Herschel's survey, some interesting contrasts are nevertheless apparent. Thus the the "nursery and spelling book" stage was occupied by botany, chemistry and mineralogy; these were only just past the "outset of their inductive career". The sciences under the heading "physics" — for example, magnetism and electricity — were further advanced because portions of their findings had been reduced to "mathematical formulae". Optics, Whewell's favourite "modern science", had made rapid progress and was close to the top of the hierarchy. At the apex, of course, were mechanics and astronomy, representing the only instances in which the combination of induction and deduction had been successfully completed; in the case of astronomy, there was no chance that its central theory could be "overturned".[22]

Now the criterion behind this classification — scope and quantification of laws — was certainly present in the *Discourse*, but Herschel did not use it to draw the same kinds of distinctions of status between the sciences. Thus while Whewell relegated chemistry and mineralogy to the "infancy" of the

[20]Herschel shared the ideals underlying Mary Somerville's *On the Connexion of the Physical Sciences* (London, 1834). For reference to his resistance to specialization, see Cannon, *op. cit.* note 5, p. 119.

[21]Whewell, *op. cit.* note 14, p. 390.

[22]*Ibid.*, pp. 390–397.

inductive project, Herschel saw them as fundamental, rather than low, in that they dealt with the basic "material constituents of the world" (290). Both these subjects concerned the most fundamental natural "substances, a knowledge of which necessarily preceded many other sciences" (143, 290–299). While Whewell omitted geology from his hierarchy, explaining that it raised additional questions, Herschel, as we have seen, gave it high praise, even illustrating the meaning of *verae causae* with examples from this "deservedly popular science", especially as practiced by Charles Lyell (144–147). Although they agreed on the marks of a mature science, Herschel and Whewell pursued different emphases in classifying the disciplines, with Herschel being more concerned to reveal their "mutual relation and dependency" (94).[23]

The most striking feature of Whewell's review was its silence on the details of Herschel's views on "*inductive logic*", which occupied the second, and largest section of the book. He regarded the *Discourse* as "the first attempt since Bacon to deliver a connected body of rules of philosophizing", but he said little about the way in which Herschel modified or reinterpreted Bacon. As Laudan and others have noticed, Herschel gave far more support to the role of hypotheses than most recent or contemporary writers, even suggesting that the source of conjectures was irrelevant if they were subsequently shown to explain phenomena separate from those which originally suggested them (164, 167, 170, 197, 203). This was a significant move from the previous orthodoxy, maintained by authors such as Reid, which rejected any hypothesis not grounded in cautious induction.[24] There was also tension between this latitude with respect to hypotheses and Herschel's apparent support for the Newtonian principle of *verae causae*, since theories such as the undulatory theory of light awarded causal status to unobservable entities (144, 203, 152–153).

In his major works Whewell subsequently developed criteria for the assessment of conjectures, and made strong criticisms of the anti-hypothetical attitudes of most Baconians. In the review, however, he did not offer any support to Herschel's liberalism; indeed, he questioned the lack of attention to Bacon's warning about the "method of *anticipation*". In discussing Dalton's atomic theory, Herschel said that it had been announced by its discoverer "on the contemplation of a few instances, without passing through the subordi-

[23]In the years after his review, Whewell began moving towards the notion of different disciplines being grounded in specific Fundamental Ideas. For some of the tensions between this emphasis and his assumptions about the unity of knowledge, see R. Yeo, "William Whewell's philosophy of knowledge and its reception", in: M. Fisch and S. Schaffer (eds), *William Whewell: A Composite Portrait* (Oxford: Oxford University Press, forthcoming). For the suggestion that Whewell and Herschel may have had different ideas about the identity of "physics", see Cannon, *op. cit.* note 5, pp. 123, 136.

[24]Laudan, *op. cit.* note 7, chap. 8; R. Olson, *Scottish Philosophy and British Physics 1750–1880* (Princeton: Princeton University Press, 1975), pp. 264–270.

nate stages of painful inductive ascent". Whewell feared that this passage
might be "liable to misinterpretation", that it might seem to promote the
notion of scientific discovery as the result of unprepared genius. To counter
this idea he cited Sedgwick — a living member of the "school of Bacon" —
who had just told the Geological Society that "the records of mankind offer
no single instance of any great physical truth anticipated by mere guesses and
conjectures". Similarly, Whewell appealed to the lessons of the past in order
to emphasize the danger of "anticipation", claiming that "the usual history of
physical theory has been the reign of one anticipation after another in
unbroken succession".[25]

It is thus possible to see that reviewing the *Discourse* was not a simple task
for Whewell, even though he considered himself "one of the most fortunate
men of the age" in being able to comment on it in the same year as Jones'
work on Rent. He believed that he was best placed to explain and advertise
the meaning of these two works by two of his closest friends. Both provided
the chance to "get *the people* into a right way of thinking about induction";[26]
but Whewell told Jones that he "did not much like the review of Herschel now
that I look at it in cold blood, and I have a strong persuasion that all the
philosophical part will repel most and puzzle the rest . . ."[27] The problem here
was one of trying to put forward a clear view of the nature of scientific
investigation when the crucial terms involved were being used in a variety of
ways. Whewell and Jones had come to this diagnosis of the situation and they
welcomed the *Discourse* as a major step towards clarifying the nature of
scientific research for a wider audience. But it was difficult for Whewell to
engage with the details of Herschel's position without undermining this.

Consequently, some of the novel features of the *Discourse* — as seen by
later writers — were not fully noticed in the review. While it is important not
to read Whewell's published views of 1837 or 1840 into his position in 1831, it
is likely that he agreed with Herschel's departure from the reigning Baconian
orthodoxy on hypotheses. Indeed, two years later in his address to the British
Association he went further than Herschel in criticizing extreme "Baconian"
empiricism by asserting the controlling role of theory in observation.[28]

[25]Whewell, *op. cit.* note 7, pp. 399–401. The message of Whewell's major works was closer to
another Baconian theme: namely that "truth comes out of error more readily than out of
confusion". See *The Philosophy of the Inductive Sciences*, 2 vols (London: Parker, 1847), vol. 2,
Bk. XI, chaps 4–5.
[26]Whewell to Herschel, 10 February 1831; Whewell to Jones, February 1831, in: Todhunter,
William Whewell, D. D. An Account of His Writings, 2 vols (London: Macmillan, 1876), vol. 2, p.
115.
[27]Whewell to Jones, 15 July 1831, in *ibid.*, p. 123.
[28]Whewell, *British Association Reports* (Cambridge, 1833), p. xx. Whewell's stress here on the
way theory informed observation contrasts with Herschel's dichotomy between observation as
passive and experiment as active (76–78). But there is nothing in the review to suggest Whewell's
later view of induction as involving the introduction of a new conception not strictly present in the
observational evidence.

However, any consideration of this would have clashed with his attempt to stabilize the meaning of Bacon's terms and the general validity of his account of the inductive process. Whewell also chose not to comment on Herschel's "admirable precepts and maxims" — his 12 rules — but told Jones that there was still a need for more, hopefully from Herschel, on "the practical rules and cautions for making experiments and collecting laws from these . . ."[29] It is possible that Whewell was anxious about reducing the method of discovery to easily grasped examples, such as the case of Wells' investigation of Dew — used by Herschel — and he hinted at this general problem in the review. When he read Mill's *System of Logic*, he suggested that Jones should "tell Herschel that he has something to answer for, in persuading people that they could so completely understand the process of discovery from a single example".[30] Perhaps more indicative of Whewell's trouble in the review of 1831 is his remark to Herschel, also by way of comment on Mill's work, that: "There is in new books of this kind a satisfaction in which you and I both may share. I mean that notions and expressions, which were new and strange when we began to write, are now familiarly referred to as part of the uncontested truth of the matter".[31] Thus in spite of the differences between himself and Mill and Herschel on epistemology, Whewell felt that some progress had been made since the early 1830s when even the basic terms employed in the discussion of scientific method were insecure. The success of the *Discourse* was that it had ensured that the meaning of terms such as induction, deduction, experiment, hypothesis, fact, theory, and verification were debated with reference to practical examples of successful science. In this sense, Mill's book, despite its critique of Whewell's idealism, was part of a shared discourse — one that Herschel, in his British Association address of 1845, earnestly sought to maintain.

The social dimension of the *Discourse* is not usually discussed by modern commentators. In his introduction, however, Fine recognizes that it was a "sustained ode to the progress of science and humankind" (xi). But the connection between this confidence and the most explicit subject of the book — the "principles and methods of scientific investigation" — deserves further attention. Towards the end of Part I Herschel made a series of significant assertions. First, that material progress in society has been dependent on the participation of the "mass of mankind" in social and moral improvement (68–69). Second, that this historical fact has implications in the intellectual sphere: namely that "knowledge can neither be adequately cultivated nor adequately enjoyed by a few", and that no body of knowledge is so complete that it may not be corrected by "passing through the minds of millions" (69). While this

[29]Whewell to Jones, 15 July 1831, *op. cit.* note 26, p. 124.
[30]Whewell to Jones, 7 April 1843, in *ibid.*, p. 314.
[31]Whewell to Herschel, 8 April 1843, in *ibid.*, p. 315.

process was most crucial when knowledge was being "applied to practical uses" (70), Herschel insisted that "the natural unencumbered good sense of mankind" was capable of assisting even the "elucidation of principles". One of the necessary conditions for this idea was the possibility of wide participation in science, and Herschel affirmed this strongly, proposing that openness and accessibility distinguished science from the arcane crafts. Furthermore, one of the recurring themes of the *Discourse* was the ability of "any person of good ordinary understanding" to contribute to scientific inquiry at some level and to understand the "*results*" of even the highest sciences (25–26).

There was a tension between this idea of science as an accessible enterprise and Herschel's emphasis on the importance of mathematics and technical knowledge. He resolved this to some extent by referring to a division of labour which allowed space for both amateurs and experts, but the claims about public appreciation of scientific thinking depended on a distinction between the *method* and the *content* of science. Although many parts of the *Discourse* showed how insecure this distinction could be, since the complexity of scientific questions often made it difficult to distil simple rules of method (97), it was crucial to Herschel's claims about the moral benefits that flowed from intellectual involvement in science. In addition to this, the notion of an identifiable scientific method played an important role in Herschel's efforts to ensure that abstract science was not valued solely for its practical outcomes, although he did produce examples of these. By defining science in terms of its special method, he was able to suggest that this systematic and experimental form of inquiry could be extended to the "more complicated conduct of our social and moral relations" (72–73).

When the young J. S. Mill reviewed the *Discourse* he welcomed this as its key message and at the same time modified its formulation. For Mill, the social significance of physical science was achieved not through "the truths which it discloses but by the process by which it attains them". Fundamental changes in society would only follow when this method was adopted by the "moral and social sciences".[32] In this respect, Herschel's *Discourse* was not only an influential statement of the scientific method but an example of how assumptions about this method could serve as a powerful rhetorical resource for wider social claims.[33]

It is also important to recognize that there was more in the *Discourse* than the analysis of methodology, although this occupied the largest section of the book. I have mentioned the classification of the sciences, but there are also

[32]Mill, 'Herschel's *Discourse*', *Examiner* 20 March 1831, p. 179.
[33]For this form of apologetic and the way it began to go beyond what Herschel contemplated, see R. Yeo, "Scientific method and the rhetoric of science in Britain, 1830–1917", in: J. A. Schuster and R. R. Yeo (eds), *The Politics and Rhetoric of Scientific Method: Historical Studies* (Dordrecht: Reidel, 1986), pp. 259–297.

other themes not discussed here, such as the tension between disinterested research and utility, and the use of history to illustrate the advance of science since Bacon. But in focusing mainly on the issue of scientific method through Whewell's review, I have tried to show how Herschel's work was not perceived as a straightforward statement of prevailing conceptions. In emphasizing practical examples of scientific reasoning while at the same time modifying the Baconian framework of contemporary discussion, Herschel was seen as offering something novel. Whewell's problem was that he had to review the *Discourse* for a general audience while he was also trying to produce something new, but different. Mill had no such problem in 1831: he was not so much interested at this stage in Herschel's rules, but in the grand implications of transferring an efficacious method from the natural to the social world. Thus if we take the reactions of its two most famous reviewers, Herschel's *Discourse* not only managed to establish a framework for the analysis of scientific method, but sustained the hope of extending its triumphs.

IV

SCIENTIFIC METHOD AND THE RHETORIC OF SCIENCE IN BRITAIN, 1830–1917

1. INTRODUCTION

The *Tribune* of October 1945 carried a contribution from George Orwell entitled 'What is Science'? While noting that this question could be answered by referring to the body of knowledge about the natural world, Orwell perceived that science had also come to denote a 'method of thought'. Alluding to the contemporary debate about curriculum content, he suggested that the demand for more science education involved 'the claim that if one has been scientifically trained one's approach to *all* subjects will be more intelligent than if one had no such training.[1] At this point, Orwell identified an important theme in the public rhetoric of science — the appeal to what Beatrice Webb called the 'cult of the scientific method'.[2] This phenomenon was enshrined in Karl Pearson's *Grammar of Science*, published in 1892, in which he declared that the 'scientific habit of mind', apart from the intellectual triumphs it had accomplished since the seventeenth century, was now called upon to play a vital historical role in a period of rapid social and political change.

I assert that the encouragement of scientific investigation and the spread of scientific knowledge by largely inculcating scientific habits of mind will lead to more efficient citizenship and so to increased social stability. Minds trained to scientific methods are less likely to be led by mere appeal to the passions or by blind emotional excitement to sanction acts which in the end may lead to social disaster.[3]

In short, Pearson claimed that scientific method was essential for good citizenship in a democracy.

The distinction between the content and method of science has been a significant element in the cultural legitimation of the scientific enterprise. In the case of the examples above, it is possible to see this theme as part of what F. M. Turner has called 'public science' — that body of rhetoric, argument and polemic by which the scientific community advanced its interest.[4] However, the appeal to scientific method in this campaign was only one aspect of the connection between methodology and the social relations of science during the nineteenth century. The aim of this chapter is to reveal the rhetorical dimension of methodological discourse, arguing that certain assumptions and statements, in both systematic works and less formal texts, were associated with debates about the nature of science, its public image, and the relationship between its various disciplines.

Discourses on methodology were important rhetorical resources for the early scientific community. At one level, it is arguable that appeals to the efficacy of experimental, inductive method were central to the formulation of distinctions between scientific and non-scientific knowledge. From different perspectives, Ian Hacking and Barbara Shapiro have shown how the classical distinction betwen knowledge (*scientia*) and opinion (*opinio*) was blurred during the seventeenth century as axiomatic, demonstrative certainty ceased to be the criterion of scientific knowledge. Although Bacon and Descartes had spoken of *certain* knowledge as the goal of science, the major practitioners of the new experimental natural philosophy were willing to accept the *probable* status of its theories about the empirical world. However, it was argued that this form of knowledge was far more secure than alternative sources of information about nature because it derived from the application of rigorous, publically repeatable, procedures which gave a *moral* certainty to its discoveries.[5] In this context, statements about the efficacy of a single scientific method became intimately associated with the definition of natural science. Furthermore, claims about this method entered into the apologetics by which men of science sought to promote the political, social and cultural value of their activites.

Joseph Ben-David has suggested that references to Bacon's inductive method provided a point of consensus for the nascent scientific community in the midst of theoretical and metaphysical disputes about science at the time of the Restoration.[6] More recently, Paul Wood has interpreted Thomas Sprat's *History of the Royal Society*, which appeared in 1667, as an apologetic text that eploited aspects of Baconian methodology in the service of a particular image of science: for example, the idea of collaborative work symbolizing the non-sectarian character of the Society.[7] It is also worth remarking that Sprat employed a dichotomy between scientific method and the 'specious *Tropes* and *Figures*' which thwarted the progress of knowledge; the clear procedures and plain style of science were set against the verbiage of other discourses. But although Sprat contrasted the 'solid Practice' of science with the 'glorious pomp of Words', and distinguished its method from the colours of '*Rhetorick*', it is arguable that the emphasis on method was itself a crucial element in the public rhetoric of the scientific community.[8]

The term 'rhetoric' is fraught with ambiguity. One definition — perhaps the most common — is epitomized by the phrase 'mere rhetoric', which refers to the persuasive use of words or literary devices. This sense of the term often carries a perjorative connotation — the implication that nothing of substance is being conveyed. There are now some approaches to this issue which challenge the rigidity of the distinction between literal and figural language in science, arguing that imagery and metaphor can have a constitutive, and not merely an expressive, role in scientific discourse.[9] The following discussion does not pursue this issue, but it does suggest that rhetoric can be understood in a more general sense as referring to what Richard Whately, a nineteenth-century authority on the subject, called 'Argumentative Composition'.[10] Recent work in both the philosophy and sociology of science views scientific activity as a process of argumentation and negotiation, rather than solely one of logical proof or justification, and this has led one writer to claim that the concept of 'rhetorical transaction' is crucial to an understanding of science.[11] Furthermore, there is evidence to indicate that apart from the putative heuristic role of scientific method, theories of methodology have been significant elements in this process of scientific debate.[12] This chapter extends the inquiry about the role of methodology by showing how it could function as a rhetorical resource for the presentation of arguments and contentions about the social relations of science. It does so by investigating how certain assumptions and statements about method were deployed in debates about the nature of the scientific enterprise in nineteenth-century Britain.

By the early nineteenth century, the discussion of scientific methodology had crystallized from the more general epistemological concerns of British philosophers. Larry Laudan has remarked that from this period 'entire books rather than prefaces or chapters were devoted exclusively to the subject.'[13] Indeed, between 1830 and 1843, the publication of major works by John Herschel, William Whewell and John Stuart Mill established the parameters of future debate in the philosophy of science. But until the later decades of the century, discussions of methodology were not confined to professional philosophers; rather, they were closely related to substantive scientific issues and, as this chapter will argue, to the social relations of science. In the first half of the century, references to epistemological and methodological issues occurred at several levels: in systematic treatises on the philosophy of science, in periodical reviews, and in the proceedings and transactions of scientific institutions and societies.[14] On the one hand, it is necessary to distinguish between these different levels, recognizing that statements about method could be shaped by perceived audience and context. On the other hand, however, when considering the social dimensions of method, it is useful to note that these levels were not entirely discrete.

John Herschel's *Preliminary Discourse on the Study of Natural Philosophy* — seen by contemporaries as a foundational text for the philosophy of science and a spur to scientific activity — was published in 1830 as a volume in Dionysius Lardner's *Cabinet Cyclopaedia*. While writing his *System of Logic*, Mill told Herschel that this *Discourse*, together with Whewell's works, was an invaluable aid in his 'attempt to methodize the process of investigating truth'.[15] But as well as contributing to formal discussions on the philosophy of science, Herschel aimed to describe the nature and scope of science for a wide audience and, in doing so, gave much attention to the character of scientific method, the hallmark, in his view, of the scientific enterprise. This emphasis on method was also a continuing theme in the addresses and speeches of the British Association for the Advancement of Science, founded in 1831, where it was linked with attempts to legitimate science, to defend it from conservative religious criticism, and to affirm its broad cultural importance. Certain themes in Herschel's *Discourse*, and in other formal writings on science, can be seen as part of this wider deployment of statements about methodology.[16]

This chapter suggests that there were three assumptions about method which were conditioned by, and served as rhetorical resources in, the social relations of science. At various levels of debate, scientific method was represented as *accessible, single* and *transferable*. These three characterizations respectively claimed that the method of science could be understood and practised by a large number of people; that there was a single method common to all branches of science; and that this method could be extrapolated from natural science to other subjects. Statements deriving from these assumptions were designed to promote the cause of science by presenting it as founded upon a well-defined and successful method, accepted by all practising researchers, accessible to laymen, and capable of being extended beyond the study of nature to the study of society. The chapter considers each of these assumptions in turn, showing how it was deployed in specific contexts: the idea of an accessible method in claims about the participation of amateurs and laymen in scientific inquiry; the notion of a single method in debates about the unity of science, the effects of specialization and the relationship between different disciplines; and the concept of a transferable method in the campaign to expand the place of natural science in education.

These assumptions about scientific method derived, to a large extent, from a Baconian framework, and it is important to recognize the influence of Bacon's work as the starting point for the philosophy of science in the early nineteenth century. His theories were subject to critical scrutiny in this period, yet Bacon was still acknowledged as the individual who most clearly comprehended the new scientific spirit and most successfully advertised its salient features. Herschel promoted this view in his *Discourse*, although he also placed an emphasis on the value of hypotheses and mathematics which was foreign to Bacon.[17] Similarly, Whewell criticized significant features of Bacon's thought but remained convinced about his historical importance.

Bacon stands far above the herd of loose and visionary speculators who, before and about his time, spoke of the establishment of new philosophies. If we must select some one philosopher as the Hero of the revolution in scientific method, beyond all doubt Francis Bacon must occupy the place of honour.[18]

Even when his methodology was abandoned as an account of scientific thinking, various features of the Baconian programme continued to exert effects. His notion of an accessible set of rules or procedures for the investigation of nature had considerable currency in debates about the educational role of science and the organization of scientific research.[19] Although certainly qualified, his idea of a single, efficacious method capable of producing certain knowledge continued in the work of Herschel and Mill; and Whewell, although rejecting the Baconian theory of induction, retained a commitment to the project of a philosophy of discovery which he saw as a continuation of Bacon's enterprise. Furthermore, all these writers on the philosophy of science were attracted to his vision of an extension of the method of science to the study of politics and society.[20] The connections between theories of method and politico-moral concerns, which informed Bacon's writings, were also visible in the ninteenth-century discourse, even in writers credited with the establishment of a distinct, academic approach to the subject. Although they clashed over epistemology and metaphysics, Whewell and Mill agreed about the wider social importance of proper conceptions of method. Whewell once spoke of the need for a 'reforming philosophy' and hoped to 'get *the people* into a right way of thinking about induction'[21]; he also entertained the idea of establishing a periodical as the organ of an inductive school embracing natural science, political economy and moral philosophy.[22] Similarly, Mill's *System of Logic* was written on the premise that 'no great improvements in the lot of mankind are possible, until a great change takes place in the fundamental constitution of their modes of thought'.[23]

But the following discussion of the place of methodology in the social relations of science should not be seen as a simple reflection of Baconianism. This chapter focuses on assumptions about the accessibility, singularity and transferability of scientific method — assumptions which betray a Baconian descent — but some of the debates investigated indicate the decline of Baconian conceptions of the nature of scientific thinking. The chapter is also concerned with the tensions and contradictions in this rhetoric, with the manner in which it could conflict, in certain situations, with the interests of particular groups within the scientific community. Thus, the following sections show how statements about methods were also implicated in disputes between experts and amateurs, between physical and organic disciplines, and between the educational claims of the natural and social sciences.

2. METHOD AND THE ACCESSIBILITY OF SCIENCE

During the early nineteenth century, individuals and institutions seeking to advance the social status of natural knowledge found it useful to present science as a cooperative pursuit embracing all classes and talents. Statements about method occupied a crucial place in this campaign: in particular, the Baconian notion of a method which rendered 'all wits and understandings ... nearly on the same level'[24] allowed an image of science as open to wide public involvement. The idea of an accessible inductive method, emphasizing factual observation rather than theoretical knowledge, appeared in the public programmes of institutions such as the Geological Society and the British Association for the Advancement of Science. William Vernon Harcourt, the first Secretary of the Association, laid much stress upon the need for massive collections of data, and indicated the scope available to the participation of amateur and part-time cultivators.[25]

In some contexts, this rhetoric of an accessible method was founded upon a contrast between observation and theory. For example, in the early years of the Geological Society, the gradual accumulation of empirical observations was presented as the guarantee of sound progress in the science; theory and hypothesis, on the other hand, were said to be dangerous. This position served the aims of the founding members of the Society who sought to distance the subject from the controversial cosmologies which had been current in the eighteenth century.[26] It also allowed them to argue that members who lacked previous knowledge were not disadvantaged, because they were not biased by a commitment to theory.[27] In this way, the concept of an egalitarian, inductive method could be linked to hostile attacks upon theory, and scientific progress construed as dependent upon collective effort rather than upon the achievements of gifted individuals.[28]

In this rhetoric, there was also a dichotomy between method and genius. In a biography of Kepler published in the *Library of Useful Knowledge*, readers were warned about 'the danger of attempting to follow his method in pursuit of truth'.[29] Kepler's undeniable achievements were seen to derive from his good fortune in 'seizing truths across the wildest and most absurd theories', and for this reason they did not provide a proper model of methodological practice.[30] In another text, also stemming from the popular education movement, the author characterized Newton as a careful observer rather than a speculative genius, and reported that the great natural philosopher had said that 'if there was any mental habit or endowment in which he excelled the generality of men, it was that of patience in the examination of the facts and phenomena of his subject'.[31] Such an attitude was compatible with the philosophy of self-help proselytized by Samuel Smiles: his redefinition of genius as 'common sense intensified' reduced all achievement, both intellectual and economic, to concepts such as hardwork, perseverance and humility. From this perspective, appropriate method was more important than inherent mental capacity.[32]

To some extent, this egalitarian view of science was sanctioned by influential general works on the scientific enterprise. In his *Reflections on the Decline of Science*, Charles Babbage suggested that the 'division of labour and of cooperation was as applicable to science' as it was to the improvement of manufactures because large areas of science involved ordinary skills of observation possessed by 'any gentleman of liberal education'.[33] Herschel supported this idea in his *Discourse*, arguing that knowledge was improved and refined when submitted to 'the pressure of minds of all descriptions'. In an explicit endorsement of the democratic image of science he urged that it 'should be divested, as far as possible, of artifical difficulties, and stripped of all such technicalities as tend to place it in the light of a craft and a mystery, inaccessible without a kind of apprenticeship.' Although science necessarily possessed a peculiar terminology, this should never give 'an appearance of superiority in its professors over the rest of mankind'. If such a situation occurred, science would be rejecting the 'light which the natural unencumbered good sense of mankind is capable of throwing on every subject, even in the elucidation of principles'.[34] Herschel returned to his theme later in the text, claiming that

There is scarcely any well-informed person, who, if he has but the will, has not also the power to add something essential to the general stock of knowledge, if he will only observe regularly and methodically some particular class of facts which may most excite his attention. . .[35]

It is not surprising that this passage was cited by those who sought to show that science was open to broad participation.[36]

The belief in the value of data collected by part-time observers was not confined to popular sciences such as geology. In 1849 the British Admiralty published a *Manual of Scientific Inquiry* which encouraged naval officers to record observations in fields such as astronomy, mineralogy, magnetism, meterology and zoology. The book carried a preface by Herschel and included chapters by eminent scientists describing appropriate methods of observation in these fields. The Memorandum from the Admiralty, directing composition of the work, said that the instructions should be 'generally plain, so that men merely of good intelligence and fair acquirement may be able to act upon them.'[37]

The assumption of an accessible scientific method also had significance outside the context of scientific institutions. Thus, in schemes of social improvement involving popular education, the controversial notion of extending knowledge to the lower classes could be justified, on utilitarian grounds, by appeals to Baconian philosophy. In his book

On the Improvement of Society, the evangelical writer, Thomas Dick, argued that

> as all science is founded on facts, and as every person possessed of the common organization of human nature is capable of observing facts . . . so there is every reason to conclude that . . . the increase of scientific observers, would ensure the rapid advancement of the different sciences, by an increase of the facts in relation to them which would thus be discovered.[38]

The contention that wide public involvement in science could actually facilitate scientific discovery also informed Henry Brougham's Society for the Diffusion of Useful Knowledge. Apart from aiming to teach science to artisans, this programme contained the notion that every man had 'a chance . . . of becoming an improver of the art he works at, and even a discoverer in the sciences connected with it'.[39]

There was, however, a danger that this emphasis on the accessibility of inductive method could be construed to imply that scientific method was simply a matter of common sense. According to some commentators, it was precisely this confusion which prevented an adequate appreciation of the systematic nature of scientific inquiry. When reviewing Herschel's *Discourse*, Mill hoped that it would succeed in demonstrating 'the superiority of science over empiricism under the name of *common sense*'.[40] Similarly, in the debate about the status of political economy, advocates of the subject found it necessary to insist that the questions it addressed could not be left to common sense, but required the adoption of systematic scientific methods.[41] Macaulay exacerbated this problem by claiming, in his essay on Bacon, that the inductive method was merely an instinctive mode of thinking, 'practised ever since the beginning of the world be every human being'.[42] This equation continued, at some levels, into the second half of the century. In a lecture at the Royal Institution in 1867, Charles Kingsley declared that scientific method needed no elaborate definition:

> It is simply the exercise of common sense. It is not a peculiar, unique, professional, or mysterious process of the understanding: but the same which all men employ, from the cradle to the grave, in forming correct conclusions.[43]

But as early as the 1830s there were signs of a reaction against the more extreme espousals of an accessible scientific method. This can be illustrated with reference to the concept of the division of labour, which had been central to eighteenth-century optimism about the promise of collaborative, inductive enterprises. As noted earlier, Herschel emphasized the importance of contributions from a wide range of observers, referring to the 'immense mass of valuable information [which might] be collected from those who, in their various lines of life, at home or abroad, stationary or in travel, would gladly avail themselves of opportunities of being useful'.[44] But throughout the *Discourse* there was a tension between this theme and the need to confirm the role of theoretical experts. Herschel appeared to be searching for a formulation of the division of scientific labour which would allow space for amateur observers without weakening the status of theory and theorists. Statements about the contributions of part-time cultivators were qualified in several ways. Firstly, Herschel argued that in the advanced branches of science, such as astronomy, 'the superior departments of theory are completely disjoined from the routine of practical observation'.[45] This relegated the amateur to a minor role which was further complicated by Herschel's stress on the need for observers to possess skills and knowledge of the various branches of science.[46] He did speak enthusiastically about the participation of the public in subjects such as meteorology and geology, but also recommended that such involvement be carefully directed by experts, suggesting the 'circulation of printed skeleton forms' which would guide the more casual observers.[47]

At the third meeting of the British Association in 1833, Whewell made a further qualification to the idea of 'combined labour' by attacking the assumption that observation and theorizing could be easily separated. He argued that facts only assume significance when viewed from a certain intellectual framework and that, therefore, 'every labourer in the field a science, however humble, must direct his labours by some theoretical views, original or adopted'.[48] This requirement, which he also defined as 'a combination of theory with facts, of general views with experimental industry',[49] severely weakened the assumptions about an accessible inductive method. And by the 1840s, even in subjects usually associated with the idea of wide involvement, the importance of theoretical experts was being reasserted. Thus, the article on geology in the seventh edition of the *Encyclopaedia Britannica* claimed that a new phase in the history of the discipline had arrived: the wild speculation of the past had been overcome, and many facts had been collected; it was now time 'to direct the labours of observation into right and fertile channels'; and those best acquainted with theory were seen as the appropriate leaders.[50]

There was, therefore, a tension between the different rhetorical uses of method. At one level, deference to Baconian induction served to distinguish science from speculative and controversial activities and, by stressing the accessibility of method, encouraged an image of natural knowledge as part of general culture. At another level, statements about method were more carefully specified so as to present an account of scientific research which justified the theoretical guidance of a governing elite. To some extent, it is possible to relate this tension to the different forms and contexts in which the rhetoric of method was employed. Thus, while appeals to an accessible, inductive method continued to be made in the second half of the century, these came to be more confined to popular texts and public lectures and addresses. However, at another level of debate which embraced both formal philosophical works on science, such as those of J. S. Mill and W. S. Jevons, and the reflections of practising scientists such as David Brewster and Justus von Liebig, the emphasis was decidedly anti-Baconian.[51] In these works, there was revision of the earlier methodological canons derived from Bacon and Newton, leading to the establishment of what Laudan has called the 'method of hypothesis', or the hypothetico-deductive method, as the new orthodoxy.[52] On this account, the *Encyclopaedia Britannica* was recording a consensus when it stated, in 1875, that:

The inductive formation of axioms by a gradually ascending scale is a route which no science has ever followed, and by which no science could ever make progress. The true scientific procedure is by hypothesis followed up and tested by verification; the most powerful instrument is the deductive method, which Bacon can hardly be said to have recognized.[53]

From the perspective of this chapter, it is important that this rejection of Baconianism should not be divorced from its intellectual and social context. The impetus for this shift in methodology cannot be seen simply in terms of an internal development within an intellectual tradition constituted by works on the nature of scientific thought from Bacon to Newton and Locke through to the Scottish philosophers such as Reid and Stewart. The qualification of Bacon's philosophy can, in some respects, be seen as an attempt to legitimize the status of the hypothetical and theoretical aspects of science in a way which countered the attitudes and expectations associated with the rhetoric of an accessible method. This can be seen in the writings of Whewell, which represent the earliest, extensive revision of Bacon's methodology.

As shown above, Whewell's public intervention in methodological debates was an attempt to qualify what he saw an an excessive emphasis on the value of empirical observation in the programme of the British Association. His subsequent major works on the history and philosophy of science were a continuation of his earlier defence of the claims of theory, and can be read as an extended vindication of the role of hypotheses and imagination in scientific thought. In his articulation of this position, Whewell had to confront assumptions about the character of inductive method which, as shown above, had currency in the rhetoric of both scientific institutions and popular societies during the early nineteenth century. Thus, in the chapter on Kepler in his *History of the Inductive Sciences*, Whewell noted that several persons,

seem to have been alarmed at the *Moral* that their readers might draw, from the tale of a Quest of Knowledge, in which the Hero, though fantastical and self-willed, and violating in his conduct, as they conceived, all right rule and sound philosophy, is rewarded with the most signal triumphs.[54]

In making this remark, he referred to the *History of Astronomy*, published by the Library of Useful Knowledge, which had difficulty in aligning the work of Kepler with its conception of 'experiment and rigorous induction' as the true method of science. Whewell used the case of Kepler to argue, as he did throughout the *Philosophy of the Inductive Sciences*, that the process which led to major scientific discovery was not 'cautious' and 'rigorous' in the sense of eschewing imaginative guesses, conjectures or metaphysical assumptions.[55]

Whewell therefore dissented from the Baconian notion of a mechanical method which levels all intellects and produces scientific truth merely through careful observation and arrangement of facts.[56] He insisted that induction only took place when observed facts were informed by an appropriate idea or conception; such ideas did not simply emerge from the facts but had to be introduced from the mind of the scientist. The knowledge established by this process could give rise to the recognition of necessary truths, a status which, in his view, was warranted by the intuition of scientific experts.[57] At this point, Whewell admitted that scientific progress depended upon the 'peculiar Sagacity which belongs to the genius of a Discoverer';[58] there was no simple rule for the acquisition of appropriate ideas and, to this extent, scientific discovery required 'some fortunate cast of intellect, rising above all rules'.[59] Thus, although he hoped to establish a philosophy of discovery, founded upon historical investigation, which would reveal the process by which science had advanced, he did not conceive this as an art of discovery or a set of mechanical rules. Whewell warned that there was no '*Popular Road*' to truth and his philosophy of science constituted a major check on any wider involvement in science.[60]

Whewell's philosophy of science also had implications for the notion of a division of labour as it appeared in arguments supporting broad participation in science. In the third edition of the *Philosophy*, he acknowledged the importance of what he called the '*social* machinery for the advancement of science', namely, the role of societies and associations in facilitating research by 'dividing the offices of observer and generalizer'.[61] But he insisted that such a segregation did not imply the separation of observation and theory. Thus, when James Spedding commended Bacon's scheme for the involvement of large numbers of people in the collection of facts, Whewell developed the comments of his earlier speech to the British Association about the dangers associated with such a division of labour. In responding to Spedding's references to the example of the *Admiralty Manual of Scientific Inquiry*, he indicated that the areas covered by this volume were those in which a workable theory already existed, thus allowing 'subordinate observers' to confirm or extend it.[62] In the case of scientific fields in which a definite theory had not emerged — such as parts of geology and biology — Whewell insisted that there could be no separation of the observer and the theorist. In this sense, he was arguing that the areas of science most commonly presented as inviting a division of labour were precisely those in which it was inappropriate.[63]

Although Whewell continued to defend the significance of Bacon's writings on science, the basic thrust of his criticism anticipated the more aggressive attacks of Brewster and von Liebig, both of whom regarded Baconian methodology as an obstacle to scientific progress. From about the 1850s, it was commonplace for scientific commentators to stress the role of genius in science in contrast with the earlier emphasis on careful empirical observation. Thus, in breaking the link between Bacon and Newton, Brewster declared that 'the impatience of genius spurns the restraints of mechanical rules, and never will submit to the plodding drudgery of inductive discipline'.[64] In different ways, writers such as John Tyndall, William Carpenter and Karl Pearson presented science as a complex intellectual construction dependent upon hypotheses, unobservable entities and theoretical assumptions. Scientific thinking was said to involve speculative leaps and intuition: Carpenter told the British Association meeting of 1872 that science was akin to poetry, and that interpretations of science were matters of individual personal judgement. Tyndall claimed that major discoveries do not 'depend upon method, but upon that genius of the investigator'.[65] However, these pronouncements about the role of imagination invariably carried the warning that this faculty could only be effective when coupled with the mental discipline which derived from scientific training.[66]

This conception of the nature of scientific thought clearly reversed the emphasis of the earlier rhetoric, supported by Baconian assumptions, which promoted the notion of an accessible method. But this shift, apparent in both formal works and public statements, should not be conceived as a uniform rejection of induction. The major systematic works on the philosophy of science, such as those of Mill and Jevons, were inductivist accounts, but they were theories about methods of verification rather than discovery. Laudan explains this shift by reference to changes in substantive scientific practice: the increasing use of hypothetical and mathematical constructs, especially in the physical sciences, undermined the Baconian concern with mechanical heuristics of discovery, and methodology became preoccupied with the elaboration of criteria for the assessment of hypotheses.[67] Furthermore, when the genesis of scientific ideas came to be viewed as the realm of unpredictable factors such as intuition and imagination, the project of formulating the conditions and mode of their action was largely abandoned. This meant that the method of science was no longer perceived as an engine or art of discovery open to the use of most people.

It is also possible that this development reflects changes in the social conditions under which science was pursued. As science became more securely institutionalized, some of the rhetorical resources provided by Baconianism became redundant. There was less need to stress the possibility of participation by amateurs or the lay public, and the pressure to indicate the connections between science and other forms of culture was no longer so intensive. This consolidation of the scientific community also meant that it did not have to be so defensive about the cultural implications of its activities. The deference to Baconianism in the early decades of the century, related, in part, to the need to advertise the sober, empirical character of scientific inquiry and to distance it from speculation and hypotheses which had controversial political or theological overtones. In contrast, the increasing emphasis on the conjectural and conceptual features of scientific thinking in the later period arguably reflected the comparative independence from religious control, and the greater confidence of a more professionalized community.[68] In this sense, methodology continued to be available as a rhetorical resource. Whereas the Baconian philosophy promoted a careful, non-conjectural inductive method, accessible to all, the emphasis on imagination combined with rigorous verification was associated with a conception of scientific method as the mark of the trained scientific expert pursuing an exciting intellectual quest which lay beyond the reach of the public.[69]

3. THE UNITY OF SCIENCE AND THE UNITY OF METHOD

In his Presidential Address to the British Association in 1850, Prince Albert referred to the increasing specialization of knowledge and feared that it would affect the 'consciousness of its unity which must pervade the whole of Science'.[70] This concern about the relationship between the various branches of science was apparent at the time of the foundation of the Association, a body conceived as a national forum for the coordination of research. Harcourt, the first secretary, claimed that the advance of science would be retarded if different fields of investigation were insulated from each other.[71] In this context, it is significant that when Whewell coined the word 'scientist' in 1834, he did so in a review of Mary Somerville's book *On the Connexion of the Physical Sciences*, a work which attempted to indicate the unity of science by illustrating the general and universal character of natural laws.[72] In searching for a generic term to describe all men of science, Whewell was not only seeking to distinguish 'scientists' from 'artists', but was also attempting to underline the common enterprise in which astronomers, chemists, geologists, and botanists were engaged. Whewell warned that excessive specialization might vitiate the unity of science so that it might come to resemble 'a great empire falling to pieces'.[73] Harcourt saw the British Association as a means of holding this 'commonwealth of Science' together, but the very fact that the Association was divided into sections, many of which were becoming increasingly specialized, suggested that a *substantive* unity of science would be difficult to demonstrate.[74]

In looking back on Somerville's work, James Clerk Maxwell remarked that 'the unity shadowed forth in . . . her book is . . . a unity of the method of science not a unity of the processes of nature'.[75] This was a salient observation, because from the 1830s there was an emphasis on method as the unifying feature of scientific inquiry. In Herschel's *Discourse*, it is possible to see this focus arising from the recognition that the increasingly mathematical character of some fields was a major factor in the growing specialization of science. This connection was tangibly expressed in the order of the sections within the British Association in which Section A, the most influential, comprised astronomy, optics, mechanics and dynamics, subjects which, as Herschel noted, were 'almost exclusively under the dominion of mathematics'.[76] In discussing the value of natural philosophy for general education, he acknowledged that a thorough comprehension of this area demanded a 'degree of knowledge of mathematics and geometry altogether unattainable by the generality of mankind'. However, he then suggested that any person of 'good ordinary understanding' could grasp 'the general train of reasoning by which any of the great truths of physics are deduced', and he indicated that there were many other areas of science in which mathematics was not crucial.[77]

Herschel's discussion of scientific method – the largest part of his *Discourse* – was presented as a general account of the kind of procedures and principles common to all scientific inquiry. This avoided the questions about the role of mathematics and the differences between disciplines; indeed, some points were illustrated by references to that 'deservedly popular science' of geology, which did not require mathematical sophistication.[78] At the most general level, scientific method was presented as the inductive philosophy outlined by Bacon. This was the 'the alpha and omega of science' which, in Herschel's interpretation, did not really exist before this method was consciously employed.[79] Most significantly, he presented method as the defining feature of the scientific endeavour, the common bond relating the cultivators of its various branches:

In the foregoing pages we have endeavoured to explain the spirit of the methods to which, since the revival of philosophy, natural science has been indebted for the great and splendid advances it has made. What we have all along most earnestly desired to impress on the student is, that natural philosophy is essentially united in all its departments, through all which one spirit reigns and one method of enquiry applies.[80]

This conception of method as the essential feature of science was a prominent theme in the rhetoric of the scientific community throughout the nineteenth century. For example, this assumption made it possible to argue that, in spite of theoretical disputes, the continuity of science was guaranteed by the consensus about method. Herschel claimed that the current opposition between Newton's corpuscular optics and the undulatory theory of light would be resolved by the proper application of scientific method.[81] In this sense, methodology was presented as an area above scientific controversy, an area in which at least general agreement could be expected. It is arguable that Herschel was sensitive to this role when, in his Presidential Address to the British Association in 1845, he sought to separate questions of method from contemporary disputes about epistemology. Alluding to the confrontation between empiricism and idealism, in which Mill and Whewell were major protagonists, Herschel emphasized what he took to be their basic agreement on 'the most essential features' of 'the inductive philosophy'.[82] In this way, method could be presented as a neutral and effective instrument of discovery.

Clearly, however, the scientific community was not unanimous on all points of methodology. For example, the debates about the wave theory of light involved significant methodological argument, with Brewster and Brougham confronting Whewell and Herschel on the role of hypotheses and unobservable entities.[83] But here it is necessary to recognise different levels of discourse about method. It was possible for detailed methodological disputes, conducted mainly in transactions and papers of limited circulation, to coexist with a consensus about the general nature of method in statements of a more public kind. This rhetoric about method was largely constituted by Baconian themes which were exploited to form both negative and positive images. Thus, the method of science was opposed to wild anticipations and conjectures, to dogmatic authority, to abstract deductive systems and to the various idols of the mind. Defined in positive terms, it was based on careful observation of nature, well constructed experiments, moving slowly and cautiously to generalizations. As such, scientific method was said to require and encourage certain attitudes and habits: humility, patience, caution and hard work. The negative image was often associated with Descartes; the positive with Newton.[84] Thomas Chalmers, the Scottish evangelical theologian, expressed the moral aspects of this contrast between methods when he stated that:

276

In the ethereal whirlpools of Descartes, we see a transgression against the humility of the philosophical character. It is the presumption of knowledge on a subject, where the total want of observation should have confined him to the modesty of ignorance. In the Newtonian system of the world, we see both humility and hardihood.[85]

Although these representations of method lacked specificity, the dichotomies they constructed were significant elements in public legitimations of science. For example, in his presidential Address to the British Association in 1839, Harcourt defended geology against conservative clerical attacks by stating that the evidence which it employed had been 'examined by all the various lights of science, and by undeniable methods of investigation'.[86] This was very much an argument from authority, based on the contention that the natural sciences possessed a method of the most reliable kind. Secondly, the notion of a single, scientific method entered into disputes involving attempts to demarcate acceptable science from unacceptable speculation. Thus, in the controversy surrounding the anonymous *Vestiges of the Natural History of Creation*, attacks upon the author's alleged ignorance of proper method served to characterize the work as pseudoscientific.[87] Even as late as 1859, when Darwin's *Origin of Species* appeared, Baconian rhetoric, similar to that of the early years of the Geological Society, was marshalled in an effort to condemn the work as speculative and conjectural. Sedgwick reviewed the book in *The Spectator* and told his readers that Darwin's theory resembled 'a vast pyramid resting on its apex'; it had not been logically derived from the facts; in short, Darwin's work had 'deserted the inductive track — the only track that leads to physical truth'. He also insinuated that this book, like the *Vestiges*, would appeal to those untrained minds who 'conclude that what is (apparently) *original*, must be a production of original *genius*'.[88] Richard Owen carried this charge further, drawing upon the Baconian notion of scientific procedure as open and accessible, and dismissing novel or controversial theories as the product of individual speculation.[89] Thus, the Baconian conception of inductive method, although increasingly rejected by the major writers on the philosophy of science, continued to function as a means of dealing with controversial theories. This emphasis on method also allowed critics to claim that their objections to a particular doctrine were based 'solely on scientific grounds' and not religious or moral considerations.[90] By attacking at the level of methodology, theories which had unwelcome ideological implications could be labelled as 'unscientific'.

These appeals to method depended upon the circulation of general and non-specific references to observation, experiment and induction. This allowed the idea of a single method, accepted by the scientific community, and capable of being appreciated by the public. However, at another level of discourse within the scientific community, the conception of the unity of method was a point of dispute in debates about the relationships between different disciplines.

The idea of a common method was seized by William Swainson, the author of the *Preliminary Discourse on the Study of Natural History*, the volume in Lardner's series which complemented Herschel's introduction to natural philosophy. Swainson wanted to ensure that the natural history disciplines were accorded scientific status. These subjects lacked the mathematical and experimental character of the physical sciences treated in Herschel's *Discourse*; and those such as zoology and botany, which Swainson discussed, belonged to a popular leisure tradition. Furthermore, they had been placed in a relatively low position in the well-defined hierarchy of sections and disciplines which crystallized in the British Association. This order was defined by the characteristics of the dominant section — the mathematically-based sciences of astronomy, mechanics and optics — and under the influence of Whewell and Herschel these disciplines, which began to assume a monopoly over the term *physical* science, were presented as the model to which all sciences should aspire.[91]

In order to defend the intellectual importance of his own subject, Swainson had to resist this tendency. He did so by asserting the links between natural history and other sciences and, perhaps in response to the hierarchical structure of the Association, referred to its position in 'the *circle* of human knowledge'.[92] 'We address ourselves', he said, 'to those who have been instructed to form more enlarged conceptions of the physical sciences; who view natural history but as a part, and consider that that part must be studied upon the same principles as any other'. He was able to argue that natural history, in spite of its non-experimental basis, shared, with natural philosophy, an important place among the inductive sciences.[93] In this sense, appeals to the idea of a common scientific method played a significant rhetorical function in the negotiations between scientific disciplines, allowing the less established subjects to claim full 'scientific' status.

But there were other situations in which this assumption of a single scientific method was more closely contested. Indeed, even at the time of Herschel's *Discourse* there was a latent tension beneath the surface of general statements about the unity of method. Although Herschel's remarks about method were taken to cover all science, his illustrations of scientific procedure were drawn largely from the advanced physical sciences. In reviewing Herschel's *Discourse*, Whewell agreed that all scientific thought consisted in a combination of induction and deduction, but emphasized that different sciences were at different stages of the 'complete cycle of ascent and descent'.[94] Thus, when compared with astronomy, chemistry was still in its infancy. In Whewell's view, a mature science was one in which the data could be expressed in the form of a 'few general and simple laws'.[95] Although he believed that, in principle, all sciences could hope to reach this point, he argued that their present stage should determine the method necessary to facilitate further progress. Thus, in some disciplines the 'Methods of Observation' would be dominant, in others, the 'Methods of Acquiring Clear Scientific Ideas'.[96] In this way, rather than acting as a counterweight to specialization, scientific method could become a focal point in disputes about the relationships between different disciplines, especially those between the physical and the organic sciences. This can be illustrated, firstly by looking at the response of Baden Powell and William Carpenter to Whewell's philosophy of science and, secondly, by investigating the connection between T. H. Huxley's methodological statements and his defence of the natural history sciences.

Whewell's philosophy of science was a search for the general features of a single philosophy of discovery. He hoped to show that 'the progress of moral, and political, and philological, and other knowledge, is governed by the same laws as that of physical science'.[97] This goal was supported by the claim that the progress of knowledge, in all these areas, depended upon a similar interaction between facts and ideas, between empirical observation and theory. But the concept of a general philosophy of knowledge was qualified by his own theory about the different fundamental ideas appropriate to various fields of science, and this stress upon the distinctiveness of scientific disciplines had implications for the notion of a single scientific method. Whewell believed that the various branches of scientific inquiry went through successive stages, characterized by different relationships between facts and ideas. However, these histories of development were not contemporaneous, and this meant that the present methods of physical astronomy, for example, could not be simply extrapolated to physiology or geology.

Whewell therefore warned that 'different sciences may be expected to advance by different modes of procedure, according to their present condition'; that different methods were appropriate at different stages of each science.[98]

When he illustrated these views in his major works, Whewell's discussion of the new sciences such as physiology and geology implied that these subjects should be conducted on principles at variance with those prescribed for the more advanced physical sciences. For example, he argued that teleological notions could play no part in the investigations of physics or astronomy, but claimed that the concept of final cause was the fundamental idea appropriate to the study of physiology; the major advances in this field had come when researchers allowed teleological assumptions to guide their thinking.[99] In criticizing this aspect of Whewell's work, Powell and Carpenter argued that his suggestion that research in physiology should be directed by speculation about purpose or final cause was tantamount to the exclusion of the subject from the domain of inductive science. In his review of Whewell's *History*, Carpenter said that if taken up, Whewell's views would retard the progress of physiology by disrupting its connections with the techniques and concepts of the physical sciences, such as chemistry. Carpenter asserted that the complexity of physiology required the training which physical science could provide and that, as an inductive science, it should be pursued by methods similar to those of physical science.[100]

Baden Powell strongly advocated the unity of science and resisted any suggestions concerning qualitative differences between the various disciplines. In 1838 he drew attention to the way in which opponents of the new science of geology attacked its conclusions by arguing that its methods of research were distinct from, and 'less strictly inductive', than those of other sciences.[101] He tried to show how the theories of the less controversial disciplines were passively accepted as certain, even though the concepts and reasoning from which they derived were more obscure than those of geology.[102] Like Carpenter, he also criticized the notion that the life sciences involved '*a new class and order of ideas* of so peculiar a kind that they must stand out as entirely exceptional cases to the general *unity of the sciences*'.[103] This attitude justified the use of concepts and methods of reasoning, in areas such as physiology and comparative anatomy, which would not be countenanced in other departments of science. Indeed, Powell's commitment to the ideal of a

280

uniformity of method and principle across the sciences was related to his attempt to bring the species question within the province of normal scientific investigation. He believed that debate on this question was being confused by the intrusion of teleogical reasoning into scientific discourse, a situation justified by Whewell's philosophy of science. Powell criticized Whewell's approach as one which isolated 'each department of science ... by a theory which would assign to each branch certain real differences of principle and peculiar fundamental ideas essentially characterizing it'.[104] At this level, therefore, statements about method did not express consensus within the scientific community, but were closely implicated in controversial areas of scientific debate. In this case, the emphasis on the unity of scientific method was part of a strategy to avoid the admission of qualitative differences between scientific disciplines.

During the second half of the century, the question of the relationship between physical and organic sciences was a matter of explicit contention. For example, representatives of the natural history disciplines felt it necessary to resist the dominance of the physical and experimental sciences, especially in science eduction.[105] The issue of method was often central to these confrontations, because distinctions between mathematical, experimental and observational modes of inquiry were associated with claims about the relative position of disciplines in the hierarchy of knowledge. This was the context in which T. H. Huxley equated scientific method with 'trained and organized common sense'.[106] By drawing this equation, Huxley tried to resist the assertion that the method of biological disciplines was inferior to that of the physical sciences.

If . . . there be no real difference between the methods of science and those of common life, it would seem, on the face of the matter, highly improbable that there should be any difference between the methods of the different sciences; nevertheless, it is constantly taken for granted that there is a very wide difference between the Physiological and other sciences in point of method.[107]

Huxley thus attempted to strengthen the status of organic sciences by claiming that they shared the method of the 'Physico-chemical and Mathematical Sciences'.[108] In this view, there was only one method of scientific investigation, which consisted of both observation and experimentation.

The context in which these claims were made illustrates the way in which the question of method had become connected with conflicts between disciplines. For example, although Carpenter agreed with Huxley's defence of physiology, he found it useful, in the case of a particular controversy, to stress the *difference* between the two kinds of methods. Writing in 1871 about the scientific attitude to psychic phenomena, he dismissed one researcher by arguing that his previous experience in observational astronomy did not prepare him for this new area of investigation:

In particular, we believe that his devotion to a branch of research which tasks the keenest powers of *observation*, has prevented him from training himself in the strict methods of *experimental* inquiry . . .[109]

Thus, in seeking to establish the priority of physiology and its techniques in the areas of psychic research, Carpenter exploited the tension between observational and experimental approaches — a tension which Huxley, for different reasons, sought to conceal.

In a number of scientific debates during the second half of the century, disputes about method, kinds of evidence, and modes of proof were central points of contention. As noted previously, the controversy over Darwin's work involved charges about his departure from sound induction and 'the true Baconian philosophy'.[110] But another aspect of these critiques illustrates the way in which the notion of a single scientific method — which Powell, Huxley and Carpenter saw as essential for the protection of the organic disciplines — could be deployed against Darwin's theory. The most forceful use of this strategy occurred in the review by William Hopkins, the Cambridge mathematician. Although he carefully avoided any moral or metaphysical objections to the *Origin of Species*, Hopkins assessed the book from the perspective of a logician rather than from that of a biologist or geologist. He lamented that 'however well the abstract philosophical rules of theorizing in physical subjects may be known, many people will still theorize as if Bacon and Newton had never lived'.[111] But for Hopkins, these methodological canons were best examplified in the most successful physical sciences, such as astronomy and optics, and his review systematically interrogated Darwin's work against criteria derived from the most established fields of science. He was happy to grant that the 'same principles of research' should apply in all sciences but insisted that the naturalist must therefore accept the standard of evidence demanded of the physicist. 'It is impossible', he said, 'to admit laxity of reasoning to the naturalist, while we insist on rigorous proof in the physicist. He who appeals to Caesar must be judged by Caesar's laws'.[112]

These disputes about methodology within the scientific community were further highlighted by the controversy between Lord Kelvin and the Darwinians over the age of the earth. The tension between the physical and organic disciplines was a central issue here, since much of the discussion exposed disagreement over modes of argument, kinds of evidence and standards of verification.[113] In these late-nineteenth-century debates therefore, methodological conflict was closely linked with clashes between increasingly specialized disciplines and rival professional interests. This made it more difficult for method to be promoted as the point of consensus in science and the means by which the layman could understand, or appreciate, scientific research. Writing in 1877, William Gladstone remarked that:

The results, in pure science, are accepted by us as facts; but on the methods by which they are reached, the mass, even of intelligent and cultivated men, are not completely informed.[114]

4. THE TRANSFERABILITY OF SCIENTIFIC METHOD

The assumption that there was a single, definable scientific method allowed claims about the possibility of extending it to other subjects. In his *Grammar of Science*, Pearson asserted that '*the unity of all science consists alone in its method, not in its material*'[115]; and he used this distinction to reinforce the claim for a scientific approach to social problems. There was, of course, a Baconian warrant for such a pro- gramme, and it was a significant theme in early-nineteenth-century writing on the philosophy of science.[116] In his *Discourse*, Herschel suggested that the success of proper methods of inquiry in the area of natural knowledge might lead to a situation in which legislation and politics were 'regarded as experimental sciences'.[117] Not unexpectedly, J. S. Mill was an early supporter of this idea, arguing that the first contribution of physical science to society would come not 'by the truths which it discloses, but by the process by which it attains them'.[118] The present weakness of moral (or social) science was due to a 'lack of proper method', and this could be taken from the natural sciences. Herschel returned to this topic in his Presidential Address to the British Association in 1845, urging that attention to scientific method was relevant to the understanding and progress of society, and suggesting that the work of Whewell and Mill on scientific thought had implications for *the principles of all general reasoning*.[119] These were early expressions of the claim to intellectual and cultural authority, beyond the realm of natural knowledge, which later become a major element in the rhetoric of public science.

Nevertheless, in the 1830s and 1840s, the extension of the method of natural science to social and political questions was still a contentious issue. Even amongst writers who defended the notion of inductive moral or social sciences, there was dispute about the precise way in which the method of physical science could be extrapolated. Herschel would have been aware that Whewell, although not unfavourable towards the scientific pretensions of political economy, was critical of hasty transfers of concepts and methods from physical to moral subjects.[120] By the nd of the century, however, the analogies between natural and social inquiries were pursued with enthusiasm.

Representatives of the natural sciences generally lent their support to the wider application of scientific procedures. Indeed, the exponents of scientific naturalism such as Huxley, Spencer and Galton campaigned in favour of this kind of transfer. Pearson repeated this aspiration, stating that 'the material of science is coextensive with the whole life, physical and mental, of the universe'.[121] In 1910, this unlimited field was formally annexed by the British Association when it established a section for Education Science which, apart from objectives in the area of pedagogic technique, was inspired by the desire to take 'a scientific view of things in every department of life. . .'.[122] This was one of the contexts in which sociology, anthropology, psychology and archaeology claimed the title of social sciences and cited the use of empirical and experimental methods as evidence of their 'scientific' status.

In addition to supporting the claim that social inquiry could be modelled on the natural sciences, assumptions about the transferability of scientific method occurred in educational debates. From the early decades of the nineteenth century, those who sought to increase the representation of natural science in university education found it necessary to conform with prevailing conceptions of liberal education. In the two ancient English universities, this curriculum was founded upon the classics and mathematics, and supported values rigidly opposed to the intrusion of 'useful' or professional knowledge. In this context, an emphasis on the method of science rather than its content allowed natural science to be presented as a form of mental discipline. Thus, when addressing the readers of the Tory *Quarterly Review*, Charles Lyell agreed 'with those who maintain that the most important part of education consists, not so much in the things taught, as in the moral and intellectual habits instilled'.[123] He was then able to argue that the physical sciences trained the mind to reason about probable evidence rather than demonstrative theorems, thus providing the best preparation for the intellectual demands of practical life.[124] As late as 1867, J. S. Mill recommended instruction in 'the Methods of the Sciences' as the best example of 'the art of thinking'.[125] In this way, a rhetorical emphasis on method made it possible to argue for the inclusion of natural science in university teaching without challenging the non-professional ethos of liberal education.

The distinction between scientific method and scientific information also entered debate about the popularization of science. Important men of science such as Powell, Lyell and Herschel pointed to the danger of an unequal distribution of knowledge throughout society and to the problem of a small scientific class appearing as the cultivators of an esoteric language.[126] However, the appropriate form of popular science education was difficult to define. There was dissatisfaction with most popular accounts of science because they stressed factual information and practical applications rather than the intellectual processes by which results were attained. But, as scientific knowledge expanded and became more complex, the possibility of explaining it to a non-scientific public became unrealistic. Out of this situation, the conception of scientific method as an educational bridge between science and the public emerged as an influential solution.[127] G. C. Lewis, for example, who was concerned about the gap between experts and laymen, recommended that:

every person should, so far as his means extend, make himself master of the methods of scientific investigation, so as to be able to judge whether in the treatment of any question, a sound and correct method has been observed.[128]

Throughout most of the century, the British Association exemplified this investment in the educational value of scientific method. At the Glasgow meeting in 1855, for example, the Duke of Argyll condemned 'cramming of facts from manuals', and suggested that 'what we want in the teaching of the young, is, not so much the mere results, as the *methods*, and, above all, the *history* of science'.[129] Pearson endorsed this theme in 1892, stating that 'it is not the facts themselves which form science, but the method in which they are dealt with'.[130] This dichotomy between the content and method of science also became the focus of a doctrine known as 'heurism', promoted by the chemist, Henry Armstrong. In the 1880s, Armstrong began to argue that current science teaching was too dogmatic, too concerned with imparting facts and theories, often by means of rote learning.[131] He advocated a shift of emphasis from content to method and outlined pedagogic techniques designed to place students 'so far as possible in the attitude of the discoverer'.[132]

Heuristic teaching attracted substantial support; but by the second decade of the twentieth century it was meeting with determined opposition. Critics complained that the system was impractical, expensive, not fully examinable and, that where implemented, it led to a situation in which science courses and students were dangerously free of content. In short, the stress on method had been excessive. This attack reached a climax in the 1917 Report of the British Association Committee on Science in Secondary Schools, headed by Richard Gregory, the editor of *Nature*. Arguing against the tradition which had culminated in Armstrong's heurism, the committee asserted that the method of scientific investigation was not separable from its matter or content:

In other words, it is not strictly true that scientific method is one and the same wherever it is employed. The physical method and the historical method, for example, have common fundamental features, but cannot be simply identified the one with the other. In short, scientific method is an abstraction which does not exist apart from its concrete embodiments; and the person who desires adequate knowledge of it must study it in all its typical manifestations.[133]

This was a virtual recantation of the assumptions which had supported the earlier rhetoric about the role of method as an educational bridge between science and the public, and the possibility of its extension to a wide range of subjects. In order to understand this reversal it is necessary to appreciate the way in which the stress on method became a strategic disadvantage in educational debates.

By the early twentieth century, the precise meaning of 'science' and 'scientific' often required clarification. Thus, when T. P. Nunn set out to write a book on *The Aim and Achievements of Scientific Method* in

1907, he felt it necessary to begin with a preface on semantics. After explaining that 'science' acquired its restricted meaning of physical or natural science with the establishment of the British Association in 1831, he warned that the word 'scientific' had now assumed a much broader denotation.

Thus, while there can be no doubt that the word "Science" refers most commonly to a certain body of knowledge, "scientific" refers more frequently to the method or procedure by which this knowledge has been established. By an easy extension of its meaning the adjective is applied to all processes by which knowledge is reached having the security which is thought to characterize "science".[134]

Because of this ambiguity, a Conference on the Humanities and Social Sciences held in 1917 was able to argue that the disciplines it represented employed scientific methods and therefore provided the mental training which the natural scientists were emphasizing. In response to such claims, the Committee on the Neglect of Science was forced to state that 'scientific method can be used in the study of literary, historical and other subjects; but applied in this way it can never be accepted as providing the same means of training as direct observation and experimental work'.[135] This was another attack upon the priority which Armstrong's heurism accorded to method over content and information; but it also derived from the recognition that this dichotomy could be used to argue that scientific method was no longer the exclusive possession of the scientific community.

If, however, it is admitted (1) that the chief business of the science teacher is to train in scientific method, and (2) that scientific method is the characteristic not of science only but of every properly conducted intellectual inquiry, the science teacher is perilously near to the surrender of his special claim to existence.[136]

The debate over science education illustrates the way in which the implications of this rhetoric could, in specific situations, conflict with the interests of the scientific community. In 1917, when various committees reported on the place of science in the curriculum, there was a shift away from the heavy emphasis on method which had emerged in the second half of the nineteenth century. The rise of the social sciences and their claim to possess a 'scientific' approach made the earlier rhetoric more problematic. But furthermore, some scientists believed that the stess on method had produced a technical and instrumental image which divorced the natural sciences too sharply from humanistic studies. Given the war and the increased awareness of the technological applications of science, there were fears in some sections of the scientific community that future funding could be too closely influenced by judgements about utility.[137]

During the twentieth century, an emphasis on method continued to be significant in definitions of science and in claims about its educational and social value. In 1917, one of the first collaborative publications in the history and philosophy of science, Charles Singer's *Studies in the History and Method of Science*, gave the concept of scientific method a foundational role, explaining that 'the need for a comprehensive study of the methods of science is now widely recognized', and that the historian could teach 'the method by which knowledge has been gained'.[138] A similar focus on method was also discernible in the later attempts of J. B. Conant and I. B. Cohen to explain science to educated, non-scientific audiences. Conant's *Harvard Case Histories in Experimental Science*, published in 1957, attempted to encourage an understanding of 'the methods of experimental science' as a way of helping the lawyer, teacher or public servant to evaluate the work of scientific experts.[139] The more extreme claims concerning the political value of scientific thought also continued to depend upon statements about method. In December 1940, for example, *Nature* carried an editorial on the 'Cultural Significance of Scientific Method' which announced that 'the fate of democracy is bound up with the spread of the scientific attitude'.[140]

5. CONCLUSION

Most recent historiography of scientific method has been concerned with the influence of particular methodologies in the work of individual scientists, or with the role of methodological doctrines in substantive scientific debates.[141] This chapter has offered a different perspective by focusing on the rhetorical aspect of discourses about method in the wider social relations of science. Three key assumptions about method were delineated and it was argued that each of these was deployed, in specific contexts, to advance certain images of science. However, as a counterpoint theme, the analysis revealed ambiguities and contradictions within this rhetoric and investigated the ways in which these related to points of tension within the scientific community. Thus the emphasis on the accessibility of method entailed arguments about the relation between theory and observation and the role of experts and amateurs; the idea of a single scientific method was involved in conflicts between disciplines; and the concept of transferability led to situations in which the dominance of the natural sciences was questioned.

The approach taken here also suggests that the historical and socio-logical perspectives which have been applied to substantive scientific texts should be extended to works on methodology. References to method occurred at various levels, and it must be recognized that the social and strategic aspects of public lectures and addresses were more obvious than those of systematic works on the philosophy of science. For example, the nature of methodological comments at a meeting of the British Association could be expected to differ from those in works such as Herschel's *Discourse* or Whewell's *Philosophy*. Nevertheless, while these works responded to issues and questions which have occurred in a variety of historical situations, they were also written in specific social and intellectual contexts, and it should not be assumed that they conveyed only an abstract discourse on methodology.

Herschel's book lent credence to assumptions about method which were associated with an account of scientific research as accessible, unified and potentially applicable to social problems. As such, it was both a contribution to the contemporary debate about the place of science in early Victorian culture and a resource for other participants in this debate. Whewell's work, on the other hand can, in some respects, be read as a reaction against these public representations of scientific method. He severely qualified the notion of an accessible, inductive procedure and was cautious about the concept of a single and trans-ferable mode of scientific thinking. But in taking this position, Whewell was also employing methodology to advance his preferred definition of science — one which stressed the importance of theoretical experts, restricted the division of scientific labour, and saw the mathematico-physical sciences as the highest form of natural knowledge. Whewell's work illustrates the way in which certain topics in the philosophy of scientific method could involve social and institutional considerations. As we have seen, the concept of a single method of science was central to debates about the relationship between physical and biological disciplines. In this case, writers of general philosophical works on science, such as Whewell and Powell, and practising scientists such as Carpenter and Huxley, were not simply formulating descriptive or prescriptive statements about scientific research: they were deploying method as a rhetorical weapon in negotiations about the place of different disciplines in the hierarchy of scientific knowledge.

This investigation does not support a simplistic notion of method-ological discourse as an ideology which uniformly promoted the inter-ests of the scientific community.[142] Instead, it suggests that the appeal to method in the apologetics of science was not identified with any single interest. The three assumptions about method which have been dis-cussed did not indicate points of unqualified or unshifting consensus, but rather areas of debate in which statements and counter-statements were made. In fact, such a picture is to be expected if we recognize that the 'scientific community' embraced various groups which perceived different priorities for both the intellectual agenda and the social relations of science. The Baconian notion of accessibility, although influential in certain contexts and for particular audiences, was qualified by those who wanted to underline the exclusive character of scientific inquiry. The idea of a unity of method, which was promoted as a coun-terweight to the fragmentation of science, played a different and more complex role in disputes between disciplines. Thus, while there were recurring themes in the rhetorical use of method, a proper grasp of its significance also demands attention to the various points of emphasis and tensions, and to the ways in which these related to different contexts and preoccupations. The subject of methodology was an argumentative resource which could be exploited in a variety of ways in debates about the definition of science and its cultural significance.

ACKNOWLEDGEMENT

I would like to thank Geoffrey Cantor, Jonathan Hodge, Menachem Fisch and John Schuster for their helpful comments on earlier drafts of this chapter. The research was undertaken as part of study leave supported by Griffith University.

Griffith University, Australia.

NOTES

[1] G. Orwell, 'What is Science?' in *Collected Essays, Journalism and Letters* (4 vols, Harmondsworth, 1970), vol. 4, pp. 26–30, at p. 27.
[2] B. Webb, *My Apprenticeship* (London, 2nd ed. nd.), p. 113. See also p. 118 for the distinction between the 'subject matter of natural science' and the 'value of physical science as a training in scientific method'.

290

[3] K. Pearson, *The Grammar of Science* (London, 1892) pp. 10—11. See also N. Lockyer, 'Scientific Administration', *Nature*, 6 October 1870, pp. 449; M. Cohen and E. Nagel, *An Introduction to Logic and Scientific Method* (London, 1934), p. 403.

[4] See F. M. Turner, 'Public Science in Britain, 1880—1919', *ISIS* **LXXI**, 1980, pp. 589—608. For a comparative analysis, see P. Buck, 'Order and Control: The Scientific Method in China and the United States', *Social Studies of Science* **V**, 1975, pp. 237—67. See also Y. Ezrahi, 'Science and the Problem of Authority in Democracy', in *Science and Social Structure*, R. K. Merton (ed.), *New York Academy of Sciences* **xxxix**, (1980), pp. 43—60.

[5] I. Hacking, *The Emergence of Probability* (Cambridge, 1975), pp. 179—180; B. Shapiro, *Probability and Certainty in Seventeenth-Century England* (Princeton, 1983), pp. 15—73.

[6] J. Ben-David, *The Scientist's Role in Society: A Comparative Study* (New York, 1971), pp. 72—74.

[7] P. Wood, 'Methodology and Apologetics: Thomas Sprat's History of the Royal Society', *British Journal for the History of Science* **XIII**, 1981, pp. 1—26.

[8] T. Sprat, *The History of the Royal Society*, J. I. Cope and H. W. Jones (eds.), (London, 1959), pp. 62, 112. Sprat claimed that the members of the Royal Society were pledged to recovering the 'primitive purity' of language in which 'men delivered so many *things*, almost in an equal number of *words*'. (p. 113) For the connection between science and the concept of a 'plain style'. see R. F. Foster, *The Seventeenth Century: Studies in the History of English Thought and Literature from Bacon to Pope* (Stanford, 1951); W. S. Howell, *Eighteenth-Century British Logic and Rhetoric* (Princeton, 1971), pp. 441—502.

[9] See A. Benjamin, G. N. Cantor, J. R. R. Christie (eds), *The Figural and the Literal: Problems of Language in the History of Philosophy, Science and Literature, 1650—1800* (Manchester, forthcoming). For earlier references, see N. Frye, *Anatomy of Criticism* (Princeton, 1957), pp. 326—337; M. Hesse, 'The Explanatory Function of Metaphor', in *International Congress for Logic, Methodology and Philosophy of Science* (Amsterdam, 1964); R. M. Young, 'Darwin's Metaphor: Does Nature Select?', *Monist* **LV**, 1971, pp. 442—503.

[10] R. Whately, *Elements of Rhetoric* (5th ed., London, 1836), p. 6.

[11] W. B. Weimar, 'Science as Rhetorical Transaction: Toward a Nonjustificational Conception of Rhetoric', *Philosophy and Rhetoric* **X**, 1977, pp. 1—29. Weimar believes that the rhetorical aspect of scientific discourse is implied by Kuhn's work (p. 24).

[12] For references to the works of Cantor, Laudan and Ruse which deal with the role of methodology in nineteenth-century scientific debates, see the editors' introduction to this volume.

[13] L. Laudan, 'Theories of Scientific Method from Plato to Mach: A Bibliographic Review', *History of Science* **VII**, 1968, p. 29.

[14] See the comment about the concern of Victorian scientists with method in G. Basalla, W. Coleman and R. Kargon (eds.), *Victorian Science: A Self Portrait* (New York, 1970), p. 399 and the examples from British Association addresses. For the importance of the Scottish contribution, see R. Olson, *Scottish Philosophy and British Physics 1750—1880* (Princeton, 1975).

[15] J. S. Mill to J. Herschel, May 1843, Herschel Papers, Royal Society, London; see J. S. Mlll, 'Herschel's *Discourse*', *Examiner* 20 March 1831, pp. 179—180; [W. Whewell],

IV

'Modern Science, Inductive Philosophy', *Quarterly Review* **XLV**, 1831, pp. 374–407, at pp. 377, 398 for its philosophical importance, and M. Faraday to J. Herschel, 10 November 1832, in *The Selected Correspondence of Michael Faraday*, L. Pearce Williams (ed.), (2 vols, Cambridge, 1971), vol. 1, p. 235, for its inspirational value. For an exposition and analysis of Herschel's philosophy of science, see C. J. Ducasse, in *Theories of Scientific Method: Renaissance through the Nineteenth Century*, R. Blake, C. Ducasse, and E. Madden (eds.), (Seattle, 1960), pp. 153–182.

[16] For attempts to contextualize some of the themes in the *Disourse*, see W. F. Cannon, 'John Herschel and the Idea of Science', *Journal of the History of Ideas* **XXII**, 1961, pp. 215–239; J. Agassi, 'Sir John Herschel's Philosophy of Success', *Historical Studies in the Physical Sciences* **I**, 1971, pp. 1–36; R. Yeo, 'Scientific Method and the Image of Science 1831–1891', in *The Parliament of Science: The British Association for the Advancement of Science, 1831–1981*, R. M. MacLeod and P. Collins (eds.), (Northwood, Middlesex, 1981), pp. 65–88. I will also be referring to the works of William Whewell and to a lesser extent to the views of Baden Powell, William Carpenter and T. H. Huxley.

[17] J. Herschel, *A Preliminary Discourse on the Study of Natural Philosophy* (London, 1830), pp. 104–114, 204. There is some confusion about the publication date because one of the title pages carries an 1831 date.

[18] W. Whewell, *The Philosophy of the Inductive Sciences* (2 vols. London, 1847), vol. 2, p. 230. See also B. Powell, *History of Natural Philosophy from the Earliest Periods to the Present Time* (London, 1830), pp. 196–211. Like Herschel's *Discourse*, this work was published in Lardner's *Cyclopaedia*. For the various levels of the debate on Baconianism see R. R. Yeo 'An Idol of the Marketplace: Baconianism in Nineteenth-Century Britain', *History of Science* **XXIII**, 1985, pp. 251–298.

[19] See for example, R. Laudan, 'Ideas and Organization in British Geology', *Isis* **LXVIII**, 1977, pp. 527–538.

[20] For the expectation of certainty from inductive procedures in the early nineteenth century, see D. Hull (ed.), *Darwin and his Critics* (Cambridge, Mass., 1973), p. 17. Bacon claimed that his inductive method 'extends not only to natural but to all sciences'. See F. Bacon, *Novum Organum*, Book 1, Aphorism 127.

[21] W. Whewell to R. Jones, February 1831, in I. Todhunter, *William Whewell D. D. An Account of his Writings* (2 vols, London, 1876), vol. 2, p. 115.

[22] W. Whewell to R. Jones, 24 April 1831, in *ibid*, p. 118.

[23] J. S. Mill, *Autobiography*, J. Sillinger (ed.), (London, Oxford, 1971), p. 142. For further connections between method and political values, see [S. Bailey], *Essays on the Pursuit of Truth* (London, 1829); [T. P. Thompson], 'Pursuit of Truth', *Westminster Review* **XI**, 1829, pp. 478–487. For the relationship between proper method and moral duty within a religious framework, see I. Watts, *The Improvement of the Mind* (London, 1741); J. Abercrombie, *Inquiries concerning the Intellectual Powers and the Investigation of Truth* (Edinburgh, 1831); and M. Faraday, 'Observations on the Inertia of Mind', in B. Jones, *The Life and Letters of Michael Faraday* (2 vols, London, 1870), vol. 1, pp. 261–279.

[24] F. Bacon, *Novum Organum*, in *The Works of Francis Bacon*, J. Spedding, R. Ellis, and D. Heath (eds.), (14 vols, London, 1857–74), vol. 3, p. 63.

[25] W. V. Harcourt, *British Association Reports* (York and Oxford, 1831–2), pp. 23–31. Further references to this collection are cited as *B. A. R.* See also pp. 10–14

for an extract from a report of the Yorkshire Philosophical Society which stresses the role of provincial observers. For the Baconian rhetoric of cooperation, see J. B. Morrell and A. Thackray, *Gentlemen of Science: Early Years of the British Association for the Advancement of Science* (Oxford, 1981), pp. 267—9.

[26] See R. Laudan, *op. cit.* (Note 19), pp. 527—638; R. Porter, *The Making of Geology: Earth Science in Britain, 1660—1815* (Cambridge, 1977), pp. 204—208. See [W. H. Fitton], 'Transactions of the Geological Society', *Edinburgh Review* **XIX**, 1811—12, pp. 207—229, at p. 207, for the description of eighteenth-century geology as 'a species of mental derangement'.

[27] [W. H. Fitton], 'Transactions of the Geological Society', *Edinburgh Review* **XXIX**, 1817—18, pp. 70—94, at p. 70.

[28] T. Dick, *On the Improvement of Society* (London, 1833), pp. 105—110.

[29] J. Drinkwater, *Lives of Eminent Persons* (London, 1833), p. 54.

[30] *Ibid.* p. 15.

[31] G. L. Craik, *The Pursuit of Knowledge Under Difficulties* (2. vols, London, 1846), vol. 1, p. 2. See also H. Martineau, *Miscellanies* (2 vols, Boston, 1836), vol. 1, pp. 72, 100, 103.

[32] S. Smiles, *Self Help* (new edition, London, 1860), pp. 46—47.

[33] C. Babbage, *Reflections on the Decline of Science in England* (London, 1830), pp. 40, 170.

[34] J. Herschel, *op. cit.* (Note 17), pp. 69—70.

[35] *Ibid.* p. 133.

[36] See H. de la Beche, *How to Observe — Geology* (London, 1833), p. 83; also J. Phillips, *A Treatise on Geology* (2 vols, London, 1839), vol. 2, pp. 279—280 for an emphasis on the 'plainness and accessibility' of geology.

[37] J. Herschel (ed.), *The Admiralty Manual of Scientific Enquiry* (London, 1849) p. iii.

[38] T. Dick, *op. cit.* (Note 28) pp. 82—83. See also *Recreations in Science* (London, 1830) for defence of empirical fact as opposed to systems.

[39] H. Brougham, quoted in J. W. Hays 'Science and Brougham's Society', *Annals of Science* **XX**, 1964, p. 231.

[40] J. S. Mill, *op. cit.* (Note 15), p. 179.

[41] R. Whately, *Introductory Lectures on Political Economy* (London, 3rd ed., 1847), pp. 55—64.

[42] [T. Macaulay], 'Francis Bacon', *Edinburgh Review* **LXV**, 1837, pp. 1—104, at p. 87.

[43] C. Kingsley, 'Science', in *Scientific Lectures and Essays* (London, 1890), p. 210.

[44] J. Herschel, *op. cit.* (Note 17), pp. 133—134.

[45] *Ibid.* p. 132.

[46] *Ibid.* Herschel also implied that new areas of reasearch relied on 'individual penetration' and required 'a union of many branches of knowledge in one person'; it was the mature areas which could benefit from division of labour, although this was often skilled labour (pp. 102—103, 131). For a similar, but more explicit formulation, see the discussion of Whewell below.

[47] *Ibid.* p. 134.

[48] W. Whewell, *B.A.R.* (Cambridge, 1833), p. xx.

[49] *Ibid.* See also pp. xxi-xxii where he indicates the danger of dogmatic theory. See also J. B. Morrell and A. Thackray, *op. cit.* (Note 25), pp. 270—271.

[50] *Encyclopaedia Britannica* (7th ed., Edinburgh, 1830—42), vol. 15, p. 173. In fact, in the public statements of the Geological Society, which contained a strong emphasis on

Baconian empiricism, it is possible to detect another methodological message which provided a definite role for the theorist. Thus, references to the importance of empirical observation 'by persons attached to various pursuits and occupations' could be coupled with the recognition that the task of generalizing and systematizing required a knowledge of other sciences and could be 'performed only by philosophers'. See *Philosophical Magazine* **XLIX**, 1817, pp. 421–429, at pp. 421–422. Rachel Laudan's analysis of methodological pronouncements in the Geological Society overlooks the way in which the Baconian tradition could be interpreted to provide this space for theorists, as distinct from the army of observers. See R. Laudan, *op. cit.* (Note 19) and the chapter by David Miller in this volume.

[51] W. S. Jevons, *The Principles of Science: A treatise on Logic and Scientific Method* (London, 2nd ed., 1877), p. 507; D. Brewster, *Memoirs of the Life, Writings and Discoveries of Sir Isaac Newton* (2 vols, Edinburgh, 1855), vol. 2, pp. 400–403; J. von Liebig, 'Lord Bacon as Natural Philosopher', *MacMillan's Magazine* **VIII**, 1863, pp. 237–249; 257–265.

[52] L. Laudan, *Science and Hypothesis: Historical Essays on Scientific Methodology* (Dordrecht and Boston, 1981), pp. 4–5. Laudan claims that the 'method of hypothesis' had been popular in the seventeenth century but was ostracized by eighteenth-century inductivism.

[53] R. Adamson, 'Bacon', *Encyclopaedia Britannica* (9th ed., Edinburgh, 1875–89), vol. 3, p. 216. For a discussion of the reactions against Baconianism, see R. Yeo. *op. cit.* (Note 18).

[54] W. Whewell, *History of the Inductive Sciences* (3 vols, London, 1837), vol. 1, pp. 410–411. For his response to the image of Newton as merely a patient investigator, see vol. 2, p. 185.

[55] W. Whewell, *op. cit.* (Note 18), vol. 2, pp. 41, 46–59.

[56] *Ibid.* vol. 2, p. 240.

[57] W. Whewell, *On the Philosophy of Discovery* (London, 1860), pp. 336–340, 347–348.

[58] W. Whewell, *op. cit.* (Note 18), vol. 2, p. 40.

[59] *Ibid.* pp. 20–21.

[60] *Ibid.* p. 366. There was a tension in Whewell's work on the question of what could be learnt about appropriate method from a study of the history of science. Compare his *History of the Inductive Sciences* (3 vols, 2nd ed., London, 1857), vol. 1, pp. viii, 4, and *Novum Organon Renovatum* (London, 1858), pp. iii–v, 141–142.

[61] *Ibid.* pp. x–xi.

[62] [W. Whewell], 'Spedding's Complete Edition of the Works of Bacon', *Edinburgh Review* **CVI**, 1857, pp. 287–322, at p. 314.

[63] W. Whewell, *op. cit.* (Note 57), pp. 154–156. He also remarked that a 'considerable degree of scientific education is needed even for the subordinate labourers in science'. *ibid.* p. 156.

[64] D. Brewster, *op. cit.* (Note 51), vol. 2, p. 405.

[65] W. B. Carpenter, *B.A.R.* (Brighton, 1872), pp. lxix–lxxxiv, at pp. lxxi–lxxiii. Carpenter was Professor of Forensic Medicine, University of London, from 1856–1879; J. Tyndall, 'Scientific Materialism', in *Fragments of Science* (2 vols, New York, 1897), vol. 2, p. 77.

[66] K. Pearson, *op. cit.* (Note 3), p. 31. J. von Liebig, 'Induction and Deduction', *Cornhill Magazine* **XII**, 1865, pp. 296–303.

[67] L. Laudan, *op. cit.* (Note 52), pp. 181—191. Nevertheless, there were still some attempts to retain a method of discovery. See G. Gore, *The Art of Discovery* (London, 1878).

[68] See F. M. Turner, 'The Victorian Conflict between Science and Religion: a professional dimension', *Isis* **LXIX**, 1978, pp. 356—376.

[69] From 1811, the *Transactions of the Geological Society* carried as a motto a quote from Bacon's *Novum Organum*; after 1860, the last phrase of the extract, which spoke of the inner shrine of science as open to all, was deleted. For a reference to this deletion, without comment on its significance, see H. B. Woodward, *The History of the Geological Society of London* (London, 1907), p. 44.

[70] Prince Albert, *B.A.R.* (Aberdeen, 1859), p. lxiii.

[71] W. V. Harcourt, *op. cit.* (Note 25), p. 28.

[72] For Somerville's conviction about the unity of the sciences, see M. Somerville, *On the Connexion of the Physical Sciences* (London, 1834), preface and p. 413. See also E. C. Patterson, *Mary Somerville and the Cultivation of Science, 1815—1840* (Boston, The Hague, 1983).

[73] [W. Whewell], 'On the Connexion of the Physical Sciences' *Quarterly Review* **LI**, 1834, pp. 54—68, at pp. 59—60.

[74] Whewell's concern about the unity of science did not prevent him from defending the sectional divisions of the British Association. See J. B. Morrell and A. Thackray, *op. cit.* (Note 25), p. 455.

[75] J. Clerk Maxwell, *Scientific Papers* W. D. Niven (ed.), (2 vols, Cambridge, 1890), vol. 2, p. 402. For an earlier reference to the need for general rules of inquiry in the context of diverse fields of knowledge, see M. Faraday, *Some Observations on the Means of Obtaining Knowledge* (London, 1817), pp. 8—9.

[76] J. Herschel, *op. cit.* (Note 17), p. 26.

[77] *Ibid.* pp. 25—26. But in his introductory text on astronomy, intended for a large audience, Herschel warned that it assumed a basic mathematical knowledge. See his *Treatise on Astronomy* (London, 1833), p. 5. For the importance of mathematics in physical science from the 1830s, see G. N. Cantor, *Optics after Newton: Theories of Light in Britain 1704—1840* (Manchester, 1983), pp. 147—150.

[78] J. Herschel, *op. cit.* (Note 17), p. 145; also pp. 281—289.

[79] *Ibid.* pp. 104—105, 114.

[80] *Ibid.* p. 219.

[81] *Ibid.* pp. 253—254, 263—264. For an extended argument on this, see J. Agassi, *op. cit.* (Note 16), pp. 1—36.

[82] J. Herschel, *B.A.R.* (Cambridge, 1845), p. xi. Whewell himself made some attempt to acknowledge some common ground between his views on method and those of Mill. See W. Whewell, *On Induction* (London, 1849), p. 5.

[83] G. N. Cantor, *op. cit.* (Note 77), pp. 177—85 and his earlier articles cited in the Introduction to this volume.

[84] For views about the moral effects of induction and deduction, see W. Whewell, *Astronomy and General Physics Considered with Reference to Natural Theology* (London, 2nd ed., 1834), pp. 323—342. For eighteenth-century views of Newton and Descartes, see p. M. Rattansi, 'Voltaire and the Enlightenment Image of Newton', in *History and Imagination: Essays in Honour of H. R. Trevor-Roper* (London, 1981), pp. 218—231.

[85] T. Chalmers, *Evidence and Authority of the Christian Revelation* (2nd ed., Edinburgh, 1815), p. 200.

[86] W. Harcourt, *B.A.R.* (Birmingham, 1839), p. 18.

[87] See R. R. Yeo, 'Science and Intellectual Authority in Mid-Nineteenth Century Britain: Robert Chambers and *Vestiges of the Natural History of Creation*', *Victorian Studies* **XXVIII**, (1984), pp. 5—31.

[88] A. Sedgwick, 'Darwin's Origin of Species', *Spectator*, 7 April 1860, pp. 334—335. This review was a revised version of one which appeared on 24 March 1860.

[89] [R. Owen], 'Darwin on the Origin of Species', *Edinburgh Review* **CXI**, pp. 487—532, at pp. 495—496.

[90] [S. Wilberforce], 'Darwin's *Origin of Species*', *Quarterly Review* **CVIII**, 1860, pp. 225—264, at p. 256. For an account of the reaction to Darwin, see A. Ellegard, 'The Darwinian Theory and Nineteenth-Century Philosophies of Science', *Journal of the History of Ideas* **XVIII**, 1957, pp. 362—393 and D. Hull (ed.), *op. cit.* (Note 20). For a contemporary defence of Darwin's method, see H. Fawcett, 'On the Method of Mr. Darwin', *B.A.R.* (Manchester, 1861), pp. 141—143.

[91] See J. B. Morrell and A. Thackray, *op. cit.* (Note 25), pp. 271—275, 482—484. For a contemporary observation on the restricted meaning of 'science', see 'The British Association', *Oxford University Magazine* **I**, 1834, pp. 401—412. It should however be noted that Whewell did not suggest that non-mathematical disciplines were non-scientific. See for example his defence of natural history and the inexactness of its classifications. Whewell, *op. cit.* (Note 18), Vol. 1, pp. 493—494.

[92] W. Swainson, *A Preliminary Discourse on the Study of Natural History* (London, 1834), p. 163. (my emphasis)

[93] *Ibid.* pp. 163—164; also p. 150. See p. 201 where he quotes Herschel's *Discourse* in support of the concept of a single method in all the inductive sciences.

[94] W. Whewell, *op. cit.* (Note 15), p. 390.

[95] [W. Whewell], 'Jones on the Distribution of Wealth and Taxation', *British Critic X*, 1831, pp. 41—61, at p. 55.

[96] W. Whewell, *op. cit.* (Note 18), vol. 2, pp. 336—337, 358.

[97] *Ibid.* Vol. 1, p. 7. For the moral and theological dimensions of Whewell's work, see R. Yeo. 'William Whewell, Natural Theology and the Philosophy of Science in Mid-Nineteenth-Century Britain', *Annals of Science* **XXXVI**, 1979, pp. 493—516.

[98] W. Whewell, *op. cit.* (Note 18), vol. 1, p. viii and *Novum Organon Renovatum* (London, 1858), p. vi.

[99] W. Whewell, *op. cit.* (Note 54), vol. 3, pp. 377—381, 400—401, 467—472. It could be argued that Whewell was not being obsurantist, since teleology played a significant role in contemporary German biology. See T. Lenoir, *The Strategy of Life: Teleology and Mechanics in Nineteenth-Century German Biology* (Dordrecht, Boston, 1982).

[100] [W. B. Carpenter], 'Physiology an Inductive Science', *British and Foreign Medical Review* V, 1838, pp. 317—342, at pp. 318, 321. See also his *Principles of General and Comparative Physiology* (2nd ed., London, 1841), p. 2.

[101] B. Powell, *The Connexion of Natural and Divine Truth* (London, 1838), p. 60. Also p. 61 for the point that the non-mathematical character of geology did not render it less 'inductive'. For a similar appeal to the unity of science in the defence of geology see [R. Grant], 'Observations on the Nature and Importance of Geology', *Edinburgh New Philosophical Journal* **1**, 1826, pp. 293—302, at pp. 293—294.

[102] B. Powell, *op. cit.* (Note 101), p. 68.

[103] B. Powell, *Essays on the Spirit of the Inductive Philosophy, the Unity of Worlds, and the Philosophy of Creation* (London, 1855), p. 63.

296

[104] *Ibid.* p. 44. Powell did not name Whewell as the target here, but in his preface he foreshadowed disagreement with Whewell (p. vi).

[105] See D. Layton, 'The Schooling of Science in England, 1854–1939', in R. Macleod and P. Collins, *op. cit.* (Note 16), p. 191. For a defence of the life sciences in this context, see W. B. Carpenter, 'The Universities and Scientific Education', *Westminster Review* XIX, ns, 1861, pp. 381–402, at pp. 388–389.

[106] T. H. Huxley, 'On the Educational Value of the Natural History Science', in *Science and Education* (London, 1893), p. 45. Although Huxley often employed the notion of an accessible method, he did not accept the Baconianism which was usually linked with that notion in the first half of the century. In the situation discussed here, his major concern was to reject any qualitative distinctions between disciplines. For his attacks on Bacon, see C. Bibby (ed.), *The Essence of T. H. Huxley* (London, 1967), pp. 42–47.

[107] T. H. Huxley, *op. cit.* (Note 106), pp. 45–46.

[108] *Ibid.* See also 'On the Study of Biology', in *Scientific Memoirs* (4 vols, London, 1902), vol. 4, p. 258 and 'The Progress of Science' in *Collected Essays* (9 vols, London, 1893–95), vol. 1, p. 60.

[109] [W. B. Carpenter], 'Spiritualism and its Recent Converts', *Quarterly Review* CXXXI, 1871, pp. 301–353, at p. 342.

[110] W. Wilberforce, *op. cit.* (Note 90), p. 249.

[111] W. Hopkins, 'Physical Theories of the Phenomena of Life Part I', *Fraser's Magazine* LXI, 1860, pp. 739–752, at p. 741.

[112] *Ibid.* pp. 739–740; also pp. 741–744.

[113] See J. D. Burchfield, *Lord Kelvin and the Age of the Earth* (New Haven, 1975), Chapter 3.

[114] W. E. Gladstone, 'On the Influence of Authority in Matters of Opinion', *Nineteenth Century* I, 1877, pp. 2–22, at p. 20. This was a review of a book by G. C. Lewis. See (Note 128) below. In 1900, the philosopher, Henry Sidgwick, spoke of 'the actual impossibility of finding a satisfactory scientific explanation of the development of scientific knowledge'. Quoted in F. M. Turner, *Between Science and Religion: The Reaction to Scientific Naturalism in Victorian England* (New Haven, 1974), p. 20.

[115] K. Pearson, *op. cit.* (Note 3), p. 15.

[116] See (Note 20) above. For the eighteenth-century European notions of a science of society, see K. Kumar, *Prophecy and Progress* (Harmondsworth, 1981) pp. 13–44.

[117] J. Herschel, *op. cit.* (Note 17), p. 73. See also Herschel, *op. cit.* (Note 82), p. xi and 'Views on Public Education', *Philosophical Magazine* VIII, 1836, pp. 432–438, at p. 434.

[118] J. S. Mill, *op. cit.* (Note 15), p. 179.

[119] J. Herschel, *op. cit.* (Note 82), p. xl.

[120] W. Whewell, *op. cit.* (Note 95), pp. 52–57 and *op. cit.* (Note 84), p. 369. See also S. Hollander, 'William Whewell and J. S. Mill on the Methodology of Political Economy', *Studies in History and Philosophy of Science* XIV, 1983, pp. 113–126 and L. Goldman, 'The Origins of British "Social Science": Political Economy, Natural Science and Statistics, 1830–1835', *The Historical Journal* XXVI, 1983, pp. 587–616.

[121] K. Pearson, *op. cit.* (Note 3), p. 18.

[122] *B.A.R.* (Liverpool, 1901), p. 863. See also D. Layton, *op. cit.* (Note 105), p. 198 and P. Collins, 'The Origins of the British Association's Education Section', *British Journal of Educational Studies* XXVII, 1979, pp. 232–34.

[123] [C. Lyell], 'State of the Universities', *Quarterly Review* **XXXVI**, 1827, pp. 216–268, at p. 221.

[124] *Ibid.* p. 222. See also (C. Lyell), 'Transactions of the Geological Society of London', *Quarterly Review* **XXXIV**, 1826, pp. 507–540, at pp. 508–509.

[125] J. S. Mill, *Inaugural Address at St. Andrews* (London, 1867), p. 8. See also M. Pattison, *Suggestions on Academic Organization* (Oxford, 1868), pp. 266–280.

[126] C. Lyell, *op. cit.* (Note 123), p. 223; B. Powell, *The Present State and Future Prospects of Mathematical and Physical Studies in the University of Oxford* (Oxford 1832), p. 27; J. Herschel, *op. cit.* (Note 17), p. 70. See also [J. Bowring], 'Scientific Eduction of the Upper Classes', *Westminster Review* **IX**, 1828, pp. 328–373.

[127] For the background to this, see R. Yeo, *op. cit.* (Note 16), pp. 72–80.

[128] G. C. Lewis, *An Essay on the Influence of Authority in Matters of Opinion* (London, 1849), pp. 112–113. See R. Yeo, *op. cit.* (Note 87), pp. 7–8, 26 for some reference to this text.

[129] J. D. Campbell (Ducke of Argyll), *B.A.R.* (Glasgow, 1855), p. lxxxi. See also *B.A.R.* (Dundee, 1867), pp. xl–xlii.

[130] K. Pearson, *op. cit.* (Note 3), p. 15 John Dewey saw this distinction as central to better science education in America. See his 'Science as Subject-Matter and as Method', *Science* **XXXI**, 1910, pp. 121–127.

[131] On Armstrong, see W. H. Brock (ed.), *H. E. Armstrong and the Teaching of Science, 1880–1930* (Cambridge, 1973) and E. W. Jenkins, *From Armstrong to Nuffield: Studies in Twentieth Century Science Education in England and Wales* (London, 1979), pp. 41–57.

[132] *B.A.R.* (Newcastle, 1889), p. 229.

[133] *B.A.R.* (Newcastle, 1917), p. 126.

[134] T. P. Nunn, *The Aim and Achievement of Scientific Method* (London, 1907), p. 2.

[135] *Committee on the Neglect of Science: Report for the Year 1916–17* (London, 1917), p. 14. For another publication in this debate, see A. C. Seward (ed.), *Science and the Nation* (Cambridge, 1917).

[136] *B.A.R. op. cit.* (Note 128), p. 133. For a later perception of this implication, see G. Orwell, *op. cit.* (Note 1), p. 12.

[137] See D. Layton, *op. cit.* (Note 105), p. 201. For an attempt to restore the links between science and humanist values, see R. Gregory, *Discovery, or the Spirit and Service of Science* (London, 1916).

[138] C. Singer (ed.), *Studies in the History and Method of Science* (2 vols, London, 1917, 1921), vol. 1, p. vi. See also the references to this issue un the Introduction to this volume.

[139] J. B. Conant (ed.), *Harvard Case Histories in Experimental Science* (Cambridge, Mass., 1957), p. vii; also his *On Understanding Science: An Historical Approach* (New Haven, 1947), pp. 5–10 and *Science and Common Sense* (London, 1951), pp. 42–48. See also I. B. Cohen, *Science, Servant of Man: A Layman's Primer for the Age of Science* (London, 1948), pp. 13–15.

[140] *Nature*, 28 December 1940, p. 817.

[141] See the Introduction to this volume.

[142] This appears to be implied by Feyerabend's notion of a mythology of method. See P. Feyerabend, *Against Method* (London, 1975), pp. 220, 300–309.

V

William Whewell, Natural Theology and the Philosophy of Science in Mid Nineteenth Century Britain

Contents

1. Introduction

Throughout his academic career, William Whewell, the Master of Trinity College, Cambridge, was the object of many witticisms. Most of these were aimed at the apparent brusqueness of his manner, but others concerned the nature of his intellectual endeavours. Sydney Smith epitomised the latter when he remarked of Whewell that 'science is his forte, and omniscience his foible'.[1] Indeed, Whewell's range of interests was extensive (if not promiscuous), and even in an age when men could still pass freely between the disciplines, the breadth of his activity was rather unusual. In their undergraduate days, John Herschel and Charles Babbage shared

---

[1] Quoted by L. Stephen, 'Whewell', *Dictionary of national biography* (21 vols, 1885–90, London), vol. 20, 1365–1374 (p. 1371). For another contemporary appreciation see J. F. W. Herschel, 'The Reverend William Whewell, D. D.', *Proceedings of the Royal Society of London*, **16** (1867–68), li-lxi.

Whewell's broader interests and discussed them at 'philosophical breakfasts'.[2] But although his colleagues appreciated the metaphysical nature of his inquiries, they often wished that Whewell had confined himself to one section of science. In fact, Whewell's practical scientific activities were limited, and as his career progressed, he became more involved in epistemology and moral philosophy, losing contact with recent scientific advances. By the middle of the century, when increasing specialization was apparent in intellectual life, Whewell retained his earlier commitment to 'universal knowledge',[3] and would still have described himself as one who twisted the results of science into metaphysical speculations.[4]

The comprehensiveness of Whewell's thought raises problems for modern scholars, just as it bemused or baffled his nineteenth-century contemporaries. In twentieth-century studies, Whewell has received attention by virtue of his dispute with J. S. Mill over the philosophy of science and, in particular, the logic of induction. There is now a significant body of literature which applies critical philosophical analysis to this debate, and Whewell's position on various issues of method and epistemology has been greatly elucidated.[5] A second area of study has been concerned with the philosophical lineage of Whewell's work, especially with the relative influence of Kantian and Platonic thought. Two extensive books, both written by non-English academics, have been devoted to this theme.[6] However, these studies rarely take any serious account of Whewell's moral and theological ideas and the intellectual context in which they arose.

Of course, the neglect of such topics does not in itself prejudice the validity of the two approaches discussed above. But there is a danger that certain areas of Whewell's work will be abstracted too sharply from the rest of his ideas. Thus, as Andrew Belsey has recently complained, there has been a tendency to treat Whewell's theological interest as an unfortunate and disturbing element in his philosophy of science.[7] Further, although it is important to delineate the intellectual influences which acted upon his thought, our understanding of Whewell would be seriously impoverished if we simply assumed that he was dominated by the desire to construct a systematic philosophical system, or that he was trying to be either a thorough Kantian or a consistent Platonist. Thus, in addition to the specialist critiques of Whewell's philosophy of science there is a need to consider the moral and theological dimensions of his epistemology. This article attempts to supply these dimensions by treating Whewell's thought within the context of nineteenth-century British natural theology.

[2] T. Forster to W. Whewell, 24 December 1841, in I. Todhunter, *William Whewell, D.D., An account of his writings* (2 vols., 1876, London), vol. 1, 6. The correspondent was probably Thomas Ignatius Maria Forster, a psychological and physiological writer. The full quote given by Todhunter is interesting: 'We have all made some advances in mere *physical* science, but in *metaphysics*, as far at least as I am concerned, I am not conscious of having advanced one single step, since the period when you and I and Herschel and Babbage used to meet at our Sunday morning's philosophical breakfasts in 1815' (pp. 5–6).

[3] Whewell to Rev. G. Morland, 15 December 1815, *ibid.*, vol. 2, 10.

[4] Whewell to J. Herschel, 1 November 1818, *ibid.*, 29. For Charles Lyell's attitude to this, see *ibid*, vol. 1, 112.

[5] See, for example, C. J. Ducasse, 'William Whewell's philosophy of scientific discovery', *Philosophical review*, **60** (1951), 56–69, 213–234; G. Buchdahl, 'Inductivist vs. deductivist approaches in the philosophy of science as illustrated by some controversies between Whewell and Mill', *The monist*, **55** (1971), 343–367; and R. E. Butts (ed.), *William Whewell's theory of scientific method* (1968, Pittsburgh).

[6] R. Blanché, *Le rationalisme de Whewell* (1935, Paris); and S. Marcucci, L' *"idealismo" scientifico di William Whewell* (1963, Pisa). See also G. C. Seward, *Die theoretische Philosophie William Whewells und der Kantische Einfluss* (1938, Tübingen).

[7] A. Belsey, 'Interpreting Whewell', *Studies in history and philosophy of science*, **5** (1974–75), 49–58.

Most recent works on the history of nineteenth-century British science refer to the subject of natural theology. Such a reference is essential because the vocabulary of natural theology constituted much of the style and rhetoric of scientific communication in the period. Natural theology proclaimed the harmony of science and religion, and in this sense it was a useful apologetic on behalf of science against conservative religious criticism. But its more general influence meant that concepts such as final cause, design, law, miracle and Providence were integral components of scientific discussion. Indeed, a number of scholars have argued that the teleological assumptions of natural theology provided the intellectual framework in which scientific thinking occurred.[8] This suggestion militates against a simple bifurcation between misguided clerical scientists and progressive secular scientists: both may have been influenced, in positive and negative ways, by the tradition of natural theology.

But apart from its more obvious connection with substantive scientific theories, natural theology may have also provided a context for significant discussion of the methodological and epistemological features of both scientific and theological knowledge. From this perspective, questions concerning the limits and use of reason, the grounds of knowledge, the role of hypotheses and imagination and empiricism versus idealism were not esoteric ones—they were part of the broad nineteenth-century British debate on man's place in nature.[9]

2. Natural theology and the philosophy of knowledge

In describing the intellectual temper of mid-nineteenth-century Victorians, Walter Houghton remarked that although they questioned many traditional values, their doubt 'never involved a denial of the mind as a valid instrument of truth'.[10] There was a belief in the existence of ultimate truths in all areas of knowledge, an abiding trust in the capacity of mind to find them, and a faith in the possibility of intellectual certitude. These generalizations certainly apply to the men of science in this period. The majority of them were convinced that careful research would reveal the underlying laws of nature, the principles which actually guided the process of creation.[11] But, as a corollary to these assumptions, there was a deep concern with questions of method and epistemology: the ability of man to attain truth was thought to depend upon the manner in which reason was employed, upon the observance of proper methods of inquiry.

The major writers of natural theology shared this interest in method and epistemology. They regarded the faculty of reason as a Divine gift which enabled man to comprehend the Divine creation. Indeed, they viewed man's place in nature as primarily defined by this capacity to understand the world: man was the

---

[8] See for example W. F. Cannon, 'The problem of miracles in the 1830's', *Victorian studies*, 4 (1960), 5–32; his 'The bases of Darwin's achievement: a revaluation', *Victorian studies*, 5 (1961), 109–134; R. M. Young, 'Darwin's metaphor: does nature select?', *The monist*, 55 (1971), 442–503; and his 'The historiographical and ideological contexts of the nineteenth-century debate on man's place in nature', in M. Teich and R. M. Young (eds.), *Changing perspectives in the history of science* (1973, London), 344–438.

[9] I have dealt with this theme in my unpublished Ph.D. thesis, 'Natural theology and the philosophy of knowledge in Britain, 1819–1869' (1977, University of Sydney). For related approaches, see A. Ellegard, 'The Darwinian theory and nineteenth-century philosophies of science', *Journal of the history of ideas*, 18 (1957), 362–393 (pp. 391–393); and R. Smith, 'The background of physiological psychology in natural philosophy', *History of science*, 11 (1973), 75–123 (pp. 97–105).

[10] W. E. Houghton, *The Victorian frame of mind, 1830–1870* (1957, London and New Haven), 13.

[11] See M. Mandelbaum, *History, man and reason. A study in nineteenth century thought* (1971, Baltimore and London), 87–88, 400.

Interpreter of Nature.[12] But this status carried a moral obligation to use the intellect in the pursuit of truth in all areas of knowledge. The duty to know the world was conceived in terms of the Biblical injunction to subdue the earth and have dominion over its creatures. Intellectual indolence and the ignorance it perpetuated were condemned as violations of God's will. Furthermore, while it was imperative that this sacrosanct gift of reason be employed rather than neglected, it was essential that it should not be abused. Like any other part of God's providence towards man, it had to be exercised with scrupulous care. The realization of man's potential for attaining truth was thought to depend upon the adherence to methods of inquiry which were not only intellectually justifiable, but morally respectable. Further, epistemological theories about the foundation and limits of reason were intimately connected with theological ideas about the appropriate relationships between man, God and nature. These assumptions meant that discussions on method and epistemology were part of a wider debate involving moral, social and religious considerations.[13]

Discussion of epistemology entered in a subtle way into the arguments of natural theology. Exponents of this subject wanted to move from the physical to the spiritual; they hoped to disclose moral purposes reflected in the material world. The argument from design epitomised this project. But in order to do this they needed to assume (and possibly demonstrate) the unity of the moral and physical spheres of creation. This was a serious problem because if design could only be perceived in the physical world it would be difficult to maintain that the Creator of nature was also responsible for man's intellectual and spiritual character. Recognizing this problem, Whewell wanted to prove that 'the Creator and Preserver of the world is also the Governor and Judge of men; that the Author of the Laws of Nature is also the Author of the Law of Duty'.[14]

In response to this problem, some writers acknowledged the need to embrace psychological evidence within the design argument.[15] They hoped to reveal examples of design in the mind so that man's mental nature could be seen as part of the Divine order manifested in the physical world. But this task was not an easy one; it involved philosophical and religious dilemmas relevant to the question of man's place in nature.

To begin with, Christian religious and moral values were invested in the Cartesian dualism of mind and matter. The mind of man, its intellectual and moral faculties, were regarded as immeasurably superior to, and qualitatively different from, the inert, lifeless matter of the physical world. This distinction supported values associated with free-will and moral responsibility.[16] But because of this ontological gulf between mind and matter, it was difficult to define the points of contact which would demonstrate the unity of creation. Furthermore, these presuppositions meant that it was impossible to accept the empiricist position which resolved the dichotomy

---

[12] This Baconian phrase appeared as the first of Whewell's Aphorisms in his *Philosophy of the inductive sciences* (2 vols., 1840, London). Herschel used it as the title of a poem, and the physiologist William B. Carpenter made it the title of his Presidential Address to the British Association in 1872.

[13] This article will deal with epistemology rather than method. But for references to the social and moral aspects of scientific method, see J. Agassi, 'Sir John Herschel's philosophy of success', *Historical studies in the physical sciences*, 1 (1969), 1–36; and R. R. Yeo (footnote 9), ch. 4.

[14] Whewell, *Astronomy and general physics considered with reference to natural theology* (2nd ed., 1834, London), 254–255.

[15] For a discussion of these attempts see Yeo (footnote 9), 111–124.

[16] For clear statement of these assumptions see G. Moore, *The power of the soul over the body* (1845, London), 9. George Moore was a popular medical writer.

between moral and physical spheres by reducing mind to the level of matter. The determinism and materialism entailed by this approach would have contradicted the Christian belief in man's unique spiritual nature. Nor was the solution advanced by German Romantics open to British natural theologians, because the extreme idealist philosophy dissolved the distinctions between man, God and nature, and challenged the authority of Revelation. Consequently there was a tension between the logical requirements of natural theology and the demands of Christian metaphysics. Writers of natural theology had to establish a relationship between the physical and moral spheres which ensured the validity of inferences from the material to the spiritual. But in accepting this task, they had to avoid the twin errors of materialism and pantheism. Therefore, they needed to affirm the unity of physical and moral spheres without implying genetic continuity; they needed to assert the harmony of mind and world without suggesting identity or reductionism.

When this problem was confronted there was often a transition from psychological to epistemological discussion. That is, it was in the course of seeking a psychological answer to the problem of the mind's relation to the world that an epistemological answer emerged as a viable solution. When illustrating the accommodation between man's psychological needs and the provisions of his environment, some writers argued that the ability of the mind to know the world was in itself proof that God had adapted one to the other. This conception of design avoided the difficulties associated with psychological discussion because it could envisage the mind as epistemologically related to, rather than genetically derived from, the world. In this sense, the idea of man as the Interpreter of Nature resolved the dichotomy between the physical and moral spheres of creation.

This use of epistemology is apparent in some of the major works of natural theology. For example, Thomas Chalmers, writing in his Bridgewater Treatise, found an impressive example of design in the harmony between subjective mental concepts and objective reality. There was a consonance between certain abstract ideas and the structure of the world. This was most obvious in the case of mathematical notions which derived solely from the mind's inner resources but subsequently proved to be accurate descriptions of the tangible world. Chalmers enlisted the support of Herschel, who had noted that abstract knowledge of conic sections preceded its application to astronomical data.[17] For Chalmers, this fact indicated the 'intervention of a Being having supremacy over all...who had adjusted the laws of matter and the properties of mind to each other'.[18] Thus the ability of the human mind to comprehend the processes of nature demonstrated the unity of the Divine creation. And, in manifesting 'the adaptation of the intellectual to the material order of things', by the success of his speculations, man was 'but filling up an essential part in the universal harmony' of the world.[19] For nineteenth-century natural theologians, man's ability to know the world was the measure of his special place in creation. Herschel reflected this opinion when he asserted that the universe was incomplete before the appearance of man. The true purpose of creation was fulfilled only when man comprehended the majesty and harmony of nature. In

---

[17] J. Herschel, *A preliminary discourse on the study of natural philosophy* (1830, London), 18–34.
[18] T. Chalmers, *On the power, wisdom and goodness of God as manifested in the adaptation of external nature to the moral and intellectual constitution of man* (2 vols., 1835, London), vol. 2, 159.
[19] B. Powell, *The connexion of natural and divine truth* (1838, London), 203.

the last lines of his undated poem, 'Man the Interpreter of Nature', Herschel exclaimed:

> Say! was the WORK wrought out! Say, was the GLORY complete?
> What could reflect, though dimly and faint, the
>   INEFFABLE PURPOSE
> Which from chaotic powers, Order and Harmony drew?
> What but the reasoning spirit, the thought and the
>   faith and the feeling?
> What, but the grateful sense, conscious of love
>   and design?
> Man sprang forth at the final behest. His
>   intelligent worship
> Filled up the void that was left. Nature at length
>   had a soul.[20]

3. The moral and theological context of Whewell's philosophy of science

Whewell's epistemological ideas also belong to this theological context. In his early notebooks and sermons, and in his Bridgewater Treatise, he attempted to use the philosophy of knowledge as evidence for natural theology. In recognizing this, one reviewer observed that 'Mr. Whewell gets a glimpse of no law, without a reverent perception of the powers, functions and endowments of the intellect which traces it; and in observing these, he is led constantly upwards to the mightier intellect, which framed man and the universe. Of all the philosophy he teaches, that which seems most emphatically his own, is the philosophy of mind'.[21] After adducing evidence of design from astronomy and physics, Whewell devoted the third section of his Bridgewater Treatise to the religious significance of man's knowledge of nature. He drew a parallel between the process of discovery in science and that involved in deciphering an unknown language. In both cases, there was a transition from discrete facts to an understanding of the relationship between them which gave meaning to the whole.[22] By grasping these general laws of nature man was comprehending 'the language in which the book of nature is written'.[23] This language was the expression of God's thoughts, and man could read it in so far as he shared some affinity with the Divine Mind.[24] Whewell therefore believed that man's ability to understand the laws (and thoughts) of God implied a threefold harmony between the mind of God, the mind of man and the laws of nature.

There were also strong moral and theological dimensions in Whewell's major works. Indeed, these can be interpreted as sophisticated responses to the need for a philosophy of science which guaranteed the principles of natural theology and the values of Christianity. He wanted to proclaim the virtue of scientific knowledge without accepting empiricist epistemology; he wanted to dissociate science from utilitarian philosophy.

---

[20] Herschel, 'Man, the interpreter of nature', in *Essays from the Edinburgh review, with addresses and other pieces* (1857, London), 737.

[21] 'Whewell's astronomy and general physics', *British magazine*, **3** (1833), 589.

[22] Whewell (footnote 14), 304–307.

[23] Whewell, notebook of 1825, quoted in Todhunter (footnote 2), vol. 1, 363.

[24] Whewell, unpublished sermon, 11 February 1827, Trinity College, Cambridge, R. 6.17.no.14. See also D. B. Wilson, 'Herschel and Whewell's version of Newtonianism', *Journal of the history of ideas*, **35** (1074), 79 97 (pp. 94–97).

Whewell's concern with this problem was apparent in early correspondence with his Cambridge colleague, Hugh James Rose.[25] Rose was not sympathetic towards physical science because he believed that it was producing a concentration upon external facts to the detriment of moral and methaphysical speculation, while at the same time encouraging a pragmatic attitude towards all knowledge.[26] In replying to this critique, Whewell attempted to affirm the moral and intellectual value of the natural sciences.[27] To some extent he was successful, for, in referring to Whewell's Bridgewater Treatise, Rose acknowledged that 'the most ... which has ever been said for these sciences, as they can affect the human mind, has been said by one whom I can never name without the strongest emotions of respect and regard'.[28] Whewell was able to promote an image of science which was free of pragmatic and irreligious connotations because he espoused an epistemology which emphasized the spiritual character of mind.

One of the intellectual contexts of Whewell's philosophy of science was the nineteenth-century reaction against Utilitarianism. The British Idealist movement is usually viewed as one which began with the English Romantics and consolidated itself in the writings of T. H. Green and F. H. Bradley.[29] But another episode in the development of this movement was the Cambridge reaction of the 1830s against the philosophy of Locke. Adam Sedgwick, Julius Hare, Frederick Maurice, Connop Thirlwall and Whewell were the major figures in this campaign.[30] These writers strove to weaken the dominant position of Locke's philosophy in the Cambridge curriculum because they regarded it as the principal intellectual foundation of a sensationalist epistemology which had infiltrated contemporary speculation on ethics, language and the philosophy of science. They saw this sensationalist (and in their view materialist) theory embodied in important and influential documents of the period. In the field of ethics, William Paley's *Principles of moral and political philosophy* espoused a utilitarian system which denied the existence of an innate moral sense or faculty.[31] Secondly, in the philosophy of language, the conventionalist theory of John Horne Tooke, which traced the meanings of all words to simple sensations and concrete objects, was quickly accepted by James Mill and other utilitarian writers as a vindication of the materialist account of mind.[32] And thirdly, in the philosophy of science, an empiricist approach which reduced all knowledge to experience was powerfully advanced in the work of J. S. Mill. For the Cambridge thinkers, all these areas were battlegrounds in their fight against Lockean

[25] Rose was a High Church theologian and editor of the *British magazine*.

[26] H. J. Rose, *The tendency of prevalent opinions about knowledge considered* (1826, Cambridge), v-vi. 3–11.

[27] Whewell to Rose. 19 November 1826 and 12 December 1826, in Todhunter (footnote 2), vol. 2, 75–79; also his unpublished sermons of 1827, Trinity College, R.6.17, nos. 13–16.

[28] Rose, *An apology for the study of divinity* (1834, London), 12; see also p. 48 for criticism of the physical sciences.

[29] See, for example, F. Copleston, *A history of philosophy* (8 vols., 1962–67, New York), vol. 8, part 1, 171–190.

[30] See H. Aarsleff, 'Locke's reputation in nineteenth century England', *The monist*, 55 (1971), 392–422 (pp. 398–399, 411–416); and J. B. Schneewind, 'Sidgwick and the Cambridge moralists', *The monist*, 58 (1974), 371–404 (pp. 371–387).

[31] For criticism of Paley see A. Sedgwick, *A discourse on the studies of the University of Cambridge* (1834, Cambridge), 57–67, 126–142; and W. Whewell (ed.), *Butler's three sermons on human nature* (1848. London), ix–x, xxvi–xxvii.

[32] On Tooke. see H. Aarsleff. *The study of language in England, 1780–1860* (1967, Princeton), chs. 2 and 3. For a critique of Tooke by another member of the Cambridge group, see J. W. Donaldson, *The new Cratylus* (1850, London).

V

epistemology and its manifestation in Utilitarianism. The underlying error in the works they rejected was a depreciation of the metaphysical aspects of knowledge and the spiritual character of the human mind.

In dedicating his *Philosophy of the inductive sciences* to Sedgwick, Whewell referred to their common enemy as the 'ultra-Lockian school'.[33] He saw his own work as part of 'that Reform of Philosophy' which had been necessary since the ascendancy of Locke's epistemology in England.[34] Whewell wanted to attack the empiricist theory which traced the origin of all knowledge, including scientific knowledge, to sensory experience. Against this view, he argued that the mind was not simply passive in the act of cognition; rather, it actively contributed ideas which gave intelligible form to sense-impressions. Moreover, he claimed that generalized knowledge only occurred when sensations were informed by mental conceptions.[35]

Central to Whewell's philosophy of science was the notion of 'Fundamental Ideas'. These metaphysical ideas, such as Space, Time and Cause, gave rise to ideal conceptions which played a major role in human thought. The fundamental ideas were inherent rather than innate to the mind; while not derived from experience, they did require experience to unfold them; but they also served to organize experience.[36] They were not objects of thought, but laws of thought.[37] The fundamental ideas were the mind's contribution to the act of cognition.

Whewell acknowledged the similarity between his fundamental ideas and Kant's *a priori* categories. Indeed, he granted that sections of his work dealing with the ideas of Space and Time 'were almost literal translations of chapters in the *Kritik der Reinen Vernunft*'.[38] But he did not regard himself as a submissive disciple of Kant, and rebuked G. H. Lewes and Henry Mansel for failing to see the novel points of his position.[39] Whewell wanted to go further than Kant by proposing additional fundamental ideas as the foundation of knowledge in the mechanical, chemical and biological sciences. And in expanding the number of fundamental ideas, he urged that the quality of necessary truth, rather than being limited to mathematics and geometry, could be expected in the physical sciences. While conceding that experience and observation alone could support contingent, but not necessary truth, he argued that certain physical facts, when informed by a fundamental idea, could be seen as necessarily true.[40]

Whewell also diverged from the original Kantian thesis by postulating the gradual emergence of *a priori* ideas. The most novel aspect of his theory was the contention that contingent truths could be apprehended as necessary truths during the historical development of a scientific discipline. While strenuously upholding the need for a philosophical distinction between contingent and necessary truth, Whewell contended that the intuition of self-evident axioms was progressive, rather than immediate. The fundamental ideas, the source of intuitive axioms, were not innate, and so necessary truth was not instantly perceived; the fundamental ideas were only disclosed through the mind's experience of the world. But in emphasising

[33] Whewell, *The philosophy of the inductive sciences* (2 vols. 2nd ed., 1847, London), vol. 1, iv.
[34] *Ibid.*
[35] *Ibid.*, 26–27, 33–37.
[36] *Ibid.*, 66–67.
[37] *Ibid.*, vol. 2, 677.
[38] Whewell, *On the philosophy of discovery* (1860, London), 335.
[39] *Ibid.*, 334—336; and his *Letter to the author of the Prolegomena Logica* (1852, Cambridge), 3–8.
[40] Whewell (footnote 33), vol. 1, 54–78.

the progressive intuition of necessary truths, and in admitting their dependence upon observation, Whewell did not imply that they were derived from experience. Although necessary truths emerged in the course of scientific investigations, they could not be proved by experience, but rather supplied the conditions for a general interpretation of experience.[41]

Whewell's philosophy of science presented the highest levels of physical knowledge as dependent upon metaphysical ideas provided by the mind. This theory offered an alternative to the empiricist school, in which Whewell detected an exclusive emphasis upon experience. Further, in his opinion, this idealist epistemology not only permitted a better account of science, but supported a more adequate view of man, because it avoided the materialist implications of an extreme sensationalist philosophy. The stress upon the role of intuitive, metaphysical concepts in knowledge ensured a spiritual conception of mind which reinforced the Christian notion of man's special place in creation. Therefore, Whewell's work made it possible to defend the validity of natural science within a philosophy of science which satisfied the requirements of a Christian natural theology.

But, while urging the importance of the metaphysical component of knowledge, Whewell did not neglect the need for empirical observation. He stressed the interdependence of ideal and empirical elements in all thought, and aimed to delineate the intimate relationships between ideas and sensations, between subjective and objective factors.[42] Whewell argued that there could be no rigid separation of theory and fact: there was 'a mask of theory over the whole face of nature'[43] He also attenuated any severe dichotomy by proclaiming that 'a true theory is a Fact; a Fact is a familiar Theory'.[44] In this manner, he was able to qualify the division which rested upon the equations of objective knowledge with physical truth and subjective knowledge with moral truth. Indeed, he believed that the progress of truth, both physical and moral, could be embraced under one law of discovery based upon the history of scientific thinking. And while drawing his data from 'the most certain and stable portions' of existing knowledge,[45] he expected that this study would offer 'some general analogies which belong to the essence of truth, and run through the whole intellectual universe'.[46] Thus he viewed his work on the philosophy of science as part of a general philosopy of knowledge. In particular, he wished to extend the analysis to moral and theological knowledge, and, in his capacity as Professor of Moral Philosophy, Whewell began to move in this direction. In the first of his *Introductory lectures on moral philosophy*, the conviction behind this broader project was clearly stated: 'Inquiries into the nature of truth, the means and methods of its discovery, and the philosophy of science, even though they set out from the study of physical science, ... cannot fail to exercise a strong and favourable influence upon our studies with regard to moral truth, moral science, and the true philosophy of human life'.[47]

---

[41] Whewell (footnote 38), 347–349. See also R. E. Butts, 'Necessary truth in Whewell's theory of science', *American philosophical quarterly*, **2** (1965), 161–181.

[42] Whewell. 'On the fundamental antithesis of philosophy'. *Transactions of the Cambridge Philosophical Society*, **8** part 2 (1849), 170–181 (read 5 February 1844).

[43] Whewell (footnote 33), vol. 1, 42.

[44] *Ibid.*, 40.

[45] *Ibid.*, 1.

[46] *Ibid.*, 3.

[47] Whewell, *Two introductory lectures on moral philosophy* (1841, Cambridge). 28.

502

Like other members of the anti-Lockean school, Whewell wanted to replace utilitarian ethical systems with a theory which held moral ideas to be irreducible, rational and intrinsic to the nature of man. Furthermore, he hoped to show that an intuitive theory of ethics was consonant with the principles of physical science. He therefore rejected the contention that there was a natural alliance between utilitarian ethics and the development of science since the seventeenth century. By underlining analogies between the progress of moral and material knowledge, Whewell sought to diminish any absolute distinctions between them.[48] The emphasis upon intuitive ideas (as well as empirical facts) in his philosophy of science, made it possible to claim that an idealist theory of ethics conformed to the general principles of human knowledge. Again, there was theological significance in this epistemology. It meant that the harmony of moral and physical truth, demanded by natural theology, could be affirmed without reductionism; both moral and physical knowledge were obtained by the one process and this intellectual unity indicated a more comprehensive unity in the mind of God.

This analysis of the nature of knowledge went beyond physical and moral science to embrace concepts which were directly relevant to theology. The idea of Final Cause, a fundamental idea like those of Space and Time, and the conception of a First or Supreme Cause, a modification of the fundamental idea of Cause, fell within the compass of Whewell's philosophy of knowledge. These ideas made it possible for man to rise from a study of nature to an awareness of God. In defending this claim, Whewell stated that; 'The Ideas which we necessarily employ in the contemplation of the world around us, afford us the only natural means of forming any conception of the Creator and Governor of the Universe'.[49]

Whewell believed that man's rational knowledge of God and his scientific knowledge of the world, depended upon the one intellectual process: that which involved the continual interaction of ideal and empirical elements. This position enabled him to avoid some of the criticisms of the design argument. Addressing this issue in his Bridgewater Treatise, Whewell was able to admit that neither the idea of design, nor the inference to Designer, were founded upon strict logical reasoning: 'It is not therefore at the end, but at the beginning of our syllogisms, not among remote conclusions, but among original principles, that we must place the truth, that such arrangements, manifestations, and proceedings as we behold about us imply a Being endowed with consciousness, design, and will, from whom they proceed'.[50] Similarly, the idea of First Cause was 'not extracted from the phenomena, but assumed in order that the phenomena may become intelligible to the mind'; as such, it was a necessary idea, like 'the ideas of Space, or Time, or Cause in general'.[51] These ideas, and the knowledge they allowed, were not *a priori*, but were unfolded from the mind by experience. And by consistently stressing this need for both ideas and experience, Whewell was able to demonstrate the affinities between theological knowledge and other forms of human belief.

In the above discussion it has been suggested that Whewell's philosophy of science can be viewed as part of a wider attempt to establish a philosophy of knowledge which confirmed Christian values associated with man's special place in

[48] *Ibid.*, 31, 42. On Whewell's moral philosophy, see J. B. Schneewind, 'Whewell's ethics', *American philosophical quarterly monograph*, **1** (1968), 108–141.

[49] Whewell (footnote 33), vol. 1, 634.

[50] Whewell (footnote 14), 344; also Whewell (footnote 33), vol. 1, 622–623.

[51] Whewell (footnote 33), vol. 1, 706.

V

nature. The main points can now be summarised. Whewell sought to dissociate science from Utilitarianism and empiricist philosophy. In his writings on ethics, language and the philosophy of science, he espoused an idealist epistemology which stressed the importance of intuitive mental concepts. By underlining the metaphysical component of knowledge, this theory supported a spiritual conception of man. And in emphasizing the role of both ideal and empirical elements in all thought, it allowed a natural theology which demonstrated the intimate relationship between physical and moral knowledge.

4. Scientific and theological reaction to Whewell's philosophy of science

Although an acceptance of idealist epistemology could be recommended on these grounds, Whewell's philosophy of knowledge evoked suspicious reaction from both scientific and theological writers in Britain. For example, in commenting upon Whewell's ethical writings, J. S. Mill warned that a reliance upon intuitive philosophy could lead to *a priori* speculation about the physical world.[52] In fact, this was the continuing problem for Whewell: the tension between his idealist epistemology and his commitment to the physical science. There was always the danger that Mill's prediction would be fulfilled. On the other hand, Whewell sought to avoid this situation by insisting upon the interdependence of thought and experience, by maintaining a balance between idealism and empiricism. However, his critics doubted the stability of this equilibrium, and, noting the Kantian affiliations of his work, feared that it might move in the direction of post-Kantian German philosophy. This possibility entailed the most serious implications for British science and religion.

In reviewing Whewell's *Philosophy*, one writer with an eye for historical irony was startled to find 'that the doctrines of Kant and Transcendental Philosophy are now promulgated from the university which educated Locke'.[53] Whewell's scientific colleagues also displayed this feeling of surprise and tension. Although many of them probably sympathized with his attack upon sensationalism, they believed that his theory, with its stress upon the creative contribution of mind, would endanger the ontological status of scientific laws and concepts. John Herschel and Richard Jones expressed this concern, and Jones hoped that their friend could be dissuaded from a position which might 'lead to scepticism as to all things exterior to us and all their relations'.[54] This view was put more bluntly by Chalmers, who said that, in spite of Whewell's opinion, he would 'persist in regarding the whole of the intermediate space between ourselves and the planet Uranus as an objective reality'.[55] In general, the critics feared that the idealist aspect of Whewell's work might produce a speculative *Naturphilosophie* similar to that advocated by Friedrich Schelling.[56]

The anxiety aroused by Whewell's work was not confined to the realm of physical science. It was argued that his acceptance of the Kantian conception of Time and

[52] J. S. Mill, 'Whewell's moral philosophy', *Westminster review*, **58** (1852), 351–353.
[53] A. de Morgan, 'The Philosophy of Inductive Sciences'. *Athenaeum*, no 672 (12 September 1840), 707.
[54] R. Jones to Herschel. 1 June 1841, Herschel Papers, vol. 10. no. 370. Library of the Royal Society, London.
[55] Chalmers, 'Morell's modern philosophy', *North British review*, **6** (1847), 307.
[56] See A. de Morgan, 'The philosophy of discovery', *Athenaeum*, No. 1694 (14 April 1860), 502; and J. D. Morell, '[Victor] Cousin', *Edinburgh review*, **93** (1851), 451. On Schelling, see B. Gower, 'Speculation in physics: the history and practice of *Naturphilosophie*', *Studies in history and philosophy of science*, **3** (1973), 301–356.

Space would undermine the 'whole fabric of human knowledge';[57] it would vitiate that natural theology which Whewell sought to maintain because it failed to guarantee a philosophical realism. If the arguments of natural theology were to have validity, the concepts of design, law and order had to refer to a reality beyond the mind; they could not be interpreted as mere subjective categories imposed upon the world. It was also quickly observed that Kant himself had rejected the argument from design.

Apart from these theological difficulties, Kant's epistemology was seen as a direct threat to Revealed Religion. The *British quarterly review*, for example, claimed that there was nothing in modern Rationalism which had not been sanctioned in the writings of Kant: the necessity for Revelation had been ignored and man was said to be self-sufficient in the knowledge of his moral duty.[58] This concern was also voiced by the *Dublin University magazine*, in its review of Whewell's philosophy of science. Again, the ultra-rationalist character of Kant's religious ideas were observed, and the reviewer concluded that 'The *moral* results of this theory—which, we need not say, are in every theory the most important results—are proved both by reasoning and experience to be such as cannot be contemplated without dread . . . '.[59] Given this judgement, Whewell was advised to think very carefully before he undertook to 'popularize the *whole* of Kant in the cloisters of Cambridge'.[60]

In the nineteenth-century debates over the philosophy of knowledge, theories were not only assessed in terms of their logical coherence, but in terms of their moral and theological consequences. It was therefore not unusual that Whewell's work should be scrutinized for these wider ramifications. Indeed, a number of writers saw favourable theological potential in his epistemology. Writing to Robert Wilberforce in 1845, Henry Manning exclaimed: 'Surely divine truth is susceptible, within the limits of revelation, of an expression and a proof as exact as the inductive sciences. Theology must be capable of a "history and philosophy" if we had a Master of Trinity to write them'.[61] Similarly, Adam Farrar proposed that 'a true philosophy of the action of the intellectual faculties in reference to religion might be obtained by transferring to it the analysis which Dr. Whewell has given of their action in reference to science'.[62]

Nevertheless, although the possibilities of extending Whewell's analysis to theology may have been exciting, there were also extreme dangers. Yet again, the Kantian element in his thought was distrusted. British commentators viewed the German idealist philosophy of the early nineteenth century as the inevitable consequence of Kant's epistemology. When considered from a religious and theological perspective, the results of this Kantian legacy were profoundly disturbing. The critics did not make subtle distinctions between the systems of

[57] 'Whewell's philosophy of the inductive sciences', *Dublin University magazine*, **17** (1841), 203.

[58] 'German philosophy and Christian theology', *British quarterly review*, **2** (1845), 310–313.

[59] (Footnote 57), 206.

[60] *Ibid.*, 201.

[61] Quoted in D. Newsome, *The parting of friends: a study of the Wilberforces and Henry Manning* (1966, London) 302.

[62] A. Farrar, *A critical history of free thought in reference to the Christian religion* (1862, London), 39. In fact, John Daniel Morell (1816–1891), a philosophical and historical writer, told Whewell that he had attempted to apply an idealist philosophy of science to theology in his *Philosophy of religion* (1849, London) (see Morell to Whewell, 20 December 1848, Trinity College, Add. ms. c. 89, no. 172). But the intuitional religion which Morell espoused was precisely the kind which British commentators rejected (See Yeo (footnote 9), 300–317).

V

Fichte, Schelling and Hegel, but they were generally appalled by the growth of an Absolute Idealism in which mind, nature and God were identified. They believed that this position had subsidised extreme intuitionism in religion and blatant *a priori* speculation in science. In Sedgwick's view, German philosophical and theological thought presented similar problems: Hegel's philosophy of science was complemented by D. F. Strauss's philosophy of religion. Whereas Hegel ignored the experiential component of science, Strauss denied the historical foundation of Christianity. Both sought to derive all knowledge from *a priori* ideas of the mind in a manner which challenged the necessity of empirical observation in science and the status of Revelation in religion.[63]

Whewell was certainly aware of the problems which an idealist epistemology could entail; and he recognized the difficulties involved in translating a foreign philosophy into English culture. Writing to Richard Jones in 1837, he admitted that 'it is hardly possible to introduce foreign metaphysics in the lump, and that we must read German writers for some other purpose than that of substituting German metaphysics for English'.[64] Again, in 1839 he told Jones that he was resolved not to adopt any German 'fancies', but that he would 'see what light their speculations will throw upon mine'.[65] But even with this cautious attitude to German thought, Whewell's attempt to establish an idealist philosophy of science in Britain was fraught with problems. As noted above, the critics doubted his ability to maintain a stable reconciliation between empiricism and idealism, and they were anxious about the religious implications of post-Kantian thought. Behind these objections was the charge that Whewell's epistemology could not guarantee the philosophical realism essential to both science and natural theology. Yet this was precisely what Whewell was striving to do: he wanted to show that 'Man is the Interpreter of Nature, Science the right interpretation'.[66] But in order to proclaim this realist philosophy of science he had to justify an epistemology which relied upon intuitive conceptions of fundamental ideas. Whewell recognized this difficulty and referred to it as 'the ultimate problem of all philosophy'.[67]

5. The Plurality of Worlds Debate: natural theology and Whewell's idealist epistemology

Whewell's struggle with this dilemma has been analyzed by Robert Butts. In discussing Whewell's theory of necessary truth, Butts suggested that he became dissatisfied with the Kantian view, which asserted that experience confirmed laws of certain forms simply because experience could not take place in any other forms. In contrast with this epistemological solution, Whewell needed a metaphysical solution capable of providing an ultimate ground in which both ideas and facts could be resolved. According to Butts, Whewell found an answer to this problem in the seventeenth-century Platonic rationalism which made God the source of both

[63] A. Sedgwick. *A discourse on the studies of the University of Cambridge with additions and a preliminary dissertation* (1850, Cambridte and London), cclxxi-cclxxiv. cccc-ccccii.
[64] Whewell to Jones, 12 July 1837, in Todhunter (footnote 2), vol. 2, 257.
[65] Whewell to Jones, 14 July 1839, in *ibid.*, 280–281.
[66] Whewell (footnote 33), vol. 2, 443.
[67] Whewell to Herschel, 11 April 1844, in *ibid.*, 676.

thought and things: the constitution of the human mind, and its fundamental ideas, corresponded to the structure of the world because both were Divine creations.[68]

It is possible to interpret Whewell's book, *Of the plurality of worlds*, as a framework for this justification of his philosophy of knowledge. Published anonymously in 1853,[69] this work has reinforced the image of Whewell as a controversialist and intellectual gadfly. Until recently, the entire mid-nineteenth-century debate which surrounded it had been neglected. However, J. H. Brooke has now argued that the conflict between Whewell and David Brewster over the question of rational life on other planets might disclose deeper cleavages within the subject of natural theology; and also that Whewell's denial of plurality was a disguised form of attack upon naturalistic theories of evolution.[70] In a sense, the following discussion is complementary to this analysis, but it concerns the philosophy of science rather than the substantive scientific theories. That is, Whewell's rejection of the plurality of worlds is seen as an important aspect of his defence of an idealist epistemology. In asserting the unique place of man in the universe, he was able to rationalize the notion of a close affinity between the mind of man and the mind of God. And by affirming this intellectual empathy, he hoped to justify an epistemology which depended upon the validity of intuitive mental concepts. He attempted to deny the plurality of worlds and vindicate his philosophy of knowledge by urging the Christian conception of man's unique relationship with God.[71]

However, in order to reject the idea of a plurality of worlds, Whewell had to qualify some of the basic assumptions of traditional natural theology. The possibility of life on other planets was linked with the principle of plenitude and the doctrine of final causes.[72] In summarizing the position of his opponents, he offered an eloquent statement of these convictions:

> It is sometimes said, that it is agreeable to the goodness of God, that all parts of the creation should swarm with life: that life is enjoyment; and that the benevolence of the Supreme Being is shewn in the diffusion of such enjoyment into every quarter of the universe. To leave a planet without inhabitants, would, it is thought, be to throw away an opportunity of producing happiness.[73]

From this perspective, the prospect of vacant stars and wasted planets was intolerable. Matter was made for life and planets without life would be devoid of

---

[68] Butts (footnote 41), 173–180. Whewell was not the only writer to support a scientific realism by an appeal to this form of the design argument. David Wilson has shown a similar connection in the thought of Sir William Thomson (Lord Kelvin), noting also that Kelvin was influenced by the Cambridge theological traditions which Whewell represented. See his 'Kelvin's scientific realism: the theological context', *The philosophical journal*, **11** (1974), 41–60.

[69] There was also a larger unpublished edition of 1853 which is now held in Trinity College Library, Cambridge. See below (footnote 91).

[70] J. H. Brooke, 'Natural theology and the plurality of worlds: observations on the Brewster–Whewell debate', *Annals of science*, **34** (1977), 221–286. For other recent comment see W. C. Heffernan, 'The singularity of our inhabited world: W. Whewell and A. R. Wallace in dissent', *Journal of the history of ideas*, **39** (1978), 81–100; and Yeo (footnote 9), ch. 6.

[71] But this was one of the contentious issues in the debate: Brewster and Chalmers, for example, did not believe that Christian doctrine was incompatible with the existence of rational beings on other planets (see Brooke (footnote 70), 237–238, 252–258). For Whewell's religious beliefs in relation to this issue, see Whewell to J. Stephen, 4 November 1853, in Todhunter (footnote 2), vol. 2, 392–394.

[72] For the philosophical background see A. O. Lovejoy, *The great chain of being* (1960, New York).

[73] [W. Whewell], *Of the plurality of worlds: an essay. Also a dialogue on the same subject* (2nd ed., 1854, London), 334.

purpose.[74] However, Whewell argued that this extreme principle of plenitude ignored the actual state of affairs on earth; it overlooked large areas of the globe which did not contain life. To the proposition that Nature does nothing in vain, he replied by suggesting that waste, in the sense of unrealized potential or abortive design, was the rule rather than the exception. The fertility of plants and animals was far greater than the resources necessary for their existence; in many cases, the majority of offspring could not survive.[75] The universe was full of the 'rudiments of things'.[76] Given this more realistic picture of nature, Whewell was not surprised that only one of the planets was a seat of life.

These observations were a direct assault upon the teleological tradition of natural theology which conceived Divine design in terms of practical adaptation. On this view, lack of discernible purpose amounted to absence of design; and, in challenging this assumption, Whewell threatened the intelligibility of the universe. Moreover, by underlining the prodigal and abortive features of creation, he disturbed the cheerful image of nature usually associated with the principle of plenitude. His evidence indicated that the 'superfecundity' of nature, often praised by natural theologians, did not ensure a plenitude of existence, but demanded the sacrifice of potential life.[77] The Malthusian theory which described this situation was embarrassing to an optimistic natural theology.[78] Whewell had no easy solution to the problem of evil, but he was able to show that the popular notion of plenitude, together with the traditional concept of final cause (that is, as teleological adaptation), could not be accepted as the sole criterion of creation.

Although Whewell's strictures conflicted with one school of natural theology (represented by Brewster), they were not alien to all exponents of this subject. In fact, the orthodox teleological conception of design had been criticized by important writers such as Baden Powell, John Tulloch and James McCosh.[79] By about 1850, there was a noticeable emphasis upon the notions of law, order, symmetry and harmony as the grandest principles of Divine creation. This was a significant development, because in the early decades of the century these concepts of design had been less favoured due to their affiliation with the morphological theories of the French paleontologist Etienne Geoffroy Saint-Hilaire. Most English scientists preferred the more strictly teleological school of Georges Cuvier, Geoffroy's opponent.[80] As late as 1845, Sedgwick felt that Geoffroy's views 'shut out all

---

[74] See for example D. Brewster, *More worlds than one: the creed of the philosopher and the hope of the Christian* (1854, London), 183–196.

[75] Whewell (footnote 73), 330–334.

[76] *Ibid.*, 331. Alfred Tennyson had made similar observations in his poem 'In memoriam', of 1850. For a detailed study see S. Gliserman, 'Early Victorian science writers and Tennyson's "In memoriam": a study in cultural exchange', *Victorian studies*, **18** (1975), 277–308, 437–460.

[77] For this concept see W. Paley, *Natural theology* (6th ed., 1803, London), 512–514.

[78] The Reverend Thomas Malthus was willing to admit the unpalatable features of Nature, and regarded physical hardship as the means, instituted by God, for the improvement of the human mind. See his *Essay on the principle of population* (1798, London), 348–371, 394–396. See also R. M. Young, 'Malthus and the evolutionists: the common context of biological and social theory', *Past and present*, **43** (1969), 109–145.

[79] See B. Powell, *Essays on the spirit of the inductive philosophy, the unity of worlds, and the philosophy of creation* (1855, London), 135–137; J. Tulloch, *Theism: the witness of reason and nature to an all-wise and beneficent creator* (1855, London), 171–173; and J. McCosh and G. Dickie, *Typical forms and special ends in Creation* (1856, Edinburgh), 1–9, 30–44.

[80] On this dispute see F. Bourdier, 'Geoffroy Saint-Hilaire versus Cuvier: the campaign for paleontological evolution (1825–1838)', in C. J. Schneer (ed.), *Toward a history of geology* (1969, Massachusetts), 36–61.

argument from *design* and all notion of a Creative Providence'.[81] Thus Whewell probably recorded the prevailing attitude when he registered an unfavourable opinion of Geoffroy and praised the work and character of Cuvier.[82]

By the time Whewell came to write his *Plurality of worlds* in 1853, this situation was changing; and, in 1857, when Baden Powell reviewed the literature of natural theology, he was pleased to note a recognition of morphological concepts as the highest indication of a Supreme Intelligence.[83] The major stimulus for this development seems to have been the work of Richard Owen, the distinguished comparative anatomist.

Owen argued that the various classes of vertebrate animals were templates of an ideal Archetype of the vertebrate skeleton. In support of this claim he produced evidence of 'homological' relationships in organic nature: that is, correspondences between organs in different animals, such as that between the arm of a man and the wing of a sparrow.[84] He cast these propositions within a neo-Platonic framework, affirming that the ideal exemplar for the vertebrate animals existed as an idea in the Divine Mind.

Owen's work had important implications for natural theology. In particular, it affected the traditional concept of teleology: his anatomical theories militated against the extreme doctrine of final cause which searched for practical purpose in every feature of animal structure. Owen offered examples of organic structures which, although serving an adaptive function in one animal, also existed in other animals without discernible purpose. Furthermore, he claimed that teleological assumptions could not account for the striking resemblances throughout the vertebrate kingdom which were not necessary for the survival of individual animals. He then argued that these aspects of creation had to be either ascribed to chance or explained in terms of a wider Unity of Plan.[85] Yet, in spite of these criticisms, Owen's position allowed for a reconciliation of teleology and morphology: evidence of final cause could be seen in the modification of the general Plan for the special needs of individual organisms.[86] In this way, Owen's scientific research improved the status of morphological concepts within natural theology.

A favourable attitude towards morphology was important for Whewell's case against the plurality of worlds: it allowed him to argue that the notion of final cause was too limited to stand as the model for all creation. By appealing to Owen's work he was able to claim that 'many parts of the structure of animals, though adapted for particular purposes, are yet framed as a portion of a system which does not seem, in its general form, to have any bearing on such purposes'.[87] Once this particular lack of

[81] Quoted in J. Clark and T. Hughes (eds.), *Life and letters of the Reverend Adam Sedgwick* (2 vols., 1890; Cambridge). vol. 2, 86. See also Whewell (footnote 33), vol. 1, 629.

[82] Whewell, *History of the inductive sciences* (3 vols., 1837, London), vol. 3, 463–478. During the 1830s Powell was one of the few English writers to praise the work of Geoffroy (see Powell (footnote 19), 128–134). However, the Scottish situation may have been different; see E. Richards, 'The German romantic concept of embryonic repetition and its role in evolutionary theory in England up to 1859' (Ph.D. thesis, 1976. University of New South Wales), 178–186.

[83] B. Powell, 'The study of the evidences of natural theology', in *Oxford essays* (1857, London), 170.

[84] R. Owen, *On the archetype and homologies of the vertebrate skeleton* (1848, London). 7. For an introduction to Owen's ideas, see R. M. Macleod, 'Evolutionism and Richard Owen, 1830–1868; an episode in Darwin's century', *Isis*, 56 (1965), 259–280.

[85] R. Owen, *On the nature of limbs* (1849, London), 29–40.

[86] *Ibid.*. 9, 84–85. See also his *Instances of the power of God in his animal creation* (1864, London), 18; and Owen to Whewell, 31 October 1937, Trinity College, Add. ms. a 210 no. 54.

[87] Whewell (footnote 73), 318.

purpose was acknowledged, it became possible to assert that the planets need not support life: their existence could be part of a wider plan independent of anthropomorphic notions of utility. Whewell contended that the grandeur of mountains, the symmetry of animal skeletons and the magnificence of the planets did not serve any practical use or function; instead, they could exist in order to fulfil laws of beauty entertained by the Divine Mind. Again, the opposition between teleology and morphology could be attenuated.

At another level, Owen's ideas suggested a case for the special place of man in nature which strengthened Whewell's epistemology. The appeal to morphology, sanctioned by Owen's research, supported an argument for man's intellectual significance: the ability of mind to comprehend the archetypes of creation confirmed man's special character; the adaptation of mind to world was the grandest form of teleology. Thus man's intellectual nature became a reason for his unique place in the universe. Furthermore, his ability to grasp the fundamental principles of the Divine plan suggested an affinity between the mind of man and the mind of God. In his *Plurality of worlds*, Whewell was able to emphasize this intellectual relationship as a major reason for denying the plurality of worlds and as an important defence of his philosophy of knowledge.

Since his sermons of 1827 and his Bridgewater Treatise of 1833, Whewell had been concerned with the analogy between Divine and human minds. But in the *Plurality of worlds* he returned to it with renewed vigour. However, his full discussion of this topic did not reach the public because Sir James Stephen, his advisor during the preparation of the book, regarded portions of the original draft as too metaphysical for a wide audience.[88] Stephen reminded Whewell that many English readers would be suspicious of metaphysics, especially German metaphysics.[89] Whewell accepted this advice and deleted the offending sections. Nevertheless, he compressed a significant part of the original discussion into one chapter of the published version without altering the substance of the argument. It is important to recognize that he did publish the essence of these metaphysical speculations, and that he considered them crucial to the debate on the plurality of worlds.[90] Furthermore, given his longstanding interest in these non-scientific topics, his discussion of them in 1853 cannot simply be interpreted as a convenient return to theology in order to solve epistemological problems.

Whewell claimed that the laws which man detected in the universe were the laws by which God had ordered the creation. Because it required intellect to delineate such laws, men were 'irresistibly led to suppose that these laws must have been present to the Divine Intellect, before they were apprehended by the human Intellect'.[91] When laws could be expressed in mathematical form Whewell was convinced that man had deciphered the language in which the Supreme Mind spoke to human minds. He also argued that the capacity of man to discern these laws indicated that his mind was of the same nature as the mind of God.[92] This conviction,

[88] Stephen was permanent under-secretary in the British Colonial Office from 1836 to 1847. In 1849 he became Professor of Modern History at Cambridge.

[89] Stephen to Whewell, 31 October 1853, Trinity College. Add. ms. a. 216 no. 137.

[90] Whewell wanted his friends to read the unedited version. See Whewell to Sedgwick, 8 June 1854, in J. M. Douglas, *The life and selections from the correspondence of William Whewell, D. D.* (1882, London), 434–435; and Whewell to J. D. Forbes, 19 February 1854, in Todhunter (footnote 2), vol. 2, 401.

[91] Whewell, *Of the plurality of worlds* (1853, printers copy containing five chapters cancelled from the published work), Trinity College, ADV.C.16.27, 250.

[92] *Ibid.*, 254–258, 267–268.

in turn, supported a belief in the special significane of man and vindicated his denial of the plurality of worlds.

Whewell freely acknowledged this position as a revival of the Platonic theory of Ideas—the notion that objects and laws of nature reflect archetypal ideas in the Divine Mind. However, he contended that this doctrine now rested upon scientific evidence as well as philosophical reasoning; he adduced the recent work of Owen as empirical confirmation of the existence of archetypal ideas which were the organizing principles behind organic nature.[93] Indeed, Owen himself had suggested this interpretation of his research by arguing that the discovery of an ideal exemplar for the vertebrate animals proved that a knowledge of that archetype and all its modifications had prior existence in the Divine Mind.[94] Whewell then claimed that man, in recognizing this archetype, was truly sharing in the original thoughts of the Deity. And, by perceiving beauty in the order and symmetry of natural laws, man was sharing in the aesthetic values of the Creator.[95]

It is therefore possible to view Whewell's thought as a return to the Cambridge Platonism of the seventeenth century; to the conviction that human reason is a reflection of a universal and Divine Reason. In one of his strongest statements of this position Whewell argued that man could 'discover truths, to which all things, existing in space and time must conform. These are conditions of existence to which the creation conforms, that is, to which the Creator conforms; and man, capable of seeing that such conditions are true and necessary, is capable, so far, of understanding some of the conditions of the Creator's workmanship. In this way, the mind of man has some community with the mind of God...'.[96]

Given this broad empathy between Divine and human minds, Whewell did not think it strange that man should exist only on one planet. Indeed, in an earlier paper he suggested that the 'Idea of Man', like other fundamental ideas, could serve to render a number of facts comprehensible, by idealising them. He was therefore prepared to speak of 'the Idea of Man as the principal Object in Creation; to whose sustenance and development the other parts of the Universe are subservient as means to an end...,.[97] This belief in the affinity of Divine and human minds also offered a solution to Whewell's philosophical problems; it provided a metaphysical basis for his epistemology. The fundamental antithesis between ideas and facts, between form and matter, between mind and world, could now be resolved. Whewell was able to claim that the ideal and empirical aspects of knowledge had their 'common source in the Deity, and in the relation borne to Him, both by our own Minds, and by the Universe'.[98]

Most reviewers of the *Plurality of worlds* were dissatisfied. For the present discussion the most relevant comments were those which focused upon the

---

[93] *Ibid.*, 271–273; also Whewell (footnote 73), 34–37, 360–363.

[94] Owen (footnote 85), 85–86.

[95] Whewell (footnote 91), 236–240, 269–271. Although Owen shared Whewell's views about the relationship between Divine and human minds, he did not think that a denial of plurality followed as a necessary conclusion (see Owen to Whewell, 20 May 1854, Trinity College, Add. Ms. a.210 no. 82). Hugh Miller took a similar line; see his *Geology versus astronomy* (1855, Glasgow), 9–11, 16–20.

[96] Whewell (footnote 73), 201–202. Stephen's reservations did not prevent Whewell from making this metaphysical statement in the published version.

[97] W. Whewell, 'Second memoir on the fundamental antithesis of philosophy', 13 November 1848, in *Transactions of the Cambridge Philosophical Society*, 8, part 5 (1849), 614–620 (p. 616).

[98] Whewell (footnote 91), 288; but see also p. 285 and (footnote 97), 617–620 for his rejection of the Absolute Idealism of German philosophy.

epistemological assumptions implicit in Whewell's statements. In particular, a number of writers objected to his remarks about the inconceivability of non-human intelligence as an argument against the existence of other rational beings.[99] James Stephen, for example, could not understand 'why the limits of my conceptions should be supposed to be also the limits of possibility or of probability'.[100] Another reviewer appreciated the wider relevance of this issue: its connection with the debate between idealism and empiricism. Writing in the *Westminster review*, this critic observed that many remarks from the anonymous author of *Plurality* betrayed more sympathy 'with the principles of the "Philosophy of the Inductive Sciences" than with those of the "System of Logic" '.[101] Indeed, in this latter work, J. S. Mill was severely critical of Whewell's views on the notion of inconceivability as a test of truth.[102]. And furthermore, the empiricist philosophy was thought to be perfectly consonant with the prospect of a different physics, geometry and logic in different worlds. Whewell could not tolerate this possibility because he was trying to defend an idealist epistemology which guaranteed the universal validity of man's fundamental ideas.[103] He attempted to justify this epistemology by affirming the intimate relationship between Divine and human minds; he rejected the plurality of worlds because he feared that it would injure this special relationship between man and God, thus qualifying the privileged status of human knowledge.[104]

Whewell's twentieth-century critics have also been dissatisfied with this theological foundation of his epistemology. However, apart from any logical weaknesses (and these are beyond the scope of this article), his philosophy of knowledge was a very significant contribution to the British form of idealism. Whewell attempted to repudiate empiricist philosophy without sacrificing the empirical dimension of physical science. At the same time, he was anxious to dissociate himself from Continental idealist thought which endangered philosophical realism. He rejected both the *a priori* science and the pantheistic religion which stemmed from German idealism and hoped to avoid these consequences by emphasizing the interdependence of intuitive and empirical elements in all knowledge. Moreover, he believed that his idealist philosophy of science could confirm a Christian conception of man's place in the universe. In the context of the intellectual debates of the 1860s, this position was significant: it allowed Whewell to defend the speculative powers of reason against positivism and the critical Kantianism of Henry Mansel.

---

[99] Whewell (footnote 73), 40–41, 125–136, 373; and (footnote 91), 328.

[100] Stephen to Whewell, 3 November 1853, Trinity College, Add.ms.a.216 no. 139. For similar criticism, see H. Smith, 'The plurality of worlds', *Oxford essays* (1855, London), 134.

[101] 'Contemporary literature', *Westminster review*, **61** (1854), 593.

[102] J. S. Mill, *Collected works* (ed. J. Robson: 17 vols., 1965–72, Toronto), vol. 7, 237–248.

[103] Whewell (footnote 73), 113–115; and (footnote 91), 279.

[104] Whewell did admit the possibility of *animal* life on other planets (see (footnote 73), 112, 118). However, he did not concede the existence of intelligent, moral beings on other worlds, even if these were *inferior* to man. To grant this would have been to allow the notion of degrees of rationality which would threaten his commitment to the absolute character of truth. I am not suggesting that Whewell's argument was flawless; indeed, it was circular. His confidence in the universal validity of human conceptions was dependent upon his previous belief in the empathy between Divine and human minds; but he then used this assumption to assert the special status of man, a status which explained his unique position in the universe and weakened the probability of a plurality of similar moral and intellectual beings. See (footnote 73), 118–136.

6. Natural theology versus nescience: Whewell versus Mansel

In his *Moral and metaphysical philosophy* of 1862, F. D. Maurice said that it was unfortunate for Kant's reputation in England that the 'transcendental' aspect of his philosophy had been more influential than the 'critical'.[105] Indeed, most writers neglected Kant's strictures regarding the possibility of metaphysical knowledge and emphasized the creative, rather than the regulative, function of *a priori* ideas. But the critical side of Kant's work was not ignored by all English philosophers. Perhaps the most notable exception was Dean Henry Mansel, author of the Bampton Lectures for 1858 on *The limits of religious thought*.[106]

In these lectures, Mansel sought to underline the limits of reason in the area of religious and theological speculation. His critique derived from an application of Sir William Hamilton's 'philosophy of the unconditioned' to theology. Unlike most British writers, Hamilton exploited the critical facet of the Kantian system and stressed the relativity of knowledge. While following Reid in affirming the mind's direct knowledge of phenomena, he argued that this could not extend to direct acquaintance with mind-in-itself or matter-in-itself. Things were only known as they were related to our experience; only knowledge of the 'conditioned' was possible; knowledge of things-in-themselves, of the 'unconditioned' or the Absolute, was impossible.[107]

Taking up the question of religious thought, Mansel argued that the mind was disqualified, by the very conditions of knowledge, from the comprehension of an Infinite Being. Following Hamilton, he defined the Infinite, or Absolute, as something existing out of all relations. Then, on the assumption that the Deity could be identified with the Absolute, he concluded that man was unable to form a clear conception of such a Being. He attempted to show that whenever the human mind ventured into speculations concerning the nature of God as Infinite Being, it inevitably involved itself in contradictions.[108] But he also claimed that the inability of mind to comprehend the nature and attributes of God did not imply his non-existence. 'In this impotence of Reason', said Mansel, 'we are compelled to take refuge in Faith, and to believe that an Infinite Being exists, though we know not how'.[109]

James Martineau, the Unitarian philosopher and theologian, was greatly alarmed by Mansel's lectures. He linked Mansel's views with those of the positivists, referring to them both as doctrines of 'Nescience', a philosophy which declared that the limits of human mental powers prevented man from reaching metaphysical or theological knowledge.[110] Martineau seized upon this distrust of the intellectual faculties as the common feature of 'a Religion which exaggerates the functions and overstrains the validity of an external authority, and a Science which deals only with objective facts, perceived or imagined'.[111] In this analysis, Mansel was associated

---

[105] F. D. Maurice, *Moral and metaphysical philosophy* (2 vols., 1886, London), vol. 2, 620.

[106] On Mansel, see K. D. Freeman, *The role of reason in religion: a study of Henry Mansel* (1969, The Hague).

[107] I have relied upon the exposition in J. Passmore, *A hundred years of philosophy* (1968, Penguin), 32. For Hamilton's discussion of his position in relation to the views of Kant, Schelling and Cousin, see his 'M. Cousin's course of philosophy', *Edinburgh review*, **50** (1829), 194–221.

[108] H. Mansel, *The limits of religious thought examined* (1859, Boston), 25, 75–80, 84–87, 89–90.

[109] *Ibid.*, 127.

[110] J. Martineau, 'Science, nescience and faith', in *Essays, reviews and addresses* (4 vols., 1890–91, London), vol. 3, 185–218 (pp. 196–202).

[111] J. Martineau, 'The restoration of belief', in *Studies in Christianity* (1879, London), 356–398 (p. 395). Martineau wrote this before Mansel's lectures appeared.

with such unlikely allies as Auguste Comte and Herbert Spencer.[112] For Martineau, the two groups were united by their depreciation of the metaphysical components of knowledge and by their fixation upon external authority of Biblical testimony or sensory experience. Both groups denied the validity of natural theology.

Writing at the close of the nineteenth century, John T. Merz suggested that the work of Mansel, together with that of Darwin, did most to bring about the separation of scientific and theological debate in Britain.[113] In the light of recent scholarship, this analysis probably exaggerates the extent to which Darwin undermined natural theology. But if Darwin weakened the arguments from design in nature, Mansel did far more; he challenged the efficacy of human reason and its ability to deal with metaphysical questions raised by science and theology. Although less obvious than the impact of Darwinism, Mansel's lectures were a serious threat to the status of natural theology as a rational foundation of religious belief. Thus, Goldwin Smith suggested that the only conclusion to be drawn from Mansel was that all works of rational religion should be discarded 'except the Bampton Lectures for 1858, which will be preserved to prove to us that Natural Theology does not exist'.[114] In a sermon of 1859, Whewell extended this criticism, warning that Mansel's doctrine must be rejected, not only 'because it makes Natural Theology impossible, but because it makes Revealed Theology equally impossible. If we *cannot* know anything about God, revelation is in vain. We cannot have anything revealed to us if we have no power of seeing what is revealed'.[115]

Mansel's critique highlighted the theological aspect of Whewell's philosophy of knowledge. In 1860, two years after the Bampton Lectures, Whewell published his *Philosophy of discovery*, and the new sections in this work can be seen as a reply to Mansel. In these sections, Whewell tried to summarize the 'Theological Result of the Philosophy of Discovery',[116] which had been foreshadowed in his *Plurality of worlds*. He considered Mansel's views in a chapter on 'The Philosophy of the Infinite', and began by complaining about the lack of a clear definition of the Absolute in the writings of Hamilton and Mansel. If they were referring to Schelling's concept of an Absolute in which both thought and things were united and identified then Whewell agreed that such a notion was incomprehensible. But he noted that they included other concepts, such as 'the Infinite', within the category of Absolute or Unconditioned. In his opinion, the notion of Infinite, although one of great abstraction, was an appropriate object of human thought: the success of mathematical inquiry which depended upon this concept was testimony to the fruitful manner in which it could be employed. Although the mind could not form an image of the Infinite, this did not exclude the concept from processes of reasoning.[117] But apart

---

[112] J. Martineau. 'Mansel's limits of religious thought', in Martineau (footnote 110), vol. 3, 117–142 (p. 134).

[113] J. T. Merz. *A history of European thought in the nineteenth century* (4 vols., 1896–1914, London and Edinburgh), vol. 2, 326.

[114] G. Smith. *Rational religion and the rationalistic objections of the Bampton Lectures for 1858* (1861, Oxford and London), 70. See also C. Remusat. *Philosophie religieuse* (1864, Paris), 69–89. Remusat was surprised to find an English theologian undermining English natural theology.

[115] Whewell. Sermon, 16 October 1859, quoted in Todhunter (footnote 2), vol. 1, 341.

[116] Whewell (footnote 38). 374. For his discussion of this theme see chs. 29–32.

[117] *Ibid.*, 320–325; also Martineau (footnote 110), vol. 3, 136–138, 212–213; and J. McCosh, *The intuitions of the mind* (1860, London), 227.

514

from these objections, Whewell was anxious about the theological implications. He observed that

> One of the consequences which is drawn by the assertors of the doctrine that we cannot know anything about Infinity, is that we cannot obtain from science any knowledge concerning God: And I have been the more desirous to show the absence of proof of this doctrine, because I conceive that science *does* give us some knowledge, though it be very little, of the nature of God.[118]

Whewell therefore agreed with Martineau's assessment of Mansel's work as a doctrine of Nescience—a theological positivism which denied the possibility of reaching philosophical knowledge of God. Against this view, he reaffirmed his commitment to natural theology: that is, to the ability of man to gain knowledge of God through the study to nature, to a belief in the harmony of physical and religious truth. Furthermore, he was able to meet the critical Kantianism of Mansel with the Christian Platonism advocated in the *Plurality of worlds*. He admitted that man's knowledge of God was limited, but his own philosophy of science led him to reject Mansel's scepticism. According to Whewell, the human mind was capable of supplying fundamental ideas which made the phenomena of nature intelligible; and, by comprehending the laws of nature, man was understanding something of the nature of God because he was sharing, to some extent, in Divine knowledge.

It should also be noted that, in both the *Plurality of worlds* and the *Philosophy of discovery*, Whewell stressed the historical dimension of human knowledge. In the course of man's intellectual history, the scope of knowledge had expanded; new sciences had been formed and old sciences extended; new fundamental ideas had emerged as the foundation of new knowledge.[119] For Whewell, this progressive understanding indicated a Providential plan and suggested that man's life on earth was one not only of moral probation but also of intellectual trial, a struggle to achieve his potential for sharing in Divine truth.[120] In short, he believed that the most profound evidence for natural theology derived not simply from science but from the philosophy of scientific knowledge: not only the study of nature, but also the study of mind, gave knowledge of God and his relationship with man. Given these convictions, he felt confident in defending the validity of natural theology against both secular and religious forms of Nescience.

7. Conclusion

In summary, this article has attempted to reveal the moral and theological dimensions of Whewell's philosophy of science and its relationship with his natural theology. Whewell believed that man had a duty to study the world, and he defended the efficacy of reason against the opposition of clerical conservatives by affirming the assumptions of natural theology. The major writers of this subject saw man's ability to know the world as the measure of his special place in the universe. Man was the Interpreter of Nature: his comprehension of the laws of nature suggested a

---

[118] Whewell (footnote 38), 325; also 376.

[119] *Ibid.*, 315–319, 350–353, 372–375, 385–399; and (footnote 91), 302–309.

[120] Whewell (footnote 73), 118–125, 137–138, 366–370 and (footnote 91), 302–309. I cannot deal fully with the historical dimension of Whewell's though. but it is interesting to consider him as sharing the Providentialist philosophy of history outlined by Julius and Augustus Hare in their *Guesses at truth* (1838); and more recently, in D. Forbes, *The liberal Anglican idea of history* (1953, Cambridge). For this suggestion, see Schneewind (footnote 30).379.

consonance between mind and world and therefore indicated the unity of the physical and moral spheres of creation. These convictions meant that questions of method and epistemology had moral connotations. In considering these topics, Whewell hoped to establish a philosophy of science which guaranteed the assumptions of natural theology and the values of Christianity: the affinity of moral and physical knowledge, and the spiritual character of the human mind. He proclaimed the religious and intellectual value of natural science while repudiating the empiricist philosophy of science promoted by the utilitarians. He stressed the role of metaphysical concepts in physical knowledge, and offered an idealist philosophy of science which might satisfy the requirements of a Christian natural theology.

However, this epistemology was criticized by religious and scientific writers in Britain because they distrusted its Continental affiliations and feared that it might engender *a priori* science and Pantheistic religion. Whewell recognized these dangers and tried to justify an idealist epistemology by grounding it within the metaphysics of a Christian Platonism. This theological metaphysics, which emphasized the close relationship between Divine and human minds, was presented in his works on *The plurality of worlds* and *The philosophy of discovery*. In rejecting the possibility of rational life on other planets, Whewell confronted important assumptions of traditional natural theology. But his idealist epistemology may have strengthened natural theology against its opponents in the second half of the century. It challenged the positivist account of science without impugning the validity of scientific knowledge; it also met Mansel's scepticism by claiming that man, in comprehending the laws of nature, was understanding something of the nature of God. In this sense, an idealist natural theology could defend the validity of metaphysical speculation in both science and theology.

Whewell's work reveals a deep concern with the moral, theological and metaphysical dimensions of the philosophy of science. During the first half of the nineteenth century this concern was shared by a significant number of British scientists and theologians. However, by the 1860s, with the increasing specialization and professionalization of intellectual life, the debate between science and religion was beginning to lose its social basis.[121] Whewell's work was perhaps the last (and most sophisticated) testament to the early Victorian belief in the unity of truth. The Metaphysical Society, founded in 1869, can be viewed as an attempt to salvage the conviction, to maintain the dialogue between scientific and theological thought. But the need to institutionalize this debate portended its disintegration. And with this perspective, it is interesting to note that Cardinal Manning, a member of this society, believed that a solution to the problem of two intellectual cultures could be found in Whewell's philosophy of science:

> And this position of Whewell is, after all, only a disinterring of the Scholastic Philosophy, fragant as fresh earth. It is St. Thomas Aquinas in a Cambridge gown . . . If the Scholastic Philosophy had never been disintegrated;

[121] I have not attempted to discuss the social and institutional contexts of the debates on natural theology and science. But see W. F. Cannon, 'Scientists and Broad Churchmen: an early Victorian intellectual network', *Journal of British studies*, **4** (1964), 65–88; S. F. Cannon, *Science in culture; the early Victorian period* (1978, New York), chs. 1 and 2; W. H. Brock and R. M. MacLeod, 'The Scientists' Declaration: reflexions on science and belief in the wake of *Essays and reviews*, 1864–5', *British journal for the history of science*, **9** (1976), 39–76; and F. M. Turner, 'The Victorian conflict between science and religion: a professional dimension', *Isis*, **69** (1978), 356–376.

if the two elements of reason and of sense, which are the conditions of all knowledge in the human subject, had not been violently sundered, and after their separation falsified by exclusive theories of the opposing schools of intuition and of sense, a great part of our discussions would have been impossible.[122]

Acknowledgements

I would like to acknowledge the comments of an anonymous referee. For access to the Whewell papers, I am grateful to the Librarian and staff of Trinity College, Cambridge.

[122] Quoted in A. W. Brown, *The metaphysical society: Victorian minds in crisis, 1869–1880* (1947, New York), 78–79. Manning made this statement in his paper entitled 'A diagnosis and a prescription', delivered to the society on 10 June 1873.

VI

The Principle of Plenitude and Natural Theology in Nineteenth-Century Britain

I

In his classic study, *The Great Chain of Being,* Arthur Lovejoy delineated a complex set of concepts and assumptions which referred to the perfection of God and the fullness of creation. In attempting to distil the basic or 'unit idea' which constituted this pattern of thought, he focused on the assumption that 'the universe is a *plenum formarum* in which the range of conceivable diversity of *kinds* of living things is exhaustively exemplified'. He called this the 'principle of plenitude'. Lovejoy argued that this idea implied two others—continuity and gradation—and that together these reflected a pre-occupation with the 'necessity of imperfection in all its possible degrees', a concern which had pervaded Western thought since Plato and gave rise to the powerful ontology known as the 'great chain of being'.[1]

Although Lovejoy's approach to the history of ideas has been criticized,[2] this need not imply that the concepts he discussed are no longer of interest to intellectual historians or historians of science. Rather, it suggests a need for historical sensitivity to the different formulations of those ideas in specific contexts. William Bynum canvassed this alternative in an essay dealing with the possible fate of the chain of being in the nineteenth century, a period which lay outside Lovejoy's survey. Lovejoy's pursuit of his theme and its manifold permutations ended in the late eighteenth century, with the thesis that the chain of being acquired a temporal dimension and was thus transformed into a chain of becoming. In making this point, Lovejoy revealed the inherent tensions in the

1 A.O. Lovejoy, *The Great Chain of Being.* New York, 1960, p. 52, 338. For a recent account which stresses the principle of plenitude as 'the very foundation of the Chain of Being', see L. Formigari, 'Chain of Being', in P. Wiener (ed.), *Dictionary of the History of Ideas,* 4 vols. New York, 1968–1973, pp. i, 324–335(334). For a recent dispute about the concept of plenitude and the category of 'unit idea', see J. Hintikka, 'Gaps in the Great Chain of Being: an exercise in the methodology of the history of ideas', *Proceedings of the American Philosophical Association,* (1975–1976), 49, 22, and M.S. Gram and R.M. Martin, 'The perils of plenitude: Hintikka Contra Lovejoy', *Journal of the History of Ideas,* (1980), 41, pp. 497–511.

2 See Lovejoy, *Essays in the History of Ideas,* Baltimore and London, 1948, pp. 1–13 and 'Reflections on the history of ideas', *Journal of the History of Ideas,* (1980), 41, pp. 487–511. For other perspectives, see Q. Skinner, 'Meaning and understanding in the history of ideas', *History and Theory* (1969), 8, pp. 1–53 (10–14) and T. Bresdorff, 'Lovejoy's idea of "idea"', *New Literary History,* (1976–1977), 8, pp. 195–211. See also C. Brinton, 'Intellectual history', In: D.L. Sills (ed.), *International Encyclopedia of the Social Sciences.* 18 vols. New York, 1968–1979, 6, pp. 462–468; H.V. White, 'The tasks of intellectual history', *The Monist,* (1969), 53, pp. 606–630, and M. Foucault, *The Archaeology of Knowledge.* London, 1972, pp. 134–140.

I would like to thank the referee (Dr J. H. Brooke) and the editor for their helpful comments on an earlier draft.

conceptions of Divine perfection which stemmed from Plato, and saw the transition he described as a 'logically inevitable outcome'.[3]

Bynum suggested that this philosophical conclusion could be enriched by closer historical analysis of the social, intellectual and national contexts which involved assumptions concerning the chain of being. His essay dealt mainly with anthropological debates concerning man's place in nature, and showed the ways in which 'the chain of being was discarded as a conceptual framework' by thinkers interested in explaining social change.[4] But this example can lead to the more general contention that Lovejoy's philosophical account of the transformation of the chain of being at the close of the eighteenth century concealed the ways in which it was in fact dismantled rather than temporalized. Lovejoy claimed that 'When the notion of the Scale or Chain of Being was translated from its static to its temporalized version, some of the related ideas inherent in the former passed over into the latter'.[5] However, the continuing connection between the various unit ideas which is implied here was not apparent in nineteenth-century discourses.[6] It may therefore be too simplistic to speak of the chain of being acquiring an historical or temporal dimension. Rather, it could be proposed that the system of ideas and assumptions known as the chain of being became fragmented and that its component ideas—continuity, gradation and plenitude—although still discernible in various debates, became divorced from each other and transformed by new intellectual pressures. This article investigates the ways in which the principle of plenitude functioned in the writings of natural theology from the work of Paley and the Bridgewater Treatises through to the mid-nineteenth-century debate on the plurality of worlds in which it was the constitutive idea, even though other unit ideas of the chain of being had been rejected. In doing so, it reveals certain qualifications of this principle and relates these to discussions of the question of species and to the concepts of final cause and design in natural theology.

II

Lovejoy defined the principle of plenitude as 'the realization of conceptual possibility in actuality',[7] and traced this assumption to speculations which led to the notion that 'the existence of all possible beings at all times is . . . an implicate of the divine nature'.[8] He detected this thinking in philosophers from Plato and Abelard to Leibniz and Spinoza, and related it to deliberations about the 'sufficient reason' for the existence of finite things.[9] This area of thought was a site of enormous tension because it involved the

3 Lovejoy, op. cit. (1), p. 326 and chs. ix–xi.
4 W. Bynum, 'The Great Chain of Being after forty years: an appraisal', *History of Science*, (1975), **13**, pp. 1–28(3), also 7, 14, 18.
5 Lovejoy, op. cit. (2), *Essays*, p. 169.
6 Bynum, op. cit. (4), p. 20. See also Foucault, *The Order of Things*, London, 1974, pp. 145–157 for a distinction between eighteenth- and nineteenth-century theories of evolution and their relation to the chain of being.
7 Lovejoy, op. cit. (1), p. 52.
8 Ibid., p.154.
9 Ibid., pp. 70–72.

question of whether, as Spinoza contended, the Deity, in his perfection, was constrained to produce all logically conceivable forms of existence so that this world which does exist, must do so by absolute, rational necessity. This conclusion conflicted with the voluntarist tradition in Christian philosophy which represented the creation as an act of God's free will, and with the ideological arguments which implied a Divine choice of the best possible world.[10] Lovejoy suggested that Leibniz attempted to avoid the universal necessity of Spinoza by proposing that 'there are species which are possible but nevertheless do not exist' because, in the concrete world, as opposed to the world of essences, only certain combinations are possible. There were 'species which never have existed and never will exist, since they are not compatible with the series of creatures which God has chosen'.[11] However, Lovejoy indicated that Leibniz continued to insist that the world which actually exists was the best possible and that, within it, no gaps of any sort could be admitted.

Thus, by the eighteenth century, in spite of these tensions, there was a conception of plenitude which stemmed from the conviction that the omnipotence and goodness of God gave rise to a world containing all possible forms of life. This notion, according to Lovejoy, supported the emphasis on diversity and fullness apparent in eighteenth-century thought and sponsored, for example, the belief in intermediate forms and missing links. 'Given the principle of plenitude, which most well-instructed persons then accepted in theory, it followed that the existence of aquatic anthropoids was more probable than their non-existence'.[12] By the end of the century, when the chain of being was projected as an historical series, this principle was interpreted to include the future realization of further possible kinds of life.

Lovejoy also offered another definition of the principle of plenitude: namely, the idea that all space, or at least all matter, should contain life. The influence of this assumption was most obvious in debates about the plurality of intelligent worlds on other planets, which became more prominent with the emergence of heliocentric cosmology in the seventeenth century. Lovejoy saw these debates as 'manifest corollaries of the principle of plenitude, when that principle was applied, not to the biological question of the number of kinds of living beings, but to the astronomical questions of the magnitude of the stellar universe and of the extent of the diffusion of life and sentiency in space'.[13] But throughout his analysis of the writings on the plurality of worlds, Lovejoy gave two slightly different accounts of the way in which this corollary was derived from the principle of plenitude. Thus, in the case of Giordano Bruno, the belief in the existence of other worlds stemmed from the necessity for the realization of the full scale of being: that is, other worlds provided the accommodation for the possible forms of life which did not exist on earth.[14] More often, however, Lovejoy emphasized the argument which moved from the supposition of God's infinite creative power to the conclusion that wherever

10 Ibid., pp. 156–158.
11 Ibid., pp. 170–171.
12 Ibid., p. 272.
13 Ibid, III. For a study which also stresses the *scientific* thinking involved, see S.J. Dick, *Plurality of Worlds: the Origins of the Extraterrestrial Life Debate from Democritus to Kant*. Cambridge, 1982.
14 Lovejoy, op. cit. (1), pp. 118–119.

266

there was matter there should be life, that having created matter in any part of the universe, the Deity would not waste the opportunity to populate it with living beings.

Thus, while the salient feature of the principle of plenitude, as Lovejoy presents it, is the emphasis on the fullness of creation and the variety of life within it, it is possible to detect two different formulations of its premises. The first, and for Lovejoy's purposes, the most important of these is the notion that postulates the actual existence of all conceivable (or possible) forms of life. I call this 'conceptual' plenitude. The second assumption, which seems to have coexisted with the first in the period studied by Lovejoy, is the notion that matter exists to sustain life and that, hence, wherever there is matter there should be life. I call this 'spatial' plenitude.

It is worth remarking that these two assumptions do not logically entail each other. The belief that all conceivable (or practically possible) forms of life exist, or have existed, does not necessarily imply that all space or matter must support life. Similarly, the contention that all matter sustains life does not require a commitment to the belief in the actual existence of all conceivable forms of life. From Lovejoy's work it appears that these two assumptions coexisted without tension until the late eighteenth century, but a study of nineteenth century debates discloses a rejection of the first (conceptual plenitude) and a continuing adherence to the second (spatial plenitude). Thus, by the 1830s, the idea of a complete and continuous chain of being, embracing all conceivable forms and gradations of life, was largely rejected in both scientific and theological discourse. However, writings on natural theology continued to celebrate the variety and fullness of creation and, in the 1850s, a lively dispute over the plurality of worlds indicated the presence of assumptions about the nexus between matter and life.

This article draws attention to the ways in which different inflections of the principle of plenitude and its associated assumptions could be used to support different scientific and theological positions. It argues that writers of natural theology perceived a need to dissociate themselves from formulations which could be deployed by proponents of the transmutation of species. Secondly, it shows how the principle of spatial plenitude was implicated in arguments about the place of teleological and morphological concepts of design in natural theology, an issue highlighted by the debate on the plurality of worlds.

III

In works of natural theology since the early eighteenth century, the extent and diversity of creation were cited as illustrations of God's power and benevolence.[15] By the nineteenth century, this was a common theme across texts written from different denominational positions. Thus, in his Bridgewater Treatise, William Buckland announced that 'the design of the Creator seems at all times to have been, to fill the waters of the sea, and cover the surface of the earth with the greatest possible amount of organized beings enjoying life.'[16] Writing in the same collection, Peter Mark Roget found it impossible to

15 See for example J. Ray, *The Wisdom of God*, new edn., London, 1827, p. 302.

16 W. Buckland, *Geology and Mineralogy Considered with Reference to Natural Theology*, 2 vols. London, 1836, i, p. 301. A bequest of the eighth Earl of Bridgewater provided funds for the illustration of 'the Power, Wisdom and Goodness of God as Manifested in the Creation'. Eight treatises, covering various scientific areas, were published between 1833 and 1836. See W.H. Brock, 'The selection of the authors of the Bridgewater Treatises', *Notes and Records of the Royal Society of London*, (1966), 21, pp. 162–179.

survey this profusion of living beings without 'a feeling of profound astonishment at the inconceivable variety of forms and constructions to which animation has been imparted by creative power.'[17] Similar sentiments could be found in more controversial writers such as Charles Bray and Robert Chambers, both of whom agreed that the Deity had clearly set out 'to diffuse existence as widely as possible, to fill up every vacant piece of space with some sentient being to be a vehicle of enjoyment'.[18] But the enthusiasm of these writers was eclipsed by Thomas Dick, the Scottish evangelical and popular author. Dick provided detailed numerical calculations to evince the 'plan of boundless and universal variety'[19] upon which the universe was constructed, and went so far as to speculate about the even more beautiful character of the earth before the Fall, when 'no barren deserts of heath and sand disfigured the rich landscape of the world'.[20]

This reference to the plenitude of life throughout the globe was prominent in William Paley's *Natural Theology* of 1802, a text often cited as the classic expression of the design argument. In one section of his argument—the discussion of the Deity's benevolence— Paley celebrated the extent and diversity of sensual pleasure enjoyed by animal life in a world in which 'the air, the earth, the water, teem with delighted existence'.[21] Having accepted this principle of plenitude as a primary feature of creation, he had recourse to it as a rationalization of phenomena which apparently threatened the concept of benevolent design. In his *Essay on the Principle of Population* of 1798, Thomas Malthus referred to the enormous rate of reproduction in 'the animal and vegetable kingdoms' as part of his argument about the imbalance between population growth and food resources.[22] However, Paley saw this 'superfecundity' of nature as the means chosen by God 'to keep the world always full',[23] and he regarded the necessary destruction which accompanied it as subservient to this larger good. In his view, those who criticized this 'superabundant multiplication' overlooked the ways in which it worked to fill all parts of the world with living things.

> Where this vast fecundity meets with a vacancy fitted to receive the species, there it operates with its whole effect; there it pours in its numbers, and replenishes the waste. We complain of what we call the exorbitant multiplication of some troublesome insects, not reflecting that large portions of nature might be left void without it. If the accounts of travellers may be depended

17 P.M. Roget, *Animal and Vegetable Physiology Considered with Reference to National Theology*, 2 vols. London, 1834, i, p. 11, 13. For a reflection on the contrast between the feminine image of 'Nature' in Roget and the masculine 'Divine Artificer' in Paley, see S. Gliserman, 'Early Victorian science writers and Tennyson's "In Memoriam": a study in cultural exchange', *Victorian Studies*, (1975), **19**, pp. 277–308(290–291). While this contrast is useful, it is important to recognize that Paley also emphasized the fecundity of nature in ways which went beyond the design argument.

18 [R. Chambers] *Vestiges of the Natural History of Creation*. London, 1844, p. 367; also C. Bray, *The Philosophy of Necessity*, 2 vols. London, 1841, i, p. 24.

19 T. Dick, *On the Improvement of Society*. London, 1833, p. 257; for his calculations leading to the conclusion that there were more than sixty thousand billion 'distinct ideas, conceptions or contrivances, in relation to the animal world', see ibid., pp. 260–262.

20 T. Dick, *The Christian Philosopher*, new edn., 2 vols. Glasgow, 1863, i, p. 109.

21 W. Paley, *Natural Theology*, London, 1802, p. 490.

22 T. Malthus, *An Essay on the Principle of Population*, London, 1798, reprinted, Harmondsworth, 1976, pp. 71–72. For the relevant contemporary discussion, both scientific and social, see B.G. Gale, 'Darwin and the Concept of a Struggle for Existence: A study in the extrascientific origins of scientific ideas', *Isis*, 1972, **63**, pp. 321–344.

23 Paley, op. cit. (21), pp. 514–516.

upon, immense tracts of forest in North America would be nearly lost to sensitive existence if it were not for *gnats* . . . Again, hosts of *mice* are reckoned amongst the plagues of the north-east part of Europe; whereas vast plains in Siberia . . . would be lifeless without them.[24]

Early nineteenth-century writings on natural theology emphasized the diversity as well as the spatial extension of the living world. Indeed, the references to this manifestation of plenitude—the feature most emphasized by Lovejoy—may have been strengthened by the increasing use of evidence from the organic sciences. Although including a chapter on astronomy, Paley stated his preference for examples drawn from the more complex formations of the animal world and the evidence of purposive adaptation which they offered.[25] Roget, whose Bridgewater Treatise dealt with animal and vegetable physiology, also began with this contrast between the mathematical order of physical science and the phenomena of the organic kingdom:

Far different is the aspect of living Nature. The spectacle here offered to our view is every where characterized by boundless variety, by inscrutable complexity, by perpetual mutation.[26]

In the works of Paley and the Bridgewater authors, this feature of nature, which Roget called the '*law of variety*',[27] reinforced the teleological argument: it allowed an appeal to countless examples of form adapted to function and environment by means of minute modification. Roget observed that 'even when the purpose to be answered is identical, the means that are employed are infinitely diversified in different instances, as if a design had existed of displaying to the astonished eyes of mortals the unbounded resources of creative power'.[28] From this perspective, the plenitude of natural forms became an example of the purposiveness of variety in God's intricately designed creation.

At this point, it is necessary to be more specific about these allusions to the principle of plenitude. Although the previous statements about the fullness and diversity of creation have clear affinities with those cited by Lovejoy, it is important to recognize that they did not involve a commitment to the notion of sufficient reason which had been associated with the idea of plenitude in the eighteenth century. In nineteenth-century natural theology, the emphasis on the extensive distribution of animal life did not derive from a conviction about the necessary realization of logically possible forms, and it did not imply the complete continuity of the chain of being. Although Paley strongly emphasized the plenitude of nature, in the sense of all possible space on earth being filled with life, he had cause to reject the implication previously drawn from the principle of sufficient reason: that is, the assumption that all conceivable forms of life must actually exist. In canvassing various 'atheistic' objections to the thesis of design in nature, he came to the suggestion that there was no warrant for the inference of a purposeful Designer from the existence of organized forms of life: on this view, as Paley recorded it, these were simply the surviving relics of the many 'possible varieties and combinations of beings which have existed; every possible variety of being hath, at one time or another, found its

24 Ibid. p. 511.
25 Ibid., pp. 409–411.
26 Roget, op. cit. (17), i, p. 9.
27 Ibid., pp. 48, 10–13.
28 Ibid., p. 48.

way into existence . . ., and those which were badly formed, perished'.[29] Paley countered this hypothesis of an original infinite number of living forms by denying the assumption of conceptual plenitude.

> Multitudes of conformations, both of vegetables and animals, may be conceived capable of existence and succession, which yet do not exist. Perhaps almost as many forms of plants might have been found in the fields, as figures of plants can be delineated upon paper. A countless variety of animals might have existed which do not exist.[30]

While earlier writers such as J. B. Robinet speculated about the likely existence of mermaids and other intermediate forms, Paley pointed to the absence of 'unicorns and mermaids, sylphs and centaurs' as evidence against the suggestion that the existing animals were merely the chance remains of an infinite series. In his view, there was no such energy operating 'as that which is here supposed, and which should be constantly pushing into existence new varieties of beings'.[31] Furthermore, he argued that the classification of plants and animals into genera and species, a system 'founded in the order which prevails in external nature', stood in contradiction to the hypothesis of 'a variety which rejects all plan.'[32] Thus, in seeking to safeguard the argument from design, Paley exposed the tension between the principle of plenitude and the notion of fixed species.

Paley did not name his atheistic opponents, and the objections he cited derived from a variety of sources extending from the Epicurians to the transformist or evolutionary theories of Buffon and Erasmus Darwin.[33] He did not distinguish between general philosophical objections and those formulated in scientific treatises,[34] rejecting them all in favour of the concept of design. By the 1830s, however, some authors of the Bridgewater Treatises had to confront doctrines of transmutation current in French biology and comparative anatomy, which had definite implications for their own scientific disciplines and their natural theology. William Whewell gave a succinct statement of the implications for teleological assumptions in 1837 when he explained that 'if we allow such a *transmutation of species,* we abandon the belief in the adaptation of the structure of

29 Paley, op. cit. (21), pp. 70–71.

30 Ibid., p. 69. This aspect of Paley's text has received little, if any, attention; one commentator has claimed that his concerns were 'pastoral rather than philosophical', that he 'paid no attention to the principle of plenitude nor to the continuity of the great chain of being'. E. Manier, *The Young Darwin and his Cultural Circle*, Dordrecht, 1978, p. 72.

31 Paley, op. cit. (21), p. 69.

32 Ibid., p. 70.

33 For a discussion of eighteenth-century theories of 'evolution' and the various meanings of the term, see P.J. Bowler, 'Evolutionism in the Enlightenment', *History of Science,* (1974), **12**, pp. 159–183 and 'The changing meaning of "Evolution" ', *Journal of the History of Ideas,* (1975), 36, pp. 95–114.

34 These two levels were, of course, related in the works of contemporaries. Erasmus Darwin made use of David Hume's exploitation of the phenomenon of generation as a possible model for creation. See F.C. Haber, in B. Glass, O. Temkin and W. Straus (eds): *Forerunners of Darwin,* Baltimore, 1959, p. 251. Darwin's *Zoonomia,* originally published between 1794 and 1796, reached a third edition in 1801, just before Paley's book appeared. It is likely that Paley had some of Darwin's arguments in mind—such as the determination of structure by use. See Paley, op. cit. (21), pp. 72–76. Although he did not mention Hume's *Dialogues concerning Natural Religion,* it does seem that he sought to counter some of the points in it—such as the problem of treating the world as logically equivalent to its constituent parts. See D.L. Le Mahieu, *The Mind of William Paley: A Philosopher and his Age.* Lincoln and London, 1976, pp. 67–68.

every creature to its destined mode of being'.[35] Transmutationist theories were discussed in British scientific circles during the 1820s, and with the publication of the second volume of Charles Lyell's *Principles of Geology* in 1832, the speculations of J. B. Lamarck were clearly presented to English readers.[36] For, although Lyell criticized Lamarck, he provided an extensive account of his views, thus giving renewed significance to the species question just before the Bridgewater Treatises began to appear.

Lyell's exposition of Lamarck revealed the ways in which the principle of plenitude and its associated concepts could be used to support the idea of transmutation. In referring to the chapter on species in the *Philosophie Zoologique*, Lyell gave the following paraphrase of Lamarck's argument:

> The more we advance in the knowledge of the different organized bodies which cover the surface of the globe, the more our embarrassment increases, to determine what ought to be regarded as a species, and still more how to limit and distinguish genera. In proportion as our collections are enriched, we see almost every void filled up, and all our lines of separation effaced . . .[37]

In this way, the idea of a plenitude of living forms could be seen to endanger the immutability of species. Secondly, Lyell showed that Lamarck's conception of a tendency to progressive improvement—one of the basic assumptions of his philosophy of nature —was closely linked with the belief in a finely graded scale or chain of being.[38] In this context, concepts and images associated with the principle of plenitude—the great variety of living forms, the progressive continuity between them, and the creative powers of nature—could serve as a powerful cultural resource for the proponents of transformist doctrines.

This is not to suggest that there were logical connections between such manifestations of plenitude and theories of evolution, but rather that this set of ideas was open to various interpretations and uses. For example, a notion of continuity could be accepted by those who defended the fixity of species. Thus, Buckland regarded fossils as supplying the links 'that appeared deficient in the grand continuous chain which connects all past and present forms of organic life'.[39] But when confronted with theories which exploited the idea of continuity as part of an argument for transmutation, writers maintaining the orthodox view qualified their meaning of the term, explaining that it

35 W. Whewell, *History of the Inductive Sciences,* 3 vols. London, 1837, iii, p. 574. Compare B. Powell, *Essays on the Spirit of the Inductive Philosophy, the Unity of Worlds, and the Philosophy of Creation.* London, 1855, p. 385, 391.

36 For the presence of transmutationist ideas in England prior to Lyell's response, see P. Corsi, 'The importance of French transformist ideas for the second volume of Lyell's *Principles of Geology', British Journal for the History of Science,* (1978), 11, pp. 221–244; A. Desmond, 'Robert E. Grant: the social predicament of a pre-Darwinian transmutationist', *Journal of the History of Biology,* (1984), 17, pp. 189–223.

37 C. Lyell, *The Principles of Geology,* 3 vols. London, 1830–1833, ii, p. 3.

38 Ibid., pp. 13–14. On the qualifications made by Lamarck to the *complete* continuity of the chain of being, and for the connection between his theory of mutability and his refusal to accept widespread extinction, see R. Burkhardt, *The Spirit of System: Lamarck and Evolutionary Biology.* Cambridge, Mass., 1977, pp. 54–4, 128–137. See also L.J. Jordanova, *Lamarck.* Oxford, 1984.

39 Quoted in N. Rupke, *The Great Chain of History: William Buckland and the English School of Geology (1814–1849),* Oxford, 1983, p. 172. For the historicization of the principle of plenitude, see also M. Rudwick, *The Meaning of Fossils.* London, 1972, pp. 154–155.

applied within, but not between, the four great divisions of the animal kingdom classified by Georges Cuvier.[40] On the other hand, in promoting Lamarck's work in Britain, Robert Grant explicitly rejected this restriction and sought to trace continuity, and increasing complexity, across Cuvier's categories.[41] He was also reluctant to acknowledge extinction, suggesting that 'many fossil species to which no originals can be found, may not be extinct, but have gradually passed into others.'[42] Consequently, opponents of transmutation focused on the idea of continuity, and John Fleming was pleased to report that 'the advocates for the existence of the "law of continuity" among created beings in their mutual relations, have experienced no small degree of pain from those *chasms* which so frequently present themselves'.[43]

The distinction between conceptual and spatial plenitude acquired strategic importance here, for while rejecting the implications of conceptual plenitude such as a perfect chain of being, writers of natural theology did assume that all habitable parts of the globe were filled with life.[44] It was this fact—and not the notion that all possible forms of life existed—which, in their view, accounted for the great variety and diversity of the organic creation. The geologist, William Conybeare, exemplified this position when criticizing Lamarck's controversial speculations:

> In the original formation of animated beings, the plan evidently to be traced throughout is this. That every place capable of supporting animal life should be so filled, and that every possible mode of sustenance should be taken advantage of; hence every possible variety of structure became necessary . . .[45]

In this formulation there was a commitment to spatial plenitude but the variety of animal forms was explained in teleological terms. There is evidence that some authors of the Bridgewater Treatises, alerted by Lyell's account of Lamarck, were sensitive to the different and often controversial interpretations of references to the fullness and variety of nature. They also attempted to qualify the principle of plenitude, and incorporated it within a teleological framework, thus protecting the concepts of purposive design and fixed species.

In discussing Paley's response to 'atheistic' objections, we saw that he explicitly rejected the principle of plenitude which implied the actual existence of all conceivable forms. Within the more explicit debate of the 1830s on the question of species, John Kidd

40 See M.P. Winsor, *Starfish, Jellyfish and the Order of Nature*. New Haven, 1976, ch. 1.

41 Desmond, op. cit. (36), pp. 205–207.

42 Ibid., p. 201.

43 [J. Fleming] 'Systems and methods in natural history', *Quarterly Review*, (1892), 41, pp. 302–327(327).

44 It is worth noting that such an assumption was not confined to treatises on natural theology. Lyell's discussion of the replacement of extinct species and the filling of any vacant 'stations' in the natural economy appears to rest on suppositions about the 'fullness' of God's creation. See Lyell, op. cit. (37), i, p. 140; ii, p. 130. For references to this and to the various relationships between the views of Lyell and Darwin, see M.J.S. Hodge, 'Darwin and the Laws of the Animate Part of the Terrestrial System (1835–1837): On the Lyellian Origins of his Zoonomical Explanatory Program', *Studies in History of Biology*, (1983), 6, pp. 1–106 (6, 23). Although the opponents of transmutation feared the connotations of conceptual plenitude, it would be wrong to see this notion as subsidizing Darwin's theory of evolution. For the suggestion that Darwin saw nature as 'full' in the sense of all ecological niches being occupied rather than in the sense of a plenitude of forms, see D. Worster, *Nature's Economy: the Roots of Ecology*. New York, 1979, pp. 157–158.

45 Quoted in Rupke, op. cit. (39), p. 174. Conybeare was writing in 1821.

and Charles Bell also regarded this as a dangerous notion. Referring to Paley's remark about the possible existence of other animal forms, Bell argued that man's ingenuity was in fact quite limited. All the fabled animals of antiquity—centaurs, satyrs, griffins—displayed no real novelty, suggesting that man's inventions were merely 'the incongruous union of things seen in nature'.[46] Furthermore, he argued that such creatures could not survive because their structures were not internally adjusted nor adapted to the external environment. 'Were such forms actually in being', he concluded, 'they must creep weakly on the ground'.[47] Kidd agreed that 'there is no ground, for supposing that nature has ever produced such an individual as a chimera or centaur', and related this concern more clearly to the species question by tackling the problem of 'lusus naturae', or monsters. The phenomenon of malformations and monstrosities was used by proponents of transformist theories as evidence of the malleability of species: some of these forms were said to occupy intermediate positions in the scale of life, thus suggesting a possible mechanism of evolutionary change. But Kidd insisted that in all known cases 'the character of the species, however obscured, is never lost';[48] that anomalous deviations were always within certain limits. He went so far as to suggest that these 'anomalous productions' may have been adapted, by the Creator, to serve special ends.[49]

It was also crucial for writers of natural theology to ensure that allusions to the plenitude and fecundity of nature did not subsidize the concept of a continually active, natural mechanism of creation. William Kirby, who devoted the introduction of his Bridgewater Treatise to an attack on Lamarck and materialism, saw this notion as part of a tendency to substitute Nature for God. He regarded Lamarck's hypothesis of spontaneous generation as part of a theoretical framework which made 'nature . . . the real creator of all the forms and beings that exist'.[50] Similarly, Kidd was anxious to avoid the suggestion that monsters resulted from a 'sportive effort of the creative powers of Nature', and in attempting to avoid this notion, explained that there was great variety *within* species and that this 'wonderful diversity' was ample manifestation of the infinite power of the Deity.[51] This qualification matched Lyell's answer to the 'doctrine of transmutation': namely, that deviations are always limited, that species have a 'real existence in nature, and that each was endowed, at the time of its creation, with the attributes and organization by which it is now distinguished'.[52] In this sense, speculation about transmutation which appealed to the diversity of organic beings was met with a reassertion of teleology: the perfect adaptation of species to their environment was the

46 C. Bell, *The Hand, its Mechanism and Vital Endowments, as Evincing Design*, 3rd edn. London, 1834, p. 308.

47 Ibid., p. 307.

48 J. Kidd, *On the Adaptation of External Nature to the Physical Condition of Man*. London, 1833, pp. 335–336.

49 Ibid., p. 338.

50 W. Kirby, *On the Power, Wisdom and Goodness of God as Manifested in the Creation of Animals, and in their History, Habits and Instincts*, 2 vols. London, 1835, i, xxiv–xxxiii. See T.A. Appel, 'Henri de Blainville and the Animal Series: a nineteenth-century chain of being', *Journal of the History of Biology*, (1980), 13, pp. 291–317 (299) for Cuvier's attack on de Blainville, Geoffroy, Lamarck and Oken for their failure to distinguish adequately between nature and the Creator.

51 Kidd, op. cit. (48), p. 335.

52 Lyell, op. cit. (37), ii, p. 65.

mode in which God had chosen to manifest and limit his designing powers. Within this framework, variety in animal forms did not reflect an independent principle or power of nature, but was explained in terms of purposive adaptation to different environments.

IV

Some recent work has drawn attention to the presence of an alternative, non-teleological conception of design in early nineteenth-century science and natural theology.[53] Within the Bridgewater Treatises this was apparent in the case of Roget's contribution, which gave a generally favourable account of morphological notions such as the 'unity of composition' or type espoused by Etienne Geoffroy Saint-Hilaire. While he concurred with the dominant teleological tradition of British natural theology in underlining the image of adaptive diversity, Roget contended that this *'law of variety'* was complemented and qualified by another principle: it was not blindly followed, but controlled by another law—'that of *conformity to a definite type'*.[54] He thus made a significant departure from the teleology of Cuvier—a key authority for many of the Bridgewater Treatises—for whom any resemblances in the structures of animals derived from the functions they performed under similar 'conditions of existence'.[55]

With Robert Knox, Martin Barry and William B. Carpenter, Roget suggested that there were certain features of animal structures—for example, vestigial organs—and certain parallels between them, which could not be explained by direct adaptation to function or environment, but could best be understood as permutations of a general plan or type.[56] But unlike Knox, whose contempt for the current vogue of natural theology caused him to speak of the 'Bilgewater Treatises',[57] he claimed that this morphological perspective strengthened the subject by allowing it to appeal to grand analogies throughout creation. He then proposed a view which incorporated both teleology and morphology, thereby anticipating the terms of reference for all future discourse within natural theology concerning the principles by which God had ordered the organic world:

> We have seen that, in constructing each of the divisions so established, Nature appears to have kept in view a certain definite type, or ideal standard, to which, amidst innumerable modifications, rendered necessary by the varying circumstances and different destinations of each species, she always shows a decided tendency to conform.[58]

53 P.J. Bowler, 'Darwinism and the argument from design: suggestions for a reevaluation', *Journal of the History of Biology*, (1977), **10**, pp. 29–43; R. Yeo, 'William Whewell, natural theology and the philosophy of science in mid-nineteenth century Britain', *Annals of Science*, (1979), **36**, pp. 493–516 (507–511); D. Ospovat, *The Development of Darwin's Theory: Natural History, Natural Theology and Natural Selection, 1838–1859*. Cambridge, 1981, pp. 7–33, 140–143; T. Lenoir, *The Strategy of Life: Teleology and Mechanics in Nineteenth Century German Biology*. Dordrecht, 1982; P.F. Rehbock, *The Philosophical Naturalists: Themes in Early Nineteenth Century British Biology*. Madison, 1983.

54 Roget, op. cit. (17), i, pp. 47–48.

55 Quoted in Ospovat, op. cit. (53), p. 34; also pp. 115–116.

56 For these ideas in British comparative anatomy during the early nineteenth century, see ibid., pp. 11–12, 17–23; Rehbock, op. cit. (53), ch. 3. For Roget's possible debt to Grant, see Desmond, op. cit. (36), pp. 215–216.

57 Quoted in Rehbock, op.cit.(53), p. 56.

58 Roget, op. cit. (17), ii, p. 627.

There were three problematical areas within natural theology in which morphological concepts could be deployed. In different ways, these related to the general problem of revealing design amidst the variety and diversity of creation, intelligent and benevolent purpose behind the plenitude of nature. Firstly, as Paley realized, in order to dispel sceptical suggestions about a plurality of designers rather than a single God, it was important to evince not only marks of design but definite evidence for a unity of design.[59] This had, of course, been one of the attractions of the chain of being, and writing in 1821, William MacLeay still spoke of the image of a progressive series as the most beautiful 'evidence of the truths of natural religion'; and claimed that 'until we can imagine ourselves acquainted with every possible production of this globe . . . naturalists can never be entitled to consider the chain of being as broken.'[60] However, the Bridgewater Treatise authors were generally suspicious of this notion and Kidd, quoting Cuvier, ruled out any attempt to 'class them [animals] so as to form a single series descending gradually from the higher to the lower'.[61] Roget also rejected the idea of a single, simple, continuous series, remarking that 'if, for the sake of illustration, we must employ a metaphor, the natural distribution of animals would appear to be represented, not by a chain, but by complicated net-work, . . . by circular or recurring arrangements'.[62] Here he mentioned MacLeay's 'quinarianism' and the concept of a 'unity of composition' which, he said, was being zealously pursued 'by many naturalists, of the highest eminence, on the continent'.[63] He believed that such perspectives, which stressed the analogies and symmetries pervading organic forms, offered better indication of a unity of design than the narrow teleological argument because they could show that apparent deviations, 'far from being arbitrary, are themselves referrable to particular laws'.[64]

Secondly, the problem of accounting for monsters or apparent aberrations from species posed difficulties for those committed to the notion of final cause. As we saw earlier, John Kidd betrayed the strains involved here when he suggested, rather weakly, that monsters might be adapted to some special, but unknown ends.[65] Furthermore, although proponents of this position emphasized the limits of deviations from normal species, they had no way of explaining the form which such deformities assumed. On the other hand, those who affirmed the idea of a morphological type as a governing principle,

59 Paley, op. cit. (21), pp. 482–487. For a contemporary recognition of the spectre of polytheism, see [G.R. Gleig], 'Lord Brougham on Natural Theology', Fraser's Magazine, (1835), 12, pp. 375–393(386).

60 W.S. MacLeay, Horae Entomologicae, London, 1819–1821, p. 169. MacLeay said that he had been told that his own theory of 'a chain of being returning into itself militates against those notions of an ascending scale of nature, which are inculcated by revelation'.(162).

61 Kidd, op. cit. (48), p. 333. Cuvier had asked: 'What law would have unnecessarily constrained the Creator to produce useless forms, merely to fill voids in a chain, which is only a speculation of the mind?' Quoted in Appel, op. cit. (50), p. 299. Adam Sedgwick still thought it necessary to attack the notion of a continuous chain of being as late as 1850. See A. Sedgwick, A Discourse on the Studies of the University of Cambridge, with additions, and a preliminary dissertation, 5th edn. London and Cambridge, 1850, pp. ccxxi–ccxxv. For an earlier rejection, see [Voltaire] The Philosophical Dictionary for the Pocket. London, 1765, pp. 72–73.

62 Roget, op. cit. (17), i, pp. 53–54.

63 Ibid., ii, p. 627.

64 Ibid., i, p. 50.

65 Kidd, op. cit. (48), p. 338.

with less stress on the criterion of adaptation, could use the phenomenon of monstrosities to their advantage. William Carpenter did this in his *Principles of General and Comparative Physiology* of 1839:

> The tendency to conformity to an ideal 'archetype' is frequently shown, in a most remarkable manner, by the occurrence of *Monstrosities*; which, though once regarded by men of science with feelings very little higher than those with which they are still looked-on by the vulgar, may now be considered as among the most interesting and suggestive of all the illustrations of 'Unity and Design;' since of these malformations, a considerable proportion are such in virtue of their closer conformity to the *general* model, those modifications of it which are characteristic of the *special* form not having been evolved.[66]

Thirdly, there was the question of man's place in nature. Although it was accepted that animal life had its own value and did not simply exist for human use—a position difficult to sustain after the disclosure of a prehistoric period—Christian religious convictions demanded a conception of man as the special object of Divine Providence. In concluding his treatise, Roget raised the question of whether man could be considered as simply another product of the plenitude which characterized the organic world. But it was impossible, he said, 'to conceive that this enormous expenditure of power . . . and this profusion of existence . . . can thus, from age to age, be prodigally lavished, without some ulterior end'.[67] Consequently, the vision of a progressive plan of creation, beginning with lower forms of life and culminating in man, had obvious attractions for writers of natural theology.[68] The problem here, of course, was that it was precisely this notion of a progressive plan, so consonant with the idea of man as the denouement of creation, which reinforced theories of transmutation. As Buckland explained in his treatise, the word 'Development' contained several levels of meaning,[69] and natural theologians were caught in the difficult position of needing to affirm organic progression and unity of design while at the same time rejecting any account of these phenomena by theories of descent. Roget realized this, and after giving a favourable account of recent morphological theories and the striking analogies on which they were based, stressed that there was still 'an impassable barrier of separation' between species, and castigated the extravagancies of 'continental physiologists', such as Lamarck.[70]

Thus, by the end of the 1830s, there were tensions associated with the problem of discerning the unifying design of a single Deity within the great diversity of living forms. Writers of natural theology were uncomfortable with earlier expositions of the principle of plenitude which celebrated this superabundance of nature as testimony to the notion that all conceivable forms of life exist. While they were enthusiastic about the image of a world in which all matter supported life, they firmly rejected the assumption that all conceivable or possible varieties of being should exist. They associated such a notion

66 W.B. Carpenter, *Principles of Comparative Physiology,* 4th edn., new title. London, 1854, p. 105. See also Powell, op. cit. (35), pp. 388–389; Roget, op. cit. (17), ii, p. 635.

67 Roget, op. cit. (17), ii, p. 640.

68 See ibid., ii, pp. 628–635 for Roget's interest in the 'law of Gradation' and embryology.

69 Buckland, op. cit. (16), i, p. 585.

70 Roget, op. cit. (17), ii, pp. 635–638. Referring to the discussion of transmutation in the *Philosopie Zoologique,* he remarked: 'If this be philosophy, it is such as might have emanated from the college of Laputa'. (638).

with atheistic or materialistic philosophies which substituted references to the energies or powers of nature—its fecundity, spontaneity or plasticity—for acknowledgements of a designing Deity. Thus, in dealing with the plenitude of nature they insisted that variety had limits, did not threaten the reality of species, and always existed for some adaptive or functional purpose. From this dominant teleological perspective, the splendid diversity of animal structures reflected the different environments in which they dwelt and the different functions they performed. However, those who favoured morphological theory argued that the concept of special adaptation to environment could not sufficiently account for the phenomena of the organic world, and claimed that in restricting itself to a narrow teleology, the subject of natural theology failed to provide the strongest evidence for design.[71]

The comparative anatomy of Richard Owen played an important part in mediating between these positions. After reading Whewell's *History*, Owen wrote saying that only the present narrow basis of observation prevented a harmonious combination of 'the transcendental and teleological views';[72] and in his major works he claimed that a strictly teleological theory could not account for the similarities which existed among animal forms. In 1849, Owen argued that 'if we reject the idea that these correspondences are manifestations of some archetypal exemplar on which it has pleased the Creator to frame certain of his living creatures, there remains only the alternative that the organic atoms have concurred fortuitously to produce such harmony'.[73] After thus raising the spectre of Democritus, Owen affirmed that the discovery of this archetype of the vertebrate skeleton provided the grandest evidence of Divine design. But, significantly, he did not totally reject the notion of final cause, and incorporated a teleological factor to explain adaptation to particular circumstances. It was this compromise which appealed to writers of natural theology, and Owen's scientific work was seen as an illustration of the interdependence of teleology and morphology.[74] But in spite of the shifts which had occurred since the 1830s, there were still tensions between these perspectives. There were two sources for these: firstly, references to unity of type or plan were not generally accepted as evidence of design in the same sense as examples of purposive adaptation;

71 For opposing positions on this issue in the 1830s, see Whewell, op. cit. (35), iii, pp. 456–478 and B. Powell, *The Connexion of Natural and Divine Truth*. London, 1838, pp. 128–135.

72 R. Owen to W. Whewell, 31 October 1837, Whewell Papers, Trinity College, Cambridge, Add.MS.a.210⁵⁴.

73 R. Owen, *On the Nature of Limbs*. London, 1849, pp. 40, 85–86. On Owen, see R.M. MacLeod, 'Evolutionism and Richard Owen, 1830–1868: an episode in Darwin's Century', *Isis*, (1965), 56, pp. 259–280. For the institutional and ideological context of Owen's work, see A. Desmond, *Archetypes and Ancestors: Paleontology in Victorian London 1850–1875*. London, 1982 and 'Richard Owen's reaction to transmutation in the 1830s', *British Journal for the History of Science*, (1985), 18, pp. 25–50.

74 For the shift in Whewell's attitude, compare op. cit. (35), iii, pp. 456–478 and *History of the Inductive Sciences*, 3rd edn. London, 1857, iii, pp. 553–562. Whewell believed that Owen offered a version of transcendental theory suitable to a British audience and referred to his 'carefully chosen language' (553). For other favourable responses, see Powell, op. cit. (35), pp. 135–137 and J. McCosh and G. Dickie, *Typical Forms and Special Ends in Creation*. Edinburgh, 1856.

and secondly, theories which emphasized morphological analogies and homologies in organic nature were often seen to be linked with theories of transmutation.[75]

V

The issue which produced the most explicit confrontation between teleological and morphological ideas of design was that of the plurality of worlds. The doctrine of intelligent life on other planets became the subject of a public controversy between David Brewster, who defended what he took to be a popular orthodoxy, and William Whewell, who challenged it. The positive side of this question was largely constituted by assumptions about the manifestations of plenitude in creation; but furthermore, these were linked to arguments about final causes. In order to deny the plurality of worlds, a notion which he regarded as an ally of speculation about transmutation, Whewell was forced to qualify teleological inferences; in so doing he appealed to morphological interpretations usually associated with evolutionary theories. Since it was central to the subject of the controversy, the principle of plenitude was deeply implicated in a dispute which set the two concepts of design against each other and exposed the weakness of natural theology in dealing with the fundamental issue of man's place in nature.[76] Because the principle of plenitude—and in particular the corollary of the plurality of worlds—had entered natural theology, any attempt to assert the Christian notion of man's special relationship to God revealed the tensions between the intellectual vocabulary of this discourse and its apologetic function.[77]

Since the seventeenth century, the idea of other intellectual worlds was also coupled with an explicit qualification of narrow anthropocentrism: not all features of God's creation, either on earth or in the heavens, were made for man's use.[78] But the habitation

75 See, for example, W. Whewell, *The Philosophy of the Inductive Sciences*, 2 vols. London, 1840, i, p. 629 for the statement that: 'The regular form of a crystal, whatever beautiful symmetry it may exhibit . . . does not prove design in the same manner in which design is proved by the provisions for the preservation and growth of the seeds of plants, and of the young of animals'. Sedgwick wrote to Louis Agassiz expressing anxiety about 'the opinions of Geoffroy St. Hilaire and his dark school', warning that these would 'shut out all argument for *design* and all notion of a Creative Providence'. A. Sedgwick to L. Agassiz, 10 April 1845, in J. Clark and T. Hughes (eds.): *Life and Letters of the Reverend Adam Sedgwick*, 2 vols. Cambridge, 1890, ii, p. 86. For the connection between morphology and transmutation in the contemporary debates, see Powell, op. cit. (35), p. 392.

76 For an extensive study of this controversy which argues that Whewell's rejection of the plurality of worlds was a disguised attack on theories of transmutation, such as that of Robert Chambers, see J.H. Brooke, 'Natural Theology and the Plurality of Worlds: observations on the Brewster–Whewell Debate', *Annals of Science*, (1977), 34, pp. 221–286. For the connection with Whewell's epistemology, see Yeo, op. cit. (53). See also W.C. Heffernan, 'The singularity of our inhabited world: William Whewell and A.R. Wallace in dissent', *Journal of the History of Ideas*, (1978), 39, pp. 81–100.

77 In advising Whewell, Sir James Stephen predicted that 'the habit of mankind, would be to disbelieve the Evangelists, rather than to disbelieve the Natural Theologians'. Stephen to Whewell, 10 November 1853, Whewell Papers, Trinity College, Cambridge, Add.MS.a.216[142]. For comments on these tensions, see Brooke, op. cit. (76), pp. 232, 236–237, and P.J. Bowler, 'Sir Francis Palagrave on Natural Theology', *Journal of the History of Ideas*, (1974), 35, pp. 144–147. For a criticism of Whewell from a Christian perspective, see: 'Of the Plurality of Worlds', *British Quarterly Review*, (1854), 20, pp. 45–48 (81–82).

78 Dick, op. cit. (13), ch. 6; also G.M. McColley, 'The seventeenth-century doctrine of a Plurality of Worlds', *Annals of Science*, (1936), 1, pp. 385–430; D.M. Knight, 'Celestial worlds discover'd', *Durham University Journal*, (1965), 27, pp. 23–29.

of these worlds was still rationalized by the concept of final cause: other planets did not exist for their own sake but in order to support life. In this way, the principle of spatial plenitude—the notion that all possibly habitable parts of the universe were filled with life—was fortified by teleological reasoning: other planets had been created for the use of intelligent populations.

Brewster's defence of the plurality of worlds took this form. In his view, the plenitude of creation was synonymous with the axiom that matter was made for life. To contemplate the prospect of uninhabited planets was to imply that they had been made in vain; if other worlds were devoid of life they would exist without purpose or final cause, a situation incompatible with the idea of a rational creation:

> In peopling such worlds with life and intelligence we assign the cause of their existence; and when the mind is once alive to this great truth, it cannot fail to realize the grand combination of infinity of life with infinity of matter.[79]

Whewell offered a perceptive summary of these assumptions:

> It is sometimes said, that it is agreeable to the goodness of God, that all parts of the creation should swarm with life: that life is enjoyment; and that the benevolence of the Supreme Being is shewn in the diffusion of such enjoyment into every quarter of the universe. To leave a planet without inhabitants, would, it is thought, be to throw away an opportunity of producing happiness.[80]

In order to counter this position he argued that the extreme principle of plenitude which it embodied ignored the actual state of affairs on Earth; it overlooked the large areas of this globe which did not contain life. But he also drew attention to the ways in which the superfecundity of nature routinely involved waste (in the sense of unrealized potential) and abortive design.

> To work in vain in the sense of producing means of life which are not used, embryos which are never vivified, germs which are not developed, is so far from being contrary to the usual proceedings of nature, that it is an operation which is constantly going on, in every part of nature.[81]

79 D. Brewster, *More Worlds than One: the Creed of the Philosopher and the Hope of the Christian*, new edn. London, 1867, p. 183; first published in 1854. See also T. Dick, *Celestial Scenery; or the Wonders of the Planetary System Displayed*. London, 1871, p. 343 for the remark that 'so far as we are able to penetrate, it appears demonstrable that matter exists chiefly, if not *solely*, for the sake and convenience of sensitive and intelligent beings'. Yet it should be noted that this strong nexus between matter and life had already been questioned by the pre-organic eras revealed by geology. Buckland admitted this in his Bridgewater Treatise, op. cit. (16), i, pp. 56–57, and Whewell capitalized on it in his argument with Brewster.

80 [W. Whewell] *Of the Plurality of Worlds: an Essay. Also a Dialogue on the Same Subject.* 2nd edn. London, 1854, p. 334.

81 Ibid., p. 330. Compare the following passage from Darwin's *Origin of Species*, 1859, ed. J.W. Burrow, Harmondsworth, 1968, p. 116, cited in Gale, op. cit. (25), p. 329: 'We behold the face of nature bright with gladness, we often see superabundance of food; we do not see, or we forget, that the birds which are idly singing round us mostly live on insects or seeds, and are thus constantly destroying life'. Hume had exploited this theme in his attack upon natural theology: 'What an immense profusion of beings, animated and organized, sensible and active! You admire this prodigious variety and fecundity. But inspect a little more narrowly ... The whole presents nothing but the idea of a blind nature, impregnated by a great vivifying principle, and pouring forth from her lap, without discernment or parental care, her maimed and abortive children!' D. Hume, *Dialogues Concerning Natural Religion*, 1779, (ed. H.D. Aiken), New York, 1948, p. 79.

Whewell's emphasis on these prodigal and abortive features of creation was a challenge to the optimistic image of nature usually associated with the principle of plenitude. It also indicated a shift in the strategy for dealing with the unwanted implications of this assumption, within writings on natural theology. While authors such as Bell and Kidd had constrained plenitude by tying it to utility in order to protect the idea of fixed species, Whewell stressed the waste which it entailed in order to assert the uniqueness of life on earth. Furthermore, whereas other natural theologians had been anxious about the affinities between the principle of plenitude and various materialistic or pantheistic philosophies, it was Whewell, in his denial of the plurality of worlds, who was seen as the effective antagonist of such tendencies. Although most notices of Whewell's book were critical, there were two reviews which praised him for reasserting a Christian view of God's power over nature. In defending Whewell, the Catholic periodical, the *Rambler,* charged Brewster with materialist leanings because he had treated matter as something that made demands upon the Deity, something that had to be filled with life; he was therefore seen as a member of the school which 'think matter is a self sufficient being, over which God has plastic power to form, but not substantial power to create or annihilate'.[82] Similarly, the *Eclectic Review* welcomed Whewell's argument as 'a timely blow to [the] Nature-worship' of Carlyle, Emerson, Combe and Chambers. In stressing examples of waste and abortive design, Whewell had shown that Nature was not perfectly formed, not a living animal, but rather 'clay in the hands of an almighty Potter';[83] and in doing so he underlined the need to worship God rather than nature, to profess Christianity not Pantheism.

These comments point to tensions between aspects of Christian doctrine and the neo-Platonic assumptions about the natural world which had entered the discourse of natural theology. Whewell felt it necessary to attack the principle of plenitude, which legitimized a plurality of worlds, in order to safeguard the Christian idea of man's unique relationship with God:

> The thoughts of Rights and Obligations, Duty and Virtue . . . are thoughts which belong to a world, a race, a body of beings, of which any one, with the capacities which such thoughts imply, is more worthy of account, than millions and millions of mollusks and bellemnites, lizards and fishes, sloths and pachyderms, diffused through myriads of worlds.[84]

But in doing so he was forced to deny many of the features of creation which had been celebrated as marks of design in natural theology.

A critique of the traditional concept of final cause was also central to Whewell's case against the plurality of worlds. In exposing the connection between plenitude and waste, he was indicating that there were aspects of creation which did not betray design in the

82 'The plurality of worlds', *Rambler,* (1854), II, 2, pp. 129–137 (135–136).
83 'The plurality of worlds', *Eclectic Review,* (1854), II, 2, pp. 513–531 (527–528).
84 [Whewell] *Of the Plurality of Worlds.* 1853, printer's copy containing chapters cancelled from published work, Whewell Papers, Trinity College, ADV.C.16.27, 245. Sedgwick appears to have thought Whewell's book might encourage Pantheism. Whewell could not see the logic of this objection and protested that: 'The proper antithesis to Pantheism is not intelligent creatures besides man, but a Divine Mind'. Whewell to Sedgwick, 8 June 1854, in J. Stair Douglas, *The Life and Selections from the Correspondence of William Whewell,* 2nd edn. London, 1882, p. 434.

sense of purposive or functional utility. Yet in the Bridgewater Treatises the affirmation of this strict teleological notion of design was regarded as essential to the doctrine of the immutability of species. As we saw earlier, Whewell himself endorsed this position in 1837 in his *History*. Like the Bridgewater authors, he too wanted to reject theories of transmutation, but believed that such theories were thought to be strengthened by the popular belief in a plurality of worlds—an association made in the anonymous *Vestiges of the Natural History of Creation* which appeared in 1844.[85] In rejecting the proposition in his own anonymous work of 1853, Whewell attempted to assert the uniqueness of intelligent life on earth,[86] and therefore had to confront the teleological thinking which postulated life as the *raison d'etre* of matter. It was in this context that he appealed to Owen's work, because it allowed him to claim that the traditional concept of final cause could not provide a comprehensive account of creation. Repeating Owen's critique of teleological explanations in comparative anatomy, Whewell stated that

> many parts of the structure of animals, though adapted for particular purposes, are yet framed as a portion of a system which does not seem, in its general form, to have any bearing on such purposes.[87]

Once this particular lack of practical purpose was acknowledged, it became possible to assert that the planets need not support life: their existence could be part of a wider plan independent of anthropomorphic notions of utility. But furthermore, the morphological concept of design which Owen's work had successfully promoted allowed Whewell to contend that there were other principles which pervaded the universe—laws of symmetry, resemblance, analogy and beauty. He remarked that

> there may be large portions of the Creation in which we cannot trace any design for the good of sentient beings; and in which we must suppose that the symmetry and beauty and variety which exist, exist on their own account, and as a manifestation of Law directed to no purpose such as we can understand.[88]

After stressing the manifestation of such general laws, he argued that these must have been known to God before being actualized, and that in discerning such laws, man shared an affinity with 'the Divine Mind of the Creator'.[89] For Whewell, this ability of man to share in Divine thoughts, including the perception of his own place in the plan of creation, provided a striking legitimation of the uniqueness of intellectual life on this planet:

85 In his own Bridgewater Treatise, *Astronomy and General Physics Considered with Reference to Natural Theology*, London, 1834, pp. 269–271, Whewell had entertained the likelihood of the plurality of worlds. For the scientific reaction to '*Vestiges*', see R. Yeo, 'Science and intellectual authority in mid-nineteenth-century Britain: Robert Chambers and *Vestiges of the Natural History of Creation*', *Victorian Studies*, (1984), 28, pp. 5–31.

86 Whewell did admit the possibility of *animal* life on other planets but not the existence of other moral, intellectual beings. Whewell, op. cit. (80), p. 112, 118.

87 Whewell, op. cit. (80), p. 318.

88 Whewell, op. cit. (84), p. 269.

89 Whewell, op. cit. (80), p. 360; also pp. 363–364.

we have, in this, a reason which may well seem to us very powerful, why, even if the Earth alone be the habitation of intelligent beings, still, the great work of Creation is not wasted.[90]

Whewell thereby rejected the principle of spatial plenitude, but, recognizing the appeal which it commanded, proceeded to supply an alternative. He suggested that the existence of life was not the only manifestation of providence; it could be equally indicated by the action of physical forces and laws which God sustained. In this sense, he claimed, even 'the remotest planet is not devoid of life, for God lives there'.[91]

Whewell's references to non-teleological features of creation went beyond references to symmetry, analogy and order. He contended that there were aesthetic values informing man's response to the sublime (and irregular) beauty of nature. Having proposed this additional, and profound, level of empathy with the Divine mind, Whewell felt confident in affirming the relationship between man and God, above the principle of plenitude, as the most significant aspect of creation. Dignity and sublimity, he suggested, 'appear especially to belong to the larger objects, which are destitute of conscious life; as the mountain, the glacier, the pine-forest, the ocean; since in this, we are, as it were, alone with God, and the only present witnesses of His mysterious working'.[92] In seeking to affirm the uniqueness of man's status, Whewell thus exorcized the assumptions about plenitude apparent in earlier works of natural theology. Whereas Paley found purpose in the abundance of mice and gnats in barren places, Whewell felt compelled to ask

> whether the dignity of the Moon would be greatly augmented if her surface were ascertained to be abundantly peopled with lizards; or whether Mount Blanc would be more sublime, if millions of frogs were known to live in the crevasses of its glaciers.[93]

VI

This article has focused on the metaphysical assumption which Lovejoy called the principle of plenitude. While not committing itself to the programme of a 'history of ideas' outlined by Lovejoy, it has attempted to study the trajectory of this assumption in a period beyond the compass of *The Great Chain of Being,* and in the particular framework of the discourse of natural theology. This subject sought to evince the marks of Divine design and power in the natural order, and in doing so, referred to the fertility and abundance of living things. However, the earlier formulations of the principle of plenitude, as presented by Lovejoy, appear to have undergone significant modulation in the early nineteenth century. I have argued that in the context of anxieties about theories of transmutation, there was a rejection of what may be called conceptual plenitude and a continuing acceptance of spatial plenitude. This sensitivity of natural theologians to

90 Ibid., p. 364. Owen had in fact inferred the possibility of different forms of life on other planets from the non-existence of possible permutations of the archetype on earth. And in contrast with his other qualifications of teleology, Owen suggested that the support of life was 'the only conceivable purpose' of the planets. See Owen, op. cit. (73), pp. 83–84.

91 Whewell, op. cit. (80), p. 366. For a similar notion, see A. Sedgwick, *Discourse on the Studies of the University of Cambridge,* 3rd edn. Cambridge, 1834, pp. 106–107.

92 Whewell, op. cit. (80), p. 365.

93 Ibid., p. 366.

ideas about the protean powers of nature and the plastic and diverse character of its forms reflected their need to defend the immutability of species, a notion closely associated with the conceptions of teleological design which they espoused. There were attempts to constrain the concept of plenitude by subordinating it to teleology, explaining the variety of organic life in terms of its adaptive purpose, and its extent in terms of the proposition that matter was made to sustain life. This was one approach to the problem of delineating the plan and purpose of the natural world. But there were problems confronting this approach, at both scientific and theological levels, which related to the features of creation associated with the principle of plenitude: for example, the phenomena of monsters and aberrations, the variety of form and function, the possibility of a plurality of designers, and the status of man amidst the profusion of organic life.

The alternative morphological theory, present in Roget's Bridgewater Treatise, dealt with such issues by embracing the phenomena of variety and diversity within transcendental concepts such as archetype and unity of composition, which went beyond the concentration on adaptive utility and final cause. Throughout the second quarter of the century there was debate about the relationship between the teleological and morphological concepts of design within natural theology. Several important writers argued for the complementary character of this relationship, but in the controversy over the plurality of worlds—a doctrine in which plenitude was the constitutive assumption— those two concepts were opposed in a dispute which involved the integrity of natural theology and its account of man's place in God's creation. Whewell's argument against the plurality of worlds involved him in a critique of both the principle of spatial plenitude and the traditional notion of design. He exposed the ways in which plenitude was not always realized, and the cases in which its manifestations were inconsistent with the notion of final cause.[94] In challenging the sufficiency of teleological accounts of creation, he appealed to the morphological theories which emphasized the symmetry and order of nature. This perspective was compatible with his attempt to affirm the unique character of man's mind which, in discerning such general laws, indicated its special relationship with the Deity. It is both ironic and significant that in seeking to promote a non-evolutionary account of man's place in nature, Whewell was forced to qualify teleological conceptions of design and to embrace morphological notions often associated with theories of transmutation. This situation underlined the potential conflict between Christian doctrine and the neo-Platonic assumptions about the world which had entered the discourse of natural theology. Although the eighteenth-century concept of plenitude had been severely qualified by early nineteenth-century writers concerned with the question of species, Whewell found it necessary to make an explicit rejection in order to assert the uniqueness of man. But in doing so, he revealed the extent to which the principle of plenitude had become implicated in serious tensions within natural theology over the key issues of design and the place of man in nature.

94 Heffernan, op. cit. (76), p. 96 remarks that 'the dissenters [the opponents of the plurality of worlds] were moved by a belief in the intricate design of creation and the sufficiency of man as an end to himself'. But in Whewell's case, as argued above, the affirmation of man's uniqueness was linked with some severe qualifications of the notion of design.

Genius, Method, and Morality: Images of Newton in Britain, 1760–1860

The Argument

Focusing on the celebrations of Newton and his work, this article investigates the use of the concept of genius and its connection with debates on the methodology of science and the morality of great discoverers. During the period studied, two areas of tension developed. Firstly, eighteenth-century ideas about the relationship between genius and method were challenged by the notion of scientific genius as transcending specifiable rules of method. Secondly, assumptions about the nexus between intellectual and moral virtue were threatened by the emerging conception of genius as marked by an extraordinary personality – on the one hand capable of breaking with established methods to achieve great discoveries, on the other, likely to transgress moral and social conventions. The assessments of Newton by nineteenth-century scientists such as Brewster, Whewell, and De Morgan were informed by these tensions.

No Newton, by silent meditation, now discovers the system of the world from the falling of an apple; but some quite other than Newton stands in his Museum, his Scientific Institution, and behind whole batteries of retorts, digesters and galvanic piles imperatively "interrogates Nature," – who, however, shows no haste to answer.

Thomas Carlyle, "Signs of the Times" (1829)

. . . a work [the *Principia*], may we not add, which would be read with delight in every planet of our system, – in every system of the universe.

David Brewster, *Memoirs of the Life, Writings, and Discoveries of Sir Isaac Newton* (1855)

Introduction

Having immersed himself in the nonmathematical section of Newton's unpublished writings contained in the Portsmouth Papers, John Maynard Keynes marked the tercentenary of the great man's birth by overturning previous images of his work and character:

258

Newton was not the first of the age of reason. He was the last of the magicians, the last of the Babylonians and Sumerians, the last great mind which looked out on the visible and intellectual world with the same eyes as those who began to build our intellectual inheritance rather less than 10,000 years ago. Isaac Newton, a posthumous child born with no father on Christmas Day, 1642, was the last wonderchild to whom the Magi could do sincere and appropriate homage. (Keynes 1972, 4:363–64)[1]

In the period from his presidency of the Royal Society in 1703 to the late eighteenth century, Newton was almost universally lauded as a man of genius and high character who had established observation and experiment as the surest access to truth about the natural world. By the time Keynes was writing in 1942, this picture had become far more complex. The crucial period for this transformation was the first half of the nineteenth century, a period in which most of the documents and archives central to our present understanding of Newton became the object of scholarship. But these were read by men more immediately influenced by Enlightenment accounts of Newton's mind and by their own convictions about the close relationship between good science, sound religious beliefs, and proper moral attitudes. Their reassessment of Newton converged with significant debates concerning the nature of science and its most successful practitioners. Although the earlier image of Newton was not entirely discarded in the nineteenth century, the fundamental issues about his intellect and character were raised. This is why Keynes found it easier to read the contents of Newton's box of papers than those, like Sir David Brewster, who had opened it before him.[2]

This paper takes two areas of debate – the relationship between genius and method, and the association of genius and morality – as the framework for a consideration of changing images of Newton from the mid–eighteenth to the mid–nineteenth century. It does not attempt a comprehensive survey but rather focuses mainly on the debates engendered by the scholarly historical interest in the heroes of the scientific revolution that began to emerge in Britain during the early decades of the nineteenth century. This will involve discussion of the views of William Whewell, David Brewster, Stephen Rigaud, Francis Baily, and Augustus De Morgan, their interpretations of earlier images of Newton, and their disputes about his scientific and personal reputation. The two major themes – the relationship of scientific genius to method and to morality – could justify treatment in separate papers; the aim here will be to deal with the significant inflections that appeared in each theme and to offer some comment on the links between them.

Born on Christmas Day 1642 – in the year of Galileo's death – Newton prepared for himself a symbolic place in both the Christian calendar and the hagiography of the Enlightenment. Although the precise origin and development of the elements

---

[1] On this theme, see also McGuire and Rattansi 1966; Webster 1982.

[2] See Cohen 1960 for a survey of views to that date and a short history of the Portsmouth Papers. For nineteenth-century scholarship prior to Brewster's use of these papers, see Rigaud 1838, 1851.

that constitute the Newtonian mythology are not always clear,[3] the celebration of his name and works throughout the eighteenth century is well known. In the poetry of Pope and Young, in the memoirs of Conduitt, Stukeley, and Pemberton, in the collected anecdotes of Spence, in the eulogy of Fontenelle, and in the various notices in dictionaries and encyclopedias of the second half of the century, Newton was represented as the epitome of intellectual sagacity. His works in astronomy and optics, although arguably different in their methodologies, were cited as the model of experimental natural philosophy. Chambers' *Cyclopaedia* of 1728 explained in its entry for "Newtonian Philosophy" that this term was applied in different senses but was largely synonymous with "*Mechanical and Mathematical* Philosophy" (Chambers 1728, 2:628). Furthermore, Newton's method, his "*Experimental* Philosophy" (ibid.), was favorably contrasted with the erroneous system of Descartes (ibid.).[4] Newton was celebrated as a scientific genius, and at least from the time of Stukeley's account of his life, there was a fascination with childhood hints of this divine gift (Stukeley 1936).[5]

Intimately connected with such acknowledgments of his intellectual abilities was an affirmation of his moral character. Conduitt established the formula for this when he attested that Newton's "whole life was one continuous series of labour, patience, charity, generosity, temperance, piety, goodness, and every other virtue, without a mixture of any known vice whatsoever." The inscription on his tomb in Westminster Abbey gave a more pithy version of this theme: "Let Mortals congratulate themselves, That so great an Ornament of Human Nature has existed."[6] This dual paean to Newton was still very much the norm at the close of the eighteenth century and was thus available for exploitation at the time when biography became widely used as a vehicle of popular instruction.

Whereas mid-eighteenth-century editions of encyclopedias carried no biographical entries, later editions often included material on the lives and personal qualities of the major scientific, philosophical, and literary figures. A special example of this genre was the text, aimed at a popular audience, which narrated the lives of men of genius, often giving special attention to their childhood. The factors behind the emergence of this form of writing and its readership are beyond the scope of this paper; but it is significant that Newton was a ubiquitous figure in such texts, personifying the intellectual and moral lessons that could be drawn from a life devoted to science. The social and moral value of science had been a strong theme in works of popular instruction during the latter half of the eighteenth century: the many editions of *The Newtonian System of Philosophy, adapted to the capacities of young gentlemen and ladies,* by Tom Telescope, presented astronomy, optics, and

---

[3] For comments on Newton's apple and other elements, see McKie and De Beer 1951–52; Gjertsen 1986.

[4] For a European account of *Newtonische Philosophie* see Zedler 1740, 24:414–15. For the construction of Newtonianism, see Guerlac 1965.

[5] For the comment that the interest in childhood was far less obvious than it later became in the romantic period, see Delaney 1969, 152–57.

[6] Cited in *British Biography* 1778, 7:154.

other subjects as a refined form of leisure and amusement (for a fine contextualiza-
tion for this book see Secord 1985). The facts and laws of nature disclosed by science
were cast within the discourse of natural theology, and their moral and theological
relevance were exhibited. Similarly, but with a focus on the character of the man of
science, biographical tracts on the geniuses, heroes, and martyrs of science sought to
show that moral virtue and intellectual achievement were closely linked in the lives of
eminent men. This kind of work was quite common in Britain during the first half of
the nineteenth century and was no doubt reinforced from the 1820s by groups such as
the Society for the Diffusion of Useful Knowledge and the movement for popular
education which it reflected (Hays 1964). One earlier example, *Buds of Genius,*
employed the device of a dialogue between an adult and a child. The chapter on
Newton is interesting because it illustrates the didactic aims of this genre. Thus the
child (Henry) asks: "Was he a *good man,* as well as a great philosopher?" Mamma:
"His temper was so mild and amiable, that scarcely any accident could disturb it. . . .
He was a truly excellent character; and his modesty was as remarkable as his genius.
In his speaking of his discoveries, he once observed: 'If I have done the public any
service this way, it is due to nothing but industry and patient thought'." Henry: "But I
think, in that opinion he was much mistaken. If I were to be take as much pains as Sir
Isaac Newton did, I do not think I should be able to make any discovery at all. You
know, Mamma, he was a *great genius*" (Anon. 1816, 27–30). Here the equation
between genius and morality is affirmed, but there is a recognition that only the latter
could be emulated by most people.

   In revealing this asymmetry, the text draws attention to the problematic nature of
the concept of genius and its various interpretations in Western thought. Most
discussions of the term occurred in literary contexts, where there was a focus on the
uneasy relationship between an idea of the spontaneous creativity of genius and the
classical rules and models of poetic composition (Tonelli and Wittkower 1968–73).
From the time of Newton, the notion of *scientific* genius marked an addition to this
discourse and became an object of poetic and didactic writing. But with it came a
heightening of previous tensions. Since so much of the rhetoric of the early scientific
community stressed the importance of the rules and methods of scientific practice –
such as those laid down by Bacon – the concept of a scientific genius required delicate
exposition. It is not surprising, then, that since the earliest celebrations of Newton's
work there have been various and shifting formulations of its implications for the
method by which ·progress in science occurs, ranging from eighteenth-century
attempts to align Bacon and Newton, to later conceptions of scientific genius as
incompatible with any specifiable rules of thought or practice. Secondly, the association of intellectual and moral virtue was extremely important in the presentation of
the scientific endeavor as supportive of orthodox religion and approved social and
ethical behavior. This was especially the case in Britain, whereas on the Continent
moral behavior could be defined in terms of an enlightened crusade of reason against
the forces of obscurantism, both political and clerical. In the case of Newton as an
exemplar of the natural philosopher, moral character was illustrated in various ways

but occupied a key position in both English and European writing. While the link between genius and moral virtue was not uncontested during the eighteenth century, it became far more contentious after the impact of romanticism and its celebration of the unique and often unconventional qualities of the inspired creative individual. In this context, the promotion of Newton as the epitome of scientific genius meant that the evidence in support of his moral character had to be carefully marshalled. This became increasingly delicate with the growing interest in the connection between genius and insanity in early-nineteenth-century medical and psychological literature.

Genius and Method

Historians of eighteenth-century literary criticism have discussed the emergence of the concept of artistic genius and its impact on aesthetic theory. The story they tell is one of considerable conflict between classical notions of rules and models of good art and the idea of spontaneous and inspired creativity which ignores accepted conventions. This opposition reached a peak in the statements of the leading figures of the Romantic movement in both Europe and Britain, but was apparent from at least the 1750s. In 1759 the English writer Edward Young identified the idea of "natural genius" with the concept of originality, and according to M. H. Abrams, almost totally abandoned "the traditional rhetorical framework of art, with its emphasis on study, example, precept, and the skillful manipulation of means to end" (1953, 199). Young employed the metaphor of organic and mechanical processes in a manner which anticipated later romantic writing:

> An *Original* may be said to be of a *vegetable* nature; it rises spontaneously from the vital root of Genius; it *grows,* it is not *made. Imitations* are often a sort of *Manufacture* wrought up by those *Mechanics, Art,* and *Labour,* out of pre-existent materials not their own.[7]

Some late-eighteenth-century writers could still be optimistic about bringing this activity of the mind within definite laws akin to those described by Newton. Thus in 1776 the Scottish philosopher James Beattie argued that it was no less absurd for a poet to "violate the *essential* rules of his art" than for a mechanic to "construct an engine of principles inconsistent with the laws of motion . . ." (quoted in Abrams 1953, 159). But when Kant addressed contemporary aesthetic issues in 1792 he made a clear contrast between genius and methodical rules:

> It [genius] cannot indicate scientifically how it brings about its product, but rather gives the rule as nature. Hence, when an author owes a product to his genius, he does not himself know how the *ideas* for it have entered his head, nor has he it in his power to invent the like at pleasure, or methodically, and communicate the

---

[7] Young 1966, 12; also p. 28 on genius vs. rules and p. 76 on Newton as an "Original."

same to others in such precepts as would put them in a position to produce similar products. (Quoted in Abrams 1953, 207)

However, it would be misleading to speak of a simple contrast between an eighteenth-century and a later, romantic, conception of genius. In fact, Young's work should be regarded not as a freakish anticipation of nineteenth-century romanticism, but as part of a new analysis of genius dating from the 1760s (Engell 1981). Two important English works – William Duff's *An Essay on Original Genius* (1767) and Alexander Gerard's *An Essay on Genius* (1774) – both followed Young in stressing the role of imagination in the activities of genius. Significantly, these writers advanced a general notion of genius encompassing both science and the arts. All three mentioned Newton as displaying the "Original Philosophic Genius" which leads to "new discoveries in science." Although both Duff and Gerard acknowledged a difference between scientific and artistic genius, they did not accept the formulations which made judgment the defining feature of scientific thought, and imagination the essence of artistic creation. Gerard spoke of *penetration* and *brightness* as the distinctive modes of genius in science and the arts respectively, but still saw "force of imagination" as crucial in the process of scientific discovery (1774, 8, 322–26). Similarly, Duff asserted that "vigorous and extensive powers of Imagination are indispensably necessary to . . . the researches of Science." Referring to the "generally imagined" view that judgment was the principal ingredient of philosophic genius, he remarked that "this opinion strikes at the foundation of our theory." Duff was careful to admit that patient observation and experiment by those "enjoying a small share of imagination" was a proven method in natural philosophy. However, a mind with an "extensive Imagination" and thus the ability to combine ideas in "a greater variety of forms" could expand the range and depth of inquiry. From this perspective, Newton was "an original Genius of the first rank," whose work combined imagination, reason, and intuition (Duff 1767, 93–94, 99, 119–20).[8]

These theoretical reflections about genius in science as well as the arts do not seem to be represented in the contemporary veneration of Newton. For although his work in astronomy and optics was poeticized (Nicolson 1946; Jones 1966), there does not appear to have been a debate, analogous to that in poetry, about the respective roles of creative imagination and conventions or rules in the attainment of his scientific results. This may be due to the fact that most writers sought to refer Newton's achievement to the efficacy of the inductive method outlined by Bacon. Thus, while the exceptional character of his thought was exalted, the notion of genius operating in a mode that could not be reduced to rules was strongly resisted. This left two apparently contradictory themes: the early acknowledgment that very few people were capable of following the demonstrations in the *Principia,* and the claim that Newton had exemplified rules of scientific practice which should be adopted.

[8] It is important to note here that these writers did not associate genius with insanity, although they did warn about the dangers of excessive imagination. On Georgian views of genius, see Porter 1987, 100–104.

The first theme was the starting point of the attempts to explain Newton's work to larger audiences – notably, the efforts of Voltaire and Pemberton. In the Preface to *The Elements of Sir Isaac Newton's Philosophy* (1738), Voltaire said that it was aimed at those with "not enough Mathematical learning to read" Newton's works; he also remarked that this philosophy "has hitherto seemed to many as unintelligible as that of the Ancients."[9] The notion of Newton's "celestial genius" as a divine gift to mortals – epitomized in the poetry of Alexander Pope and Edward Young – was often accompanied by comments on the gulf between his patterns of thought and those of other men.[10] Thus in his eulogy, Fontenelle speculated that: "If those Beings that are superior to Man have likewise a progression in knowledge, they fly whilst we creep, and leap over those *mediums* by which we proceed slowly and with difficulty, from one Truth to another that has a relation to it." He went on to suggest that if Leibniz did take the calculus from Newton, "he at least resembled Prometheus in the fable, who stole fire from the Gods to impart it to Mankind" (Fontenelle 1728, 4, 7).

But in spite of this perception, most works of this kind affirmed the possibility of deriving both philosophical and practical precepts from Newton's achievement. At the most general level Newton was said to have shown the folly of proceeding solely by conjecture, a method allegedly employed by Descartes in his erroneous theory of vortices. In his *View of Sir Isaac Newton's Philosophy,* Henry Pemberton regarded it as a work that broke with the previous tendency in natural philosophy "to form conjectures." In this respect he saw it as a vindication of Bacon's theory of knowledge and produced a formulation that combined the potential tension between genius and method. After describing Bacon's doctrine of errors, Pemberton remarked that "what surprising advancements in the knowledge of nature may be made by pursuing the true course in philosophical inquiries; when these searches are conducted by a genius equal to so divine a work, will be best understood by considering Sir Isaac Newton's discoveries . . ." (1728, 2–13). In one of the earliest accounts of Newton's four rules of philosophizing, Pemberton explicitly aligned his methodology with that of Bacon. This alliance survived throughout the eighteenth century, with Newton's much-quoted injunction against the role of hypotheses being used to legitimate the continuity between his position and that of Baconian induction. This was "strongly endorsed" by the Scottish philosopher Thomas Reid, who claimed that "in the third book of his *Principia,* and in his *Optics,* [Newton] had the rules of the *Novum Organum* constantly in his eye" (1863, 711–13).[11] But when the antihypothetical doctrine came under challenge toward the end of the century, the example of Newton was exploited for another purpose. Dugald Stewart argued against Reid, claiming

---

[9] See also Algarotti 1737, 1:ii, for the dedication to Fontenelle whose *Plurality of Worlds* "first softened the savage Nature of Philosophy, and called it from the solitary Closets and Libraries of the Learned, to introduce it into the Circles and Toilets of Ladies."

[10] Marquis de l'Hôpital, quoted in "Newton," *Biographia Britannica* 1760, 5:3229. The theme of Newton's relationship with God and mankind was first employed by Edmund Halley in his "Ode on the *Principia.*" For a reading of this, see Albury 1978.

[11] See also Laudan 1981, 86–110; Olson 1975. For the continuation of this tradition of linking Bacon and Newton, see Wilson 1985.

that there had been misinterpretation of Bacon and that the Newtonian achievement depended on the use of hypotheses.

> The votaries of hypotheses, said Dr. Reid, "have often been challenged to show one useful discovery in the works of nature that was ever made in that way." In reply to this challenge, it is sufficient, on the present occasion, to mention the theory of Gravitation, and the Copernican system. (Stewart 1854–60, 3:299)[12]

These debates about the relationship between Bacon and Newton did not always involve direct reference to the opposition of genius and method. But given Bacon's notion of a method that rendered "all wits and understandings . . . nearly on the same level" (1857–64, 3:63), such a contrast was always implicit; and in the historical writings of Joseph Priestley it was an explicit theme.

In his *History of the Present State of Electricity* of 1767, Priestley argued for the accessibility of experimental philosophy, and in doing so, confronted the existing celebration of Newton's genius. Writing only forty years after Newton's death, Priestley made the bold contention that recent discoveries of electrical phenomena were about to extend "the bounds of natural science" beyond the Newtonian vision. "New worlds," he said, "may open to our view, and the glory of the great Sir Isaac Newton himself, and all his contemporaries, be eclipsed, by a new set of philosophers . . ." ([1767] 1775, xi). Behind this confidence lay Priestley's belief that discoveries of electrical phenomena had been made more by "accident" than by "human genius"; and extending this to a more general program he called for the compilation of histories for each branch of science and the encouragement of wider participation. This was feasible, he argued, because "Many modest and ingenious persons may be engaged to attempt philosophical investigations when they see, that it requires no more sagacity to find new truths, than they themselves are masters of" (ibid., 2:166, 1:xviii).[13] Priestley defended this egalitarian approach to knowledge by attacking what he viewed as the excessive worship of heroic genius in science. Such a view he claimed, was reinforced by the way in which discoveries were presented, *synthetically*, as an inevitable series of smooth steps. But discoveries were actually made by an "*analytic method*" involving far more false moves, guesswork, and accident. Addressing the case of Newton, Priestley suggested that the great philosopher had presented his work as a set of powerful logical deductions that evinced a single act of genius:

> But if, when a man publishes discoveries, he, either through design, or through habit, omit the intermediate steps by which he himself arrived at them; it is no wonder that his speculations confound others, and that the generality of mankind stand amazed at his reach of thought.

---

[12] For the revision of Bacon's methodology, see Yeo 1985. On Stewart, see Rashid 1985.
[13] See also McEvoy 1979. For the relation between his philosophy of knowledge and his theology, see McEvoy and McGuire 1975.

A closer examination of the experiments, both successful and unsuccessful, made by Newton in the course of his inquiries would, he believed, provide a better under-standing of scientific progress, although it would qualify the usual images of his genius (ibid., 2:166–69). This would lead to "an opinion of the greater equality of mankind in point of genius [which] would be of real service in the present age" (quoted in Schaffer 1986, 408).

Thus, as the romantic theories of poetic and artistic genius were making an impact both on the Continent and in Britain, Priestley affirmed the idea of a general rational process accessible, in principle, to all men. In order to contextualize Priestley's reaction to the notion of scientific genius, it is worth noting that at this time Denis Diderot was also struggling with the conflict between the Enlightenment assumption of a "standard of general Reason" and his own conviction that genius consisted of extraordinary faculties and personality. Although he found some difficulty in includ-ing what he called the *génie de la physique expérimentale* within the concept of the intuitive and inspirational artist, Diderot suggested that the former involved an unusual combination of observation, reflection, and experience. In his view this meant that scientific or experimental genius was a gift of Nature and not something that could be acquired or taught.[14] It was this concept of a special and unpredictable genius in science that Priestley resisted. In doing so, he gave a strong statement of the idea of science as an egalitarian practice, accessible to all who followed specifiable methods of observation and experiment. By the 1830s this perspective was coming under critical review and a new image of scientific genius, with Newton as its central example, was under construction.

During the early nineteenth century in Britain the scientific past became an object of sustained scholarship.[15] The reasons for this development cannot be closely analyzed here, but several factors can be noted. At the most general level, it was part of the growth of historical consciousness manifested in the new intensity with which the geological, archaeological, and medieval past were studied. With the increasing specialization of science – reflected in the establishment of societies focusing on a particular discipline – and the demise of the more general natural philosophy, it was also a period in which men of science were interested in the continuities and discontinuities between their practices and beliefs and those of the acknowledged founders of modern science. Such concerns also informed the heightened concern with methodology in the early decades of the nineteenth century: historical inquiry was relevant to debate about the method by which past scientists achieved their successes.

As indicated by the number of antiquarian societies and the sale of historical novels, the past was also popular. The most attractive medium for the dissemination

---

[14] See Dieckmann 1941, 156, 171–74. For the ideal of solitude and its links with the image of the "savant" in France at the time of the Revolution, see Outram 1978. On the Jacobin ideology of a democratic science in opposition to the perceived elitism of the Académie Royale des Sciences, see Gillispie 1959 and Hahn 1971, 138–39, 153.

[15] See Munby 1968 on the new interest in archival material, and, more generally, Levine 1986.

of the past – political, military, religious – was biography; and it is not surprising that the heroes of the scientific past should feature along with other examples of achievement and virtue. Indeed, since the scientific community was anxious to promote its cultural status, biography promised a wide audience, and by the 1830s the names of Galileo, Kepler, Boyle, and Newton began to appear in collections such as the *Lives of Eminent Persons,* published in the Library of Useful Knowledge in 1833, or the more expensive *Portraits of Illustrious Personages* issued in twelve volumes in 1835. Significantly, Brewster's biography of Newton, the first large-scale study of Newton since the entry by J. B. Biot in the *Biographie Universelle* of 1821, appeared as a volume in Murray's Family Library in 1831. Similarly, De Morgan's alternative sketch was volume nine of the Cabinet Portrait Gallery of British Worthies, published in 1846.[16] The relevance of this kind of medium to the debate on Newton's moral character will be treated in the next section; but before coming to that we need to resume discussion of the interpretation of scientific genius in these nineteenth-century studies of Newton.

While Priestley sought to record the discoveries of great natural philosophers in order to learn and disseminate their methods of inquiry, Brewster presented Newton as an object of marvel. His biography was meant to be instructive, but since Newton had eclipsed "the sages and heroes of antiquity" in "the ascendancy of his genius," the message of simple emulation recommended by Priestley was, in Brewster's view, completely inappropriate. Indeed, he concluded his book with an attack on the tradition that ascribed Newton's achievements to Bacon's rules of method. For Brewster, this association could only "tend to depose Newton from the high priesthood of nature . . ." (1831, 1, 330–31). After offering various historical arguments against this position, he proceeded to claim that the Baconian method was not the one by which major discoveries were made, and that a proper recognition of why this was so would make clear, once and for all, "the absurdity of attempting to fetter discovery by any artificial rules." Brewster did not renounce the attempt to discern "the general character" of the process by which great minds acted, but he contended that "the history of science does not furnish us with much information on this head, and if it is to be found at all, it must be gleaned from the biographies of eminent men." This marked a deliberate break with natural histories of scientific progress favored by Priestley, and with the assumption that the accumulation of facts would assist it. For Brewster, major discoveries consisted in the detection of "some deep-seated affinity which baffles ordinary research"; and he insisted that the perception of such relationships took place in a manner "the very reverse of the method of induction" (ibid., 332–36).

Underlying this position was Brewster's commitment to genius, rather than method, as the appropriate category for approaching fundamental developments in science. "The impatience of genius," he explained, "spurns the restraints of mechanical rules, and never will submit to the plodding drudgery of inductive discipline."

---

[16] For an interesting comparison of the biographies of Brewster and De Morgan, see Theerman 1985.

Thus in speaking of the "peculiar character of [Newton's] genius," he stressed fertility of imagination and exuberance of invention. Whereas Priestley claimed that a detailed knowledge of the process by which Newton came to his conclusions would reduce the references to inimitable genius and reveal science as something accessible to average capacities, Brewster acknowledged that Newton "kept back his discoveries till they were nearly perfected." Moreover, he argued that if the successive steps of his inquiries were known, we would be more amazed than ever about the imaginative leaps and conjectures they contained (ibid., 336–37). In this way, Brewster not only made a major assault upon Baconianism, but established the concept of imaginative genius and the attendant notion of an extraordinary personality as crucial elements in an alternative philosophy of scientific discovery.

During the 1830s, the legacy of the Enlightenment assumptions about a nonelitist theory of knowledge was represented in many popular publications on science. The strategy of reducing genius to method or common sense was widely disseminated throughout the mid–nineteenth century in the writings of Samuel Smiles. In his best-selling *Self Help,* Smiles announced that "in the pursuit of even the highest branches of human inquiry, the commoner qualities are found the most useful"; genius was "common sense intensified" (Smiles 1894, 94–95, 317). Newton was cited as an example here, and Smiles inverted the romantic notion of the childhood genius by stressing the fact that he was a dull boy. The earlier and more detailed biographies in the Library of Useful Knowledge of 1833 continued Priestley's criticism of the idea of scientific genius as something transcending accessible rules and procedures. Thus the volume on Galileo presented him as an exponent of the safe method of experiment and induction recommended by Bacon. But from this perspective, the work of Kepler was embarassing, because it had to be admitted that "this extraordinary man pursued, almost invariably, the hypothetical method."

His life was passed in speculating on the results of a few principles assumed by him, from very precarious analogies. . . . We nevertheless find that he did, in spite of this unphilosophical method, arrive at discoveries which have served as guides to some of the most valuable truths of modern science. ([Drinkwater] 1833, 1)

In attempting to salvage a moral from this case, the writer indicated that Kepler always abandoned hypotheses if they did not fit with facts; but this was seen as small compensation for his persistent efforts in "seizing truths across the wildest and most absurd theories" (ibid., 2, 15). Whereas Brewster saw the record of Kepler's conjectures as the mark of a scientific genius, this author regarded it as evidence of "the many years of wasted labour" in his career, and warned about "the danger of attempting to follow his method in the pursuit of truth" (ibid., 54).[17]

While Whewell was working on the history and philosophy of inductive discovery in the 1830s, he regarded this position as an obstacle. Referring to the biography of Kepler in the Library of Useful Knowledge he remarked that:

---

[17] For the wider contexts and implications of the ideas of egalitarian method, see Yeo 1986.

several persons seem to have been alarmed at the *Moral* that their readers might draw, from the tale of a Quest of Knowledge, in which the Hero, though fantastical and self-willed, and violating in his conduct, as they conceived, all right rule and sound philosophy, is rewarded with the most signal triumphs. (Whewell 1837, 1:410–11)

He used the case of Kepler to assert that the process which led to major scientific discoveries involved imagination and speculation, not the "cautious" approach endorsed by some followers of Bacon (Whewell 1847, 2:44–49; Yeo 1985, 272–77). Whewell dealt with Newton's much-quoted remarks about industry and patient thought as the key to his discoveries by suggesting that this was a fair account of the "mental *effort*" involved in science, but distinguished this from "the natural *powers* of men's minds [which] are not on that account the less different" (1847, 2:140). Whewell therefore agreed with Brewster, but he made more use of Kepler than of Newton when illustrating this conjectural aspect of science, partly because, as Priestley and Brewster also noted, Kepler had left a far more extensive account of his guesses and hypotheses, but also because he situated Newton's work more carefully against a historical background.

Whewell was more precise about Newton's original and singular contribution. Thus he indicated that several contemporaries of Newton, especially Huygens, Hooke, and Halley, had come to "consider the motion of the planets around the sun as a mechanical question, to be solved by a reference to the laws of motion, and by the use of mathematics." And while Newton produced the first full demonstration of the inverse square law applied to the motion of planets, his most original perception was the "identification of the force which retains the moon in her orbit with the force of gravity by which bodies fall at the earth's surface" (ibid., 2:115–6, 121). Thus it was only after Newton that the term "gravity," although used earlier, unequivocally came to have its modern meaning. In this way Whewell produced an account of scientific discovery as an historical process in which individual contributions from men of genius were set within the intellectual context in which they emerged.

Augustus De Morgan also emphasized the monumental character of Newton's achievement and its incompatibility with Baconian philosophy of science (De Morgan [1915] 1954, 1:75–84 for his critique of Baconianism). But he too made some qualifications to Brewster's position. De Morgan referred to "sagacity" – a word employed by Whewell – rather than genius, and dismissed Newton's attribution of his success to "patience and perseverance" as worthless (1914,[18] 50). But he also attempted to correct the contrasting image of Newton as the solitary and inspired recluse. Writing in 1837, he observed that: "the Newton of the world at large sat down under a tree, saw an apple fall, and after an intense reverie, the length of which is not stated, got up, with the theory of gravitation well planned, if not fit to print." He

---

[18] This edition brings together most of De Morgan's writings on Newton, including the biography published in *The Cabinet Portrait Gallery of British Worthies* in 1846, and the long review of Brewster's work in the *North British Review*. Subsequent references to these works will be to this edition.

confessed that it was painful to state that this was a myth constructed by Newton himself, and that it seriously misrepresented the process of scientific discovery. In this respect De Morgan believed that Newton shared "the fault of discoverers generally," by presenting conclusions as the result of an elaborate train of deduction and concealing the errors, guesses, and trials which made his own work seem "more like a book-keeping operation, than the poetical process of the fable" (1837, 242–43).

Like Whewell, De Morgan sought to relate Newton's work to that which preceded it, but admitted that the exact chronology of his contributions was often difficult to determine because, as De Morgan put it, "a discovery of Newton was of a twofold character – he made it, and then others had to find out that he had made it" (1914, 18, 38). He was also concerned to overthrow earlier notions, conveyed by Stukeley, of Newton as an untutored genius who could justifiably spurn the work of both Euclid and Descartes; for De Morgan, Newton's genius was shaped by Cambridge: "the lad carried to the University as much conceit as the man brought away of learning and judgement." On this view, therefore, the image of an independent and isolated genius, which Brewster appeared to endorse as part of his passionate anti-Baconianism, was seriously mistaken. De Morgan replaced it with a conception of great scientific work as influenced, both positively and negatively, by the institutions in which it took place. Thus while Brewster welcomed Newton's move to the Mint as a sign of state recognition of science, De Morgan could only express dismay: "And where should a high-priest of science have lived and died? At the Mint?" (ibid., 8–12, 178–79).[19]

Whewell and De Morgan agreed with Brewster's emphasis on genius over method, but they began to qualify some of his more extreme celebrations, and in doing so, historicized the concept of genius and placed it within more general patterns of discovery. While all three writers agreed about the inadequacies of an extreme inductivist view of scientific method, Whewell and De Morgan limited the poetic conception of Newton's genius promoted by Brewster. Indeed, this difference was more general and Brewster had noted it in Whewell's *History*. For although he supported the stress on the role of imagination in Kepler's work, Brewster suspected that Whewell still sought to bring the thought of "heaven-born genius" within a systematic philosophy of discovery (Brewster 1837, 150). In Whewell, this took the form of a careful analysis of the conceptual elements of a scientific discovery which, although emphasizing the importance of individual genius, also considered precursors and the various conditions necessary for the acceptance and elaboration of a new theory by the scientific community. For example, he argued that the work of "great geniuses" might expire with them in the absence of a body of scientists capable of explaining and incorporating their discoveries (Whewell 1837, 2:459–61).[20] De Morgan was more explicit in his rejection of the image of Newton as the solitary,

---

[19] For the idea of Newton's *early* manifestation of genius, see Stukeley 1936, 53–55. For a recent critique of the view that Cambridge in Newton's day was an "intellectual wasteland," see Gascoigne 1985.

[20] He also noted that there were still some Catholic states where Newton's work was heresy; see Whewell 1830, 413.

inspired genius. While attacking Bacon's method, he affirmed the importance of the cooperative and institutional aspects of science, arguing that the myths about Newton's mode of work, and his own disposition to regard his results as private knowledge, were threats to the progress of science. Thus in focusing on the role of genius and its implications for the philosophy of science, these writers went beyond both the earlier references to Newton's divine genius and the Enlightenment reactions against it. In doing so, they became involved, more seriously than their predecessors, in debates about the personality and moral character of Newton.

Genius and Morality

In a privately printed anti-Newtonian tract of 1846, William Peters remarked that "to talk about Sir Isaac Newton" may not consign a person to a madhouse, but "we have *medical advice* that *once in,* it will certainly help *to keep him there*" (Peters 1846, xi; see De Morgan [1915] 1954, 2:11–12 for comment on Peters' anti-Newtonian works). Although attacks upon Newton's system made little impact in this period, doubts about his character did indeed elicit anxious responses. This section attempts to consider the ways in which various issues involving perceptions of Newton's sanity, religious beliefs, and moral behavior were discussed by important scientific figures in nineteenth-century Britain. In treating this topic it is suggested that these questions about Newton's life, many of which had been raised in the previous century, were given new significance because of the association of genius with the concept of an extraordinary personality. We have seen in the previous section that writers on Newton interpreted his scientific discoveries as stemming from a peculiar combination of mental aptitudes and resources which enabled him to make the kind of intellectual leaps that could not be incorporated within any set of rules or method. Thus Brewster observed that his process of thought was "more characteristic of poetical than of philosophical genius" (1831, 329). The problem here was that discussions of Newton's mind, and detailed research into his habits and social behavior, coincided with renewed interest in the connections between genius and psychological disposition. While poets and artists – following the notoriety of the romantic school – were the more usual examples of this correlation, the material available on Newton was uncomfortable for those seeking to affirm him as the epitome of the connection between genius and virtue (Wittkower and Wittkower 1963). Thus in considering the manner in which men such as Brewster, Whewell, Rigaud, and De Morgan dealt with data concerning Newton's character, we have to realize that contemporary medical and psychological literature was beginning to treat the phenomenon of genius as linked with a form of personality disorder. Such a view was not confined to specialist discourse: Charles Lamb wrote against it in "The Sanity of True Genius" of 1826, contending that "it is impossible for the mind to conceive of a mad Shakespeare" (Lamb 1935, 167–69). Eminent men of science wanted to make a similar avowal on the part of Newton; but to do so they had to confront and disarm a range of anecdotes and speculations dating back to the period of his childhood. In some measure they failed to do this – for example, in 1891 J. F. Nisbet published *The Insanity of Genius,* in which Newton, along with other scientists and intellectuals, was regarded as manifesting an identifiable "neuropathic" disorder. And of course, in Frank Manuel's more recent *Portrait of Isaac Newton,* the possible connections between his personality and intellectual style are explored.[21]

Before dealing with this dimension of the nineteenth-century writings on Newton, it is necessary to explain why the question of Newton's character was so crucial. There is evidence from debates of the 1830s that this was not merely a secondary issue but one which contained serious implications for assumptions about science. The sense of urgency is apparent in the statement from Stephen Rigaud to Whewell concerning the unfavorable picture of Newton in Francis Baily's *An Account of the Reverend John Flamsteed* – a collection of unpublished letters relating to the dispute between Newton and Flamsteed in 1694–95 and again after 1704. Pleading with Whewell to enter the debate, Rigaud said that "if Newton's character is lowered, the character of England is lowered, and the cause of Religion is injured" (Rigaud to Whewell, 25 January 1836, Whewell Papers Add. MS a 211 no. 79, Trinity College, Cambridge. See Baily 1834, 462–66 for an announcement of the recovered Flamsteed correspondence).

This reaction was not only due to the national pride invested in Newton's name; it related to a wider set of assumptions and rhetoric about the relationship between virtue – moral, social and religious – and outstanding scientific achievement. There were several levels to this web of convictions. Firstly, as suggested above, the emphasis on genius as a category in analyses of scientific progress was associated with a focus on the character of such persons. The dominant messages in early nineteenth-century biographies of great men in the history of science was twofold: the unlikelihood of matching their intellectual feats, and the desirability of emulating their moral habits and character. Thus the chapter on Newton in *The Boyhood of Great Men* observed that "many of the great qualities which excite our admiration in the career of this great philosopher may be imitated by those who cannot hope to vie with him in the splendour of his genius . . ." (Edgar 1853, 173. See also Anon. 1816. See also the preface to *British Biography* 1773, 1:iii). Authors confident that Newton supplied that appropriate model could, for example, remark that "the beauties of his moral character were, if possible, more admirable than the powers of his mind, and his piety was not less genuine than his philosophy" (Lodge 1835, 10:5). There was a tension here, however, because on the one hand men of scientific genius such as Newton were being revered for the manner in which they transcended traditional views and methods, while on the other they were applauded as exemplars of conventional rules of moral conduct.

---

[21] See Nisbet 1891, 222–24, for this observation: "From . . . his father's early death, it may be concluded that the neuropathic influence in Newton's case was on the paternal side." He also noted (p. vii) that Lamb had been in a mental asylum. See also Manuel 1968. For the nineteenth-century theories on genius and madness, see Becker 1978.

There was even a further complexity insofar as the possession of sound moral and religious attitudes was often adduced as a reason for, or at least as a widely noted accompaniment of, their scientific success. Thus the common contrast in British writing between the achievement of Newton and the failure of Descartes was invariably linked with a comparison of their moral characters. Thomas Chalmers, the Scottish Evangelical who wrote on science and natural theology, presented Descartes' views as the product of presumption and immodesty, while citing the Newtonian system as one of "humility and hardihood" (Chalmers 1815, 200). There were various permutations on this theme, but the common factor was an equation between rash speculation and lack of an appropriate moral demeanor in the scientific study of God's creation. Perhaps the most theorized version of this dichotomy was that drawn by Whewell in his Bridgewater treatise of 1833. Here he argued that those who made major advances in science – such as Copernicus, Kepler, and Newton – always evinced sound moral and religious dispositions, because the inductive process of discovery involved them in a painful passage from the disorder of facts to the perception of a law.

> This step so much resembles the mode in which one intelligent being understands and apprehends the conceptions of another, that we cannot be surprised if those persons in whose minds such a process has taken place, have been the most ready to acknowledge the existence and operation of a superintending intelligence, whose ordinances it was their employment to study. (Whewell 1834, 307)

Whewell suggested that the "*great* discoverers" came in contact with the mind of the Deity, and went on to show that, historically, such men were indeed strengthened in their religious convictions by their scientific inquiries. On the other hand, deductive thinkers, who possessed a "derivative knowledge of the laws of nature," did not partake of the inductive ascent through which "genius . . . divines the general laws of nature" (ibid., 308, 329). Accordingly, he was not surprised that men such as D'Alembert, Lagrange, and Laplace, who brilliantly unfolded the consequence of the theory of gravity, had a less vivid conception of a "supreme Intelligence" than Newton, who "conceived and established the law itself," thus adding to our knowledge "something which was not contained in any truth previously known, nor deducible from it by any course of mere reasoning." In short, Whewell argued that there were inductive and deductive habits of mind, and that the former engendered, or was at least correlated with, sound religious attitudes. In contrast, if taken to extremes, the deductive cast of mind could not accept the possibility of the laws of nature being inconsonant with a particular deductive system. At this point such a manner of thinking could lead to arrogant notions about the self-sufficiency of certain mechanical principles (ibid., 328–40).[22] In his *History,* Whewell singled out Descartes and sought to make a connection between weak moral character and the

---

[22] See Babbage 1838 for a response to Whewell's attack upon deductive thinkers, and Yeo 1979 for Whewell's views. For further references to the links between intellectual and moral attacks on men of science, see Fox 1974 and Outram 1980, 35, 42.

intellectual presumption associated with the attempt to establish a system of the universe from deductive principles. After insisting that Descartes was not a true precursor of Newton, Whewell remarked that "in the whole of his philosophy, he appears to deserve the character of being both rash and cowardly ..." (Whewell 1837, 2:106).

It is now possible to see that the question of Newton's state of mind following an apparent mental collapse in 1693 was discussed during the early nineteenth century against the background of two different sets of ideas about the character of scientific genius. Firstly, there was the association of genius with insanity – an ancient notion, but one undergoing explicit theorization in contemporary psychology. Secondly, there was the set of assumptions about the dual presence of genius and virtue in great men whose biographies constituted powerful didactic resources. This found a more specific formulation in writings on the history of science and natural theology in which sound moral and religious habits and beliefs were ascribed to great discoverers and were seen as guiding their successful inquiries. In what follows here there is no attempt to bring forth new material concerning Newton's state of mind; rather, the focus is on the manner in which this debate assumed new significance during the nineteenth century as it was informed by these two conflicting perspectives on the "character" of genius.

Most of the major biographical sources have something to say about Newton's personality and social habits. These comments were of two kinds: the eulogies to his moral and pious character, and a collection of remarks about his occasional idiosyncratic behavior. While the first served as a resource for moral didacticism, the latter became part of a genre concerning the eccentric nature of men of genius. Our interest here is in the way in which this second dimension became the basis for the speculation about the nexus between genius and insanity.

Among the first-hand accounts of Newton's behavior, such as those of Humphrey Newton, Henry Pemberton, and William Stukeley, there are several references to unusual habits. On the one hand there were those illustrating the regularity of his sleeping habits and the obsession with healthy diet; on the other hand there were those pertaining to instances of absentmindedness (Stukeley 1936, 60–66; Westfall 1980, 191–94; Manuel 1968).[23] There do not appear to have been any comments on the relationship between these, but there was a tendency to see his eccentricity as an example of the behavior observed in other men of genius. Stukeley regarded Newton's indifference to the normal patterns of eating and drinking as a consequence of the "abstract reasonings" which allowed him "an anticipation of part of those divine joys in our future state of being." But although he stressed Newton's "very serious and composed frame of mind," Stukeley added that he "was easily made to smile, if not laugh"; and that "he used a good many sayings, bordering on joke and with" (1936, 57).

---

[23] The best of these anecdotes is probably apocryphal. As told by De Morgan it suggests that Newton had two holes in his study door – a small one for the use of his small cat, and a large one for the big cat. (De Morgan [1915] 1954, 52). For another version, see P. B. Shelley to his father, 6 February 1810, in Shelley 1964, 1:51.

It is tempting to see these remarks of Stukeley as a valiant attempt to save Newton from rumors about the effects of a reported depression during 1693 on his subsequent mental capacities and his personality. But although we now know that contemporaries such as Locke, Pepys, and Millington knew of this episode, it is not clear when it began to emerge as the major theme it later became. However, there is a passage in Pemberton's work which does show an appeal to eccentricity as a defense against more serious insinuations regarding Newton's state of mind: "Though his memory was very much decayed [in old age], I found he perfectly understood his own writings, contrary to what I had frequently heard in discourse from many persons." Pemberton explained that this opinion probably stemmed from the fact that Newton was not always capable of speaking on subjects on which he might be expected to have authority. "But ... it may be observed, that great geniuses are frequently liable to be absent, not only in relation to common life, but with regard to some of the parts of science they are best informed of" (Pemberton 1728, preface). This manner of dealing with allegations concerning Newton's state of mind was repeated throughout the. century, and some writers even found it possible to use the story of the loss of important papers in a fire caused by his dog, Diamond, as an indication of his tranquility under stress.[24] But by the nineteenth century, anecdotes about eccentric behavior and reports concerning mental depression converged in serious discussion about the link between insanity and genius being exemplified in the case of Newton. Thus, while earlier biographical accounts could either ignore this issue or dismiss it like Pemberton,[25] later publications had to confront it. Indeed one text of 1853 referred to Newton's "temporary attacks of insanity" as if the fact was taken for granted (Buckley 1853, 226).[26]

Without much doubt, the credit for putting this topic firmly on the agenda belongs to J. B. Biot. In his entry on Newton in the *Biographie Universelle* of 1821, a work translated into English for the Library of Useful Knowledge, Biot produced new evidence based on a letter written by Huygens relaying a report that "the illustrious geometer, Isaac Newton, had become insane...." "From these details," he concluded, "it would appear that the mind of this great man was affected, either by excess of exertion or through grief at seeing the results of its efforts destroyed." He then contended that "the fact of a derangement" explained why, after this date, Newton "*never more* gave to the world a *new* work in any branch of science...." Biot thus swept away earlier British references to eccentricity and poor memory to make mental disorder a major factor in his account of Newton's later intellectual career. But, significantly, he went further and ascribed Newton's theological writings to this derangement. While acknowledging that the tendency to combine science and theology was common in seventeenth-century England, Biot was relieved to be able

---

[24] One nineteenth-century source suggested an article in Dobley's Annual Register of 1776 as the stimulus for the emergence of this issue. See Anon. 1851, 101.

[25] For the uses of Pemberton, see *British Biography* 1778, 7:156; Hutton 1815, 2:99.

[26] Other accounts of this period offered various physiological and psychological explanations of Newton's condition. See Taylor 1875–89; Powell 1856, 525, for reference to speculation about "effects of interrupted sleep on the mental faculties."

to refer Newton's obsessions with biblical chronology to a psychological condition. This explained how a mathematical mind could indulge in such conjecture and why the attention to religious writings increased in the latter part of his life (Biot 1829, 25–37).

When Brewster wrote the next major biography of Newton in 1831 he had to confront Biot's position. He regarded the account of the episode of the lost papers and its effect on Newton as wholly novel, remarking that "no English biographer had ever alluded to such an event." In explaining why this issue required a response, Brewster indicated how it affected contemporary assumptions about the relations between genius and virtue.

> The unbroken equanimity of Newton's mind, the purity of his moral character, his temperate and abstemious life, his ardent and unaffected piety, and the weakness of his imaginative powers, all indicated a mind which was not likely to be overset be any affliction to which it could be exposed. (Brewster 1831, 224–25)[27]

Apart from this, Brewster was particularly enraged by the suggestion that Newton's deep religious concerns were largely a product of mental illness. Here he regarded Biot as continuing the skeptical crusade of Laplace who, he alleged, attempted to show that the great scientist's theological inquiries did not begin until this late and troubled period of his life. An investigation of this episode was therefore demanded as a duty to "the memory of the great man, to the feelings of his countrymen, and to the interests of Christianity itself . . ."(ibid., 227, 270–71).

Brewster's answer to this call took him beyond the short biography of 1831 and culminated in his two-volume *Memoirs of the Life, Writings and Discoveries of Sir Isaac Newton* of 1855. But even in the early biography, his treatment of this issue exposed even more damaging evidence about Newton's state of mind during 1693 from his own letters to Locke and Pepys. Brewster attempted to present the condition as a minor one and argued that Newton could hardly be said to be insane because during this period he wrote the well-known letters to Richard Bentley on natural theology. The ensuing debate exacerbated the issue. In the *Foreign Quarterly Review* of 1833, Thomas Galloway reviewed Biot, Brewster, and Biot's reply to the charge that his interpretation was motivated by anti-religious sentiment. Writing for an English audience, he took Biot's side, arguing that Brewster's case was vitiated by an erroneous chronology – the first and principal letter to Bentley was probably written *before* the onset of the illness.[28] Moreover, he cited Biot's comment that the ability to compose a letter on natural theology could not "invalidate the testimony of positive documents as to the temporary aberration of his reason." Biot continued in

---

[27] Note that here Brewster cites a weak imagination as a barrier against mental breakdown, whereas in discussing Newton's genius he refers to inventive powers (ibid., 227, 270–71). Brewster claimed that Laplace had asked Professor Gautier of Geneva to seek evidence of a post-1693 date for Newton's theological interests (ibid., 17).

[28] [Galloway] 1833, 14–20, claimed that Brewster erred in failing to note that the English calendar at that time dated the legal year from 25 March; thus an event between 1 January and 25 March "was dated a whole year *earlier* in England than on the continent."

terms which set one of the influential frameworks for future approaches to Newton's personality: "Such is the frightful condition of man. Genius and madness may exist in his mind side by side simultaneously" ([Galloway] 1833, 17).[29]

After this dispute between Biot and Brewster, the treatment of Newton's character became informed by queries about the pressures on the mind of a scientific genius. But the debate moved away from the particular issue of Newton's mental state at specific historical moments to the more general theme of the relationship between genius and personality. Baden Powell alluded to this in 1834 and offered a clear statement of the comparatively recent notion that intellectual genius was predictably associated with "marked peculiarities of ... character." After summarizing the accounts of Newton's idiosyncratic temperament and behavior, he concluded that "such peculiarities were, by some of those unknown links and mysterious sympathies which connect the phenomena of man's moral and intellectual nature, intimately wound up with the operations of Newton's mighty genius ..." (Powell 1834, 359).[30] Even in Whewell's *History,* which carefully avoided a rehearsal of the earlier arguments between Brewster and Biot, there is a clear trace of this theme:

> Often, lost in meditation, he knew not what he did, and his mind appeared to have quite forgotten its connexion with the body.... Even with his transcendent powers, to do what he did, was almost irreconcilable with the common conditions of human life; and required the utmost devotion of thought, energy of effort, and steadiness of will, – the strongest character, as well as the highest endowments, which belong to man. (Whewell 1837, 2:141)

This description is an eloquent reflection of a transition from the view that genius coexists with virtue to the acknowledgment that it has to resist vice. From this perspective, Newton's strong character was something his defenders had to demonstrate. Whewell went on to note that his contemporaries "uniformly represented him as candid and humble, mild and good"; but even in the 1830s, this was no longer an uncontested position. Galloway thought the evidence suggested that Newton "was constitutionally of a sombre and retiring disposition" (1833, 20–21). And when De Morgan wrote a short biography in 1846 he announced dissent from the tendency "to regard him as an exhibition of goodness all but perfect, and judgement unimpeachable." Later, when reviewing Brewster's *Memoirs,* he concluded that "the *moral intellect* of Newton – not his moral *intention,* but his power of judging – underwent a gradual deterioration from the time when he settled in London" (De Morgan 1914, 4, 37, 153). Powell also returned to this theme in his review of Brewster. While rejecting the parallel which some writers drew between Bacon and Newton – namely that both

---

[29] Biot later reviewed Brewster's large biography. See Biot 1855, 589–606. Although the "romantic" concept of genius was applied to Newton, the English romantic poets criticized the Newtonian world view. See Ault 1975.

[30] In his biography De Morgan remarked that "the great fault, or rather, misfortune, of Newton's character was one of temperament" (De Morgan 1914, 37).

combined high intellect with moral meanness – Powell acknowledged the troubled nature of Newton's personality:

> The truth is that the intellect which had most deeply sounded and explored the mysteries of external nature was at times perplexed and obscured by the mysteries and infirmities of its own constitution, and in embracing the system of the universe Newton at times lost possession of himself. (Powell 1834, 534)

Thus by 1855 it was not only possible temporary insanity that stood in the way of the union of genius and virtue in the person of Newton, but a series of additional issues indicative of intellectual and moral failings: anti-Trinitarianism, alchemy, the disputes with Flamsteed and Leibniz, and the question of his complicity in an immoral relationship between Catherine Barton, his niece, and Charles Montague. Ironically, it was partly Brewster's honesty in the face of material available to him in the Portsmouth Papers that either put these issues into circulation or reinforced previous suspicions (Westfall 1965, 1:cx–xlv; Christie 1984, 53–58).

There is insufficient space in the present version of this paper to discuss these issues in any detail, but some summary observations can be made. The investigation of these topics seems to have stemmed from the conviction, expressed by Rigaud, that "everything is worth knowing which is connected with such a man." Such inquiries, some of them remaining private, were conducted within a framework informed by the assumption that, as Whewell put it, "great talents are naturally associated with virtue" (1837, 141). The different issues raised various concerns. Newton's anti-Trinitarianism had long been suspected but it posed difficulties for writers such as Brewster and Whewell, who stressed the theological orthodoxy of the major scientific figures of the seventeenth century.[31] There had been less discussion of Newton's alchemical interests and until Brewster's confrontation of the full extent of this in Newton's papers, the reports which existed were either dismissed or ignored. Whewell's private papers, for example, show that he knew of articles in the *Gentleman's Magazine* during the late eighteenth century which mentioned the influence of Jacob Boehme, the German mystic, on Newton's scientific thought (Wormhoudt 1949).[32] Both Whewell and Brewster, in their writing on the history of science, treated alchemical speculation as an intellectual and moral weakness, so the revelation of Newton's involvement was extremely damaging. The question of Newton's knowledge of a possible illicit sexual relationship between his niece and his influential friend Lord Halifax was also researched by Rigaud and De Morgan, and this is why Brewster had to deal with it in his large biography (De Morgan 1885). Needless to say, it directly touched Newton's social and moral character. So too did the dispute with Flamsteed. Partly due to the extensive documentation of this episode in Baily's edition of Flamsteed's correspondence, it produced champions of Flamsteed, such as

---

[31] For various responses to this issue, see Acton 1833.

[32] For an eighteenth-century reference to Newton's mystical interests, see Spence 1964, 61–62, 69. Stukeley tried to stress Newton's interest in chemistry over alchemy (1936, 60). For a recent assessment, see Figala 1977.

278

Baily and De Morgan, and defenders of Newton, such as Rigaud, Whewell, and Brewster. The debate is significant for the way it reveals an appeal to assessments of character, both intellectual and moral, as a means of attributing blame. Thus Flamsteed's alleged ignorance of the theoretical value of Newton's work, his paranoid disposition, his belief in judicial astrology, and his known consultation in 1665 with Vincent Greatrakes, a famous magical healer in Ireland, were all cited as reasons for doubting his account. In this way, a dispute that raised important questions about the relationships between observers and theorists, and the issue of private and public scientific property, was largely conducted in a framework dominated by the assumption of a coincidence between intellectual and moral virtue.[33] By 1885 De Morgan spoke of the "desirable myth that [Newton's] goodness was paralleled only by his intellect," and noted that in his second biography Brewster could not repel the doubters and the impugners from "the temples in which science worships its founders."

Conclusion

This paper has analyzed the use of the concept of genius in interpretations of Newton in the eighteenth and nineteenth centuries. Noting the early celebrations of both Newton's intellectual achievements and his moral attributes, it has drawn attention to two areas of tension. By the close of the eighteenth century there was discussion about the conflict between the idea of genius and the notion of a specifiable and accessible scientific method – a debate which had occurred earlier in literary criticism. This contrast was also deployed in early nineteenth-century Baconian rhetoric about the possibility of wide participation in science. But this last position was rejected by writers such as Brewster, Whewell, and De Morgan who saw extreme Baconianism as an inadequate account of the creative processes involved in major scientific discovery. In their view, Newton stood as the archetypal example of a scientific genius, transcending any simple rules and methods to grasp new laws of nature. However, this stress on the uncommon intellectual capacities and processes of such men of genius appears to have converged with a new emphasis on the idea of an extraordinary personality, one likely to transgress social and moral conventions. Such a conjunction was reinforced by contemporary psychological speculation, thus transforming earlier interest in anecdotes about eccentricities of great men into a more theorized conception of a nexus between genius and insanity. In turn, this development posed difficulties for the exploitation of Newton as a moral exemplar in biographical writing. By the middle of the nineteenth century, Newton was a far more complex figure than the celestial or divine genius lauded by his contemporaries. While still exalted as the apex of scientific achievement, his genius was seen as more human in kind, and could not be dissociated from the evidence of passions,

---

[33] For a summary of the dispute between Newton and Flamsteed, see Westfall 1980, 541–50, 655–66. For the contemporary debate, see Baily 1835 and Whewell 1836. This debate and the issue of Newton's image in the nineteenth century raise questions about authority in science which lie beyond the scope of this article.

lapses, and delusions which cast severe doubts on the previous convictions about the affinity of intellectual and moral virtue. In short, this paper has been concerned not only with the idea of genius in changing images of Newton, but with the transformation of this concept under the pressure of archival inquiry and psychological theory. This was a shift from the idea of genius as a divine gift bestowed on individuals to that of genius as constituted by an exceptional personality. The result was the loss of a number of assumptions by which Newton's science was tied to a theological and moral framework and, arguably, the emergence of a more valid historical image of Newton, the seventeenth-century natural philosopher.

Acknowledgments

This article derives from a paper delivered at a workshop commemorating the 300th anniversary of the publication of Newton's *Principia* held at Tel Aviv University and at the Van Leer Jerusalem Institute, 27-30 April 1987. I would like to thank the participants for their helpful comments. Research for the article was supported by a Griffith University research grant.

References

Abrams, M. H., 1953. *The Mirror and the Lamp: Romantic Theory and the Critical Tradition.* New York: Oxford University Press.

Acton, H., 1833. *Religious Opinions and Example of Milton, Locke and Newton.* London.

Albury, W. R., 1978. "Halley's Ode on the *Principia* of Newton and the Epicurean Revival in England," *Journal of the History of Ideas* 39:24–43.

Algarotti, F., 1737. *Sir Isaac Newton's Philosophy Explained for the Use of Ladies,* trans. E. Carter, 2 vols. London.

Anon., 1816. *Buds of Genius: Or Some Account of the Early Lives of Celebrated Characters who were Remarkable in Their Childhood.* London.

Anon., 1851. *The Life of Sir Isaac Newton.* London: Religious Tract Society.

Ault, D., 1975. *Visionary Physics: Blake's Response to Newton.* Chicago: The University of Chicago Press.

Babbage, C., 1838. *Ninth Bridgewater Treatise: a Fragment,* 2d ed. London: J. Murray.

Bacon, F., 1857–64. "Novum Organum," in *The Works of Francis Bacon,* ed. J. Spedding, R. Ellis, and D. Heath, 14 vols. London.

Baily, F., 1834. "A short Account of some MSS Letters . . . ," in *British Association for the Advancement of Science Reports.* London.

——— , 1835. *An Account of the Reverend John Flamsteed, the First Astronomer Royal.* London: The Admiralty.

Becker, G., 1978. *The Mad Genius Controversy: A Study in the Sociology of Deviance*. Beverly Hills: Sage.

Biographia Britannica, or, the Lives of the Most Eminent Persons, 1760. 6 vols. London: W. Innys.

Biot, J. B., 1829. *Life of Sir Isaac Newton*, trans. H. Brougham. Library of Useful Knowledge. London: Baldwin and Craddock.

——, 1855. "Memoirs of the Life, Writings and Discoveries of Sir Isaac Newton," *Journal des Savants* (October):589–606.

Brewster, D., 1831. *The Life of Sir Isaac Newton*. London: J. Murray.

——, 1837. "Whewell's *History of the Inductive Sciences*," *Edinburgh Review* 66:110–51.

——, [1855] 1965. *Memoirs of the Life, Writings, and Discoveries of Sir Isaac Newton*, 2 vols. New York: Johnson Reprint Corporation.

British Biography; or an Accurate and Impartial Account of the Lives and Writings of Eminent Persons in Great Britain, 10 vols., 1773–80. London.

Buckley, T. W., 1853. *The Dawnings of Genius Exemplified in the Early Lives of Distinguished Men*. London: Routledge.

Carlyle, T., 1971. *Thomas Carlyle. Selected Writings*, ed. A. Shelston. Harmondsworth: Penguin.

Chalmers, T., 1815. *Evidence and Authority of the Christian Revelation*, 2d ed. Edinburgh.

Chambers, E., 1728. *Cyclopaedia*, 2 vols. London.

Christie, J. R. R., 1984. "Sir David Brewster as Historian of Science," in *Martyr of Science: Sir David Brewster 1781–1868*, ed. A. D. Morrison-Low, and J. J. R. Christie. Edinburgh: Royal Scottish Museum.

Clagett, M., ed., 1959. *Critical Problems in the History of Science*. Madison: University of Wisconsin Press.

Cohen, I. B., 1960. "Newton in the Light of Recent Scholarship," *Isis* 51:489–514.

Delaney, P., 1969. *Autobiography in the Seventeenth Century*. London: Routledge.

De Morgan, A., 1837. "Theory of Probabilities – Part 11," *Dublin Review* 3:237–48.

——, 1855. "Sir David Brewster's *Life of Newton*," *North British Review* 23:307–38.

——, 1885. *Newton: His Friend and His Niece*, ed. by his wife [S. E. De Morgan] and A. C. Ranyard. London: E. Stock.

——, 1914. "Newton," in *Essays on the Life and Work of Newton*, ed. P. Jourdain. Chicago: Open Court.

——, [1915] 1954. *A Budget of Paradoxes*, 2d ed., 2 vols. New York: Dover.

Dieckmann, H., 1941. "Diderot's Conception of Genius," *Journal of the History of Ideas* 2:151–82.

[Drinkwater, J.], 1833. "The Life of Kepler," in *Lives of Eminent Persons*. London: Society for the Diffusion of Useful Knowledge.

Duff, W., 1767. *An Essay on Original Genius*. London.

Edgar, J. G., 1853. *The Boyhood of Great Men*. London: G. Barclay.

Engell, J., 1981. *The Creative Imagination*. Cambridge, Mass.: Harvard University Press.

Figala, K., 1977. "Newton as Alchemist," *History of Science* 15:102–37.

Fontenelle, B. Le Bovier de, 1728. *The Elogium of Sir Isaac Newton*. London: J. Tomsom.

Fox, R., 1974. "The Rise and Fall of Laplacian Physics," *Historical Studies in the Physical Sciences* 4:89–136.

[Galloway, T.], 1833. "French and English Biographies of Newton," *Foreign Quarterly Review* 12:1–27.

Gascoigne, J., 1985. "The Universities and the Scientific Revolution," *History of Science* 23:39–434.

Gerard, A., 1774. *An Essay on Genius*. London: W. Strahan.

Gillispie, C. C., 1959. "The *Encyclopédie* and the Jacobin Philosophy of Science," in Clagett 1959, 255–89.

Gjertsen, D., 1986. *The Newton Handbook*. London: Routledge.

Guerlac, H., 1965. "Where the Statue Stood: Divergent Loyalties to Newton in the Eighteenth Century," in Wasserman 1965, 317–34.

Hahn, R., 1971. *The Anatomy of a Scientific Institution: The Paris Academy of Sciences, 1666–1803*. Berkeley: University of California Press.

Harman, P., ed., 1985. *Wranglers and Physicists: Studies in Cambridge Physics in the Nineteenth Century*. Manchester: Manchester University Press.

Hays, J. N., 1964. "Science and Brougham's Society," *Annals of Science* 20:227–41.

Hutton, C., [1795–96] 1815. *A Mathematical and Philosophical Dictionary*, new ed., 2 vols. London.

Jones, W. P., 1966. *The Rhetoric of Science: A Study of Scientific Ideas and Imagery in Eighteenth-Century English Poetry*. London.

Keynes, J. M., 1972. "Newton, the Man," in *Collected Works*, 14 vols. London: Macmillan.

Lamb, C., 1935. *Complete Works and Letters of Charles Lamb*. New York: The Modern Library.

Laudan, L., 1981. "Thomas Reid and the Newtonian Turn of British Methodological Thought," in his *Science and Hypothesis: Historical Essays on Scientific Methodology*. Dordrecht: Reidel.

Levine, P., 1986. *The Amateur and the Professional. Antiquarians, Historians, and Archaeologists in Victorian England, 1838–1886*. Cambridge: Cambridge University Press.

Lodge, E., 1835. *Portraits of Illustrious Personages*, 12 vols. London: Harding.

McEvoy, J. G., 1979. "Electricity, Knowledge and the Nature of Progress in Priestley's Thought," *British Journal for the History of Science* 12:1–30.

McEvoy, J. G., and J. E. McGuire, 1975. "God and Nature: Priestley's Way of Rational Dissent," *Historical Studies in the Physical Sciences* 6:325–404.

McGuire, J. E., and P. M. Rattansi, 1966. "Newton and the 'Pipes of Pan'," *Notes and Records of the Royal Society* 21:108–43.

McKie, D., and G. R. De Beer, 1951–52. "Newton's Apple," *Notes and Records of the Royal Society* 9:46–54.

Manuel, F., 1968. *A Portrait of Isaac Newton*. Boston: Harvard University Press.

Munby, A. N. L., 1968. *The History of Science in England: The First Phase, 1833–1845*. Berkeley: University of California Press.

Nicolson, M. H., 1946. *Newton Demands the Muse*. Princeton: Princeton University Press.

Nisbet, J. F., 1891. *The Insanity of Genius*. London.

Olson, R., 1975. *Scottish Philosophy and British Physics 1750–1880*. Princeton: Princeton University Press.

Outram, D., 1978. "The Language of Natural Power: The 'Eloges' of Georges Cuvier and the Public Language of Nineteenth-Century Science," *History of Science* 16:153–58.

——, 1980. "Politics and Vocation: French Science, 1793–1830," *British Journal for the History of Science* 13:27–43.

Pemberton, H., 1728. *View of Sir Isaac Newton's Philosophy*. London.

Peters, W., 1846. *Newton Rescued from the Precipitancy of his Followers: a few practical reasons suggested by fact and supported by Scripture, for questioning the soundness of the Newtonian Theory of the Universe*. London: privately printed.

Porter, R., 1987. *Mind-Forg'd Manacles. A History of Madness in England from the Restoration to the Regency*. London: Athlone Press.

Powell, B., 1834. *An Historical View of the Progress of the Physical and Mathematical Sciences*. London: Longman.

——, 1856. "Sir Isaac Newton," *Edinburgh Review* 103:499–535.

Priestley, J., [1767] 1775. *The History and Present State of Electricity*, 4th ed. London.

Rashid, S., 1985. "Dugald Stewart, 'Baconian' Methodology, and Political Economy," *Journal of the History of Ideas* 46:245–57.

Reid, T., 1863. *Works*. Edinburgh.

Rigaud, S. P., 1838. *Historical Essay on the First Publication of Sir Isaac Newton's Principia*. Oxford.

——, 1851. *Correspondence of Scientific Men of the Seventeenth Century*, 2 vols. Oxford.

Schaffer, S., 1986. "Scientific Discoveries and the End of Natural Philosophy," *Social Studies of Science* 16:387–420.

Schuster, J. A., and R. R. Yeo, eds., 1986. *The Politics and Rhetoric of Scientific Method: Historical Studies*. Dordrecht: Reidel.

Secord, J., 1985. "Newton in the Nursery: Tom Telescope and the Philosophy of Tops and Balls," *History of Science* 23:127–51.

Shelley, P. B., 1964. *The Letters of Percy Bysshe Shelley*, ed. F. L. Jones, 2 vols. Oxford: Clarendon Press.

Smiles, S., 1894. *Self Help*. London.

VII

Images of Newton in Britain, 1760–1860 283

Spence, J., 1964. *Anecdotes, Observations, and Characters ... from the Conversations of Mr. Pope, and other eminent persons of his time,* ed. S. Simpson. London: Centaur Press.

Stewart, D., 1854–60. *The Collected Works of Dugald Stewart,* ed. W. Hamilton, 11 vols. Edinburgh.

Stukeley, W., 1936. *Memoirs of Sir Isaac Newton's Life. Being some account of his family and chiefly of the junior part of his life,* ed. A. Hastings White. London: Taylor and Francis.

Taylor, H. M., 1875–89. "Newton," in *Encyclopaedia Britannica,* 9th ed., 25 vols., 17:444. Edinburgh: A. and C. Black.

Theerman, P., 1985. "Unaccustomed Role: The Scientist as Historical Biographer – Two Nineteenth-Century Portrayals of Newton," *Biography* 8:145–62.

Tonelli, G., and R. Wittkower, 1968–73. "Genius," in Wiener 1968–73, 2:293–302.

Turnor, E., 1806. *Collections for the History of Grantham. Containing Authentic Memoirs of Sir Isaac Newton.* London: W. Miller.

Voltaire, 1738. *The Elements of Sir Isaac Newton's Philosophy,* trans. J. Hanna. London: S. Austen.

Wasserman, E. R., ed., 1965. *Aspects of the Eighteenth Century.* Baltimore: Johns Hopkins Press.

Webster, C., 1982. *From Paracelsus to Newton.* Cambridge: Cambridge University Press.

Westfall, R. S., 1965. "Introduction" to Brewster 1965.

—, 1980. *Never at Rest. A Biography of Isaac Newton.* Cambridge: Cambridge University Press.

Whewell, W., 1830. "Lyell's *Principles of Geology,*" *Quarterly Review* 43:411–69.

—, 1834. *Astronomy and General Physics considered with reference to Natural Theology,* 2d ed. London: W. Pickering.

—, 1836. *Newton and Flamsteed. Remarks on an article in number cix of the Quarterly Review,* 2d ed. Cambridge: J. Deighton.

—, 1837. *History of the Inductive Sciences,* 3 vols. London: J. Parker.

—, 1847. *The Philosophy of the Inductive Sciences,* 2d ed., 2 vols. London.

Wiener, P., ed., 1968–73. *Dictionary of the History of Ideas,* 4 vols. New York: Scribner.

Wilson, D., 1985. "The Educational Matrix ... ," in Harman 1985, 23–28.

Wittkower, M., and R. Wittkower, 1963. *Born under Saturn: the character and conduct of artists: a documented history from Antiquity to the French Revolution.* London.

Wormhoudt, A., 1949. "Newton's Natural Philosophy in the Behemensitic Works of William Law," *Journal of the History of Ideas* 10:411–29.

Yeo, R., 1979. "William Whewell, Natural Theology and the Philosophy of Science in Mid-Nineteenth-Century Britain," *Annals of Science* 36: 493–516.

—, 1985. "An Idol of the Market-Place: Baconianism in Nineteenth-Century Britain," *History of Science* 23:251–98.

— , 1986. "Scientific Method and the Rhetoric of Science in Britain, 1830–1917," in Schuster and Yeo 1986, 259–98.

Young, E., [1759] 1966. *Conjectures on Original Composition*. Leeds: Scholar Press.

Zedler, J. H., 1740. *Grosses Universal Lexicon*, 64 vols. Leipzig.

VIII

William Whewell on the History of Science

INTRODUCTION

Over the last twenty years, the work of William Whewell has received serious attention, after an earlier period of unwarranted neglect. While his famous dispute with J.S. Mill over method and epistemology previously dominated approaches to Whewell's writings, it is now possible to see this controversy as part of a wider set of debates involving natural theology, education and moral philosophy – all subjects to which Whewell contributed.[1] But in spite of this recovery of Whewell as a philosopher of science, and an improved understanding of the intellectual and institutional context in which he worked, his role as an historian of science has remained comparatively neglected. In a sense this situation mirrors the uncertain reputation in Whewell's own day of his *History of the Inductive Sciences* (1837); for although that work was seen as a project of considerable importance, and more accessible than *The Philosophy of the Inductive Sciences* (1840), the response of the scientific community to it was ambivalent. "Of my History of science", he remarked, "the principal notice taken by men of science has been of a hostile kind; and I do not think that any practical cultivators of special sciences will feel any deference for a person who has presumed to speculate about them all".[2] Almost a century later, George Sarton found the work unsatisfactory, suggesting that it "is not a history of science as we understand it today, but a juxtaposition of various special histories, which is something very different; it represents a lower stage of integration".[3]

In both his major publications, Whewell was concerned with the historical development of the sciences – "the most certain and stable portions of knowledge which we already possess".[4] His research on the two projects overlapped, and in the opening remarks of both works he stressed their interdependence. A comprehensive view of Whewell's writing on the history of science would thus require reference to these two major works and to their several editions. Such a full account cannot be attempted here, but it is possible to analyze some key themes in Whewell's work against the background of other contemporary historical accounts of science.

Sarton's comment can be a useful starting point. Firstly, it is the case that most early-nineteenth-century histories of science did focus on particular disciplines and did not attempt the kind of synthesis which Sarton had in mind. But although it was organized in terms of the various branches of science, Whewell's *History* qualified and complicated some of the assumptions apparent in the histories of individual disciplines. This point will be discussed in the following four sections of this article.

Secondly, while Sarton's insinuation that Whewell's work lacked an overall framework is unfounded, it is true that he emphasized the distinctive conceptual foundations of the various disciplines.[5] However, this emphasis was more apparent in the *Philosophy*, where Whewell elaborated an epistemology that sought to combine elements from both Kantian idealism and Baconian empiricism. This topic has been discussed in recent literature and cannot be addressed here;[6] but it is important to recognize that this central feature of Whewell's epistemology led him to qualify notions of historical progress towards the unity of science, such as those promoted in the writings of Baden Powell. This point will be treated in the final section before the conclusion of this article.

HISTORIES OF DISCIPLINES

Scholars dealing with the history of historiography – the changing form and content of writing about the past – have noted significant shifts in the period from about 1800 to 1830. This development is sometimes described as 'the rise of historical consciousness', a phrase that perhaps begs more questions than it answers. But the case for a major development in historiography rests on two important grounds: firstly, that in this time there was a new critical use of sources, especially primary, archival documents; and secondly, that there was a greater sensitivity to the social and cultural differences between various historical periods. These early decades of the nineteenth century also saw the emergence of new disciplines such as geology, archaeology, anthropology and the study of language, all having strong historical frameworks.[7]

With reference to these developments, A.N.L. Munby remarked that "there is nothing mysterious about the sudden growth of the study of early science. Indeed, it would have been odd if it had escaped unstudied".[8] Munby noted that from the 1820s there was a strong interest among men of science in the earlier periods of scientific activity, particularly in the revolutionary achievements of the seventeenth century and the great names associated with them. Serious investigation of the scientific past was now based on systematic research into the surviving primary records; and by the middle of the century there were important biographies of Galileo, Kepler and Newton, scholarly editions of Bacon's works, and collections of letters and papers on seventeenth-century science.[9] Until recently, these developments were not themselves objects of study, but it is now recognized that historical representations of the past by nineteenth-century scientists were significant elements in contemporary debates both about particular theories and about the nature of the scientific enterprise. Thus while there were efforts to catalogue the primary documents of past science, there were also strong pressures to use this past in the service of present commitments.[10]

The most common vehicles for historical accounts of science were books, encyclopaedia entries or addresses dealing with particular disciplines. The crystallization of specialized subject areas and their institutionalization in separate scientific societies created the opportunity, and the need, for narratives which located the genealogy of present activities. By the first decade of the nineteenth century the preference for histories of single disciplines, or areas, rather than surveys of general science was already established. Thus in his *History of the Royal Society* of 1812, Thomas Thomson explained that "to give the reader a greater interest in the sciences ... it was thought necessary to begin the history of every science as nearly as possible at its origin, and to give a rapid sketch of its progress ...".[11] Textbooks on particular branches of science, encyclopaedias and elogia on the achievements of departed scientists all deployed historical accounts of past science in order to highlight the state of present knowledge. Given the significant, and often radical developments in several areas of science from the 1790s, these historical prefaces or chapters did not simply chronicle a list of discoveries, but sought to make a case for certain theories and methods associated with a particular conception of the subject. While there was a concern to anchor the present discipline in the past there was also a tendency to relegate some parts of it to a dark stage of pre-history from which the new science has escaped. At one extreme, Xavier Bichat (1771-1802), a proponent of physiology as a new, independent domain, regarded the past as largely irrelevant: "Our era no longer needs historical monuments. The history of science must be put to one side; it is science itself which we must present".[12] But history was too valuable as a polemical resource for Bichat's indifference to take hold. More usually, defenders of new disciplines or approaches provided a history which showed the errors of the past in order to illustrate the pitfalls from which the favoured doctrine, or hero, had rescued scientific knowledge.

A common theme of such apologetic histories was the emancipation of the discipline from various misleading preoccupations before its subsequent path to modern truth. In his *History of Chemistry* Thomson stated that "Chemistry ... sprang originally from delusion and superstition, and was at its commencement exactly on a level with magic and astrology ... It was not till about the middle of the eighteenth century that it was able to free itself from these delusions, and to venture abroad in all the dignity of a useful science".[13] Similar sketches of the history of geology were offered by members of the Geological Society of London, founded in 1807, who saw themselves as breaking away from earlier speculative cosmogonies in order to establish a new solid, empirical foundation for the study of the earth.[14] A more elaborate version of this perspective was promoted by Charles Lyell in the introduction to his *Principles of Geology* (1830). In an influential survey of ideas about the earth from the appearance of man until the early nineteenth century, Lyell constructed an historical drama in which his approach to the subject figured as the means by which geology was liberated from the superstitious, anthropomorphic and theological prejudices of the past. But he also warned against the legacy of assumptions from more primitive stages of human thought. Even in the present, he argued, geology suffered from the cosmogonies of earlier periods: "The superstitions of a savage tribe are transmitted through all the progressive stages of society till they exert a powerful influence on the mind of the philosopher". Modern geology had to clarify its "distinct nature and legitimate objects" by distinguishing itself from cosmogony and theology. Here again he found a warrant in history, because other subjects had also been more clearly defined and separated in the period since the "limits of history, poetry and mythology were ill-defined in the infancy of civilization".[15]

Another related impetus for a concern with the history of science was the belief that a record of past discoveries could reveal the method by which science progressed. Those involved in the promotion of novel approaches or new disciplines commonly referred to Bacon's notion of a survey of existing knowledge in order to uncover the causes of both failure and success.[16] As noted above, histories of particular branches of science could be used to locate the errors of previous practitioners and suggest explanations such as the influence of mysticism or rash speculation. Similarly, they were cited as keys to the most appropriate mode of pursuing scientific inquiry: the point at which the narrative located the emancipation of the science from past obscurantism also marked the discovery of the true scientific method. In the case of geology, as seen by members of the Geological Society, this occurred at a comparatively recent date and was in fact a rediscovery of the method of sober induction outlined by Bacon but ignored by earlier dogmatic and *a priori* theorists.[17] For Lyell, the more panoramic historical prelude to the *Principles* vindicated his own uniformitarian approach to geology. Thus while Lyell warned about the legacy of primitive perceptions, he also underlined a gradual but definite movement of the human mind away from fear and confusion to a rational understanding of the world around it in terms of the continuity of nature and the "undeviating uniformity of secondary causes". Lyell's history therefore revealed a psychological or anthropological warrant for his views on proper method in geology, and past scientists were judged in terms of their approximation to it. This exercise effectively yielded only one winner – Lyell himself – although Hutton was acknowledged.[18]

This brief discussion of histories of science in the early nineteenth century highlights their role in legitimating particular disciplines and methods. Such an apologetic function was less obvious in the general surveys of science which usually dealt with the more established branches of the mathematical and physical sciences. These were the subject of works such as John Playfair's *Dissertation* published in a preliminary volume to the *Encyclopaedia Britannica,* and Baden Powell's *An Historical View of the Progress of the Physical and Mathematical Sciences,* which was aimed at the wider readership of Lardner's *Cabinet Cyclopaedia.*[19] But these works shared with the disciplinary histories what has come to be called a 'Whig' perspective on the history of science.[20] Past science was largely approached with present theories as criteria for evaluating earlier thinkers and theories. Secondly, the past was usually divided into two fairly discrete periods – one of failure and one of success – with the passage from darkness to light being ascribed to the discovery or adoption of the correct method by celebrated heroes. Finally, significant points of transition such as that associated with the influence of correct method, or the appearance of a major theory, were often referred to as 'revolutions'; the subsequent development of the science was then conceived as a steady accumulation of empirical data under the compass of a general theory. In this way, histories of science incorporated a discourse about continuity and discontinuity which was central to contemporary political historiography.

This admittedly limited survey provides a set of themes through which Whewell's *History* can be discussed: first, the Whig notion of progress – in this case the gradual, cumulative progress of scientific knowledge towards nineteenth-century doctrines; second, the idea of historical study as a means of revealing the true scientific method; and third, the concept of revolution in science and its implications for Whig historiography.

PROGRESS

Introducing the *History,* Whewell said that the *"progress* of knowledge" was the "main action of our drama".[21] There is no doubt that a focus on the emergence of general theories successfully explaining empirical phenomena was a strong feature of his work. Nor is it difficult to find some similarity between Whewell's partisanship on behalf of the wave theory of light in his treatment of the development of optics and the Whiggish histories of disciplines such as geology.[22] But in contrast to the Whig view, Whewell's perspective was not one of continual progress away from past obscurantism and towards modern truth; rather, it suggested long "stationary periods".[23] The history of science was certainly not a manifestation of an improvement in mental capacities or the "March of Mind", as the satirist, Thomas Peacock, dubbed the triumphalist assumptions of his early-nineteenth-century contemporaries. And unlike eighteenth-century writers such as Priestley and Condorcet, who celebrated recent achievements as emancipations from past errors, Whewell made a case for the serious study of failures in science, arguing that they disclosed important clues about successful scientific discovery. His *History* was not a mere record of discoveries but an analysis of the dialectic between facts and theories, sensations and ideas, which he later elaborated in his *Philosophy* as the "fundamental antithesis" of all knowledge.[24] Since the lack of balance or fit between facts and ideas was the major reason for scientific error, modern science was also potentially susceptible to failures.[25] In writing of the mistakes and achievements of past science, Whewell, unlike the rationalist historians of the Enlightenment, displayed some sensitivity to cultural relativism, recognizing, for example, that social factors were relevant to the kind of science practised in particular times and places. Geoffrey Cantor attributes this feature of Whewell's historiography to his association with what Duncan Forbes has called the "Liberal Anglican" historians – English writers such as Julius and Augustus Hare, Connop Thirwall and Thomas Arnold – who were influenced by German thought and by the example of the great historian, Bartold Niebuhr. Their work questioned the assumptions of eighteenth-century rationalism and its optimism about the inevitable progress of reason.[26] When applied to the history of thought, including science, this perspective cautioned against a dismissive attitude to the past and regarded present progress as dependent upon past achievements.

This point is apparent in Whewell's attitude to what was often castigated by other writers as the baneful influence of past thought. For example, both Lyell and Powell described a movement from a period of intellectual darkness to the illumination of modern doctrines. For Lyell, as explained above, this movement was a liberation of the new science of geology. Powell's account of the effect of past beliefs was less colourful and not so explicitly intended to legitimate a particular discipline; but it was strongly informed by Auguste Comte's three stages of intellectual progress – the theological, metaphysical and positive (scientific). Powell's attraction to this aspect of Positivism was partly related to its strategic value in his efforts to clarify the boundaries between science and theology: for example, it allowed him to present teleological thinking as a vestige of an earlier, less scientific state.[27] But in addition to this, Powell conceived of the development of the various sciences as a movement "all in one direction", away from metaphysical notions to "the recognition of regulated causes, law and order ... ". Agreeing with Lyell, Powell saw geology as the subject which had "exhibited more of those changes from mysticism towards rationalism than any other branch". More generally, he pronounced that "we shall never see a recession from the more natural towards the more mysterious"; "all future progress must be in the same direction ...".[28]

As suggested earlier, Whewell's historiography did not endorse a simple contrast between modern science and past obscurantism. Confident about the progress of science, he was interested in the reasons for past failures and did not dismiss erroneous doctrines as foolish. As Yehuda Elkana has noted, when Whewell discussed failures in science, such as those of the Greeks, he spoke of the ideas involved as "vague, indeed, but not, therefore, unmeaning".[29] He had less positive things to say about the stationary period of the Middle Ages. This he considered a time in which the prevailing intellectual tendencies – characterized by the verbalism and dogmatism of the Scholastics – were unfavourable to science. The thinkers of this period were also possessed by the spirit of Mysticism, which either pursued arcane facts or wild generalizations. Thus "their physical science became Magic, their Astronomy became Astrology, the study of the Composition of bodies became Alchemy ...".[30] But even here Whewell saw some value: mystical pursuits signified a striving towards "something above the mere objects of sense and appetite", towards speculative truth. And later, when discussing Kepler, he said that the mysticism of the great natural philosopher was able to act as a stimulant in his thinking; although he was quick to add that this effect was possible because Kepler possessed clear scientific ideas.[31]

Another feature of this stationary period, the "commentatorial spirit", also impeded scientific progress because it encouraged exegesis rather than experiment: "criticism took the place of induction; and instead of great discoverers we had learned men".[32] But like mysticism, this tendency manifested a concern with ideas, and Whewell believed that this "speculative tendency" stimulated the abstract reflections by which "scientific Ideas" were distinguished from "common Notions".[33] He stressed that this activity could not be divorced from observation and experiment – as it had been during the Middle Ages – but equally insisted on the necessity of the discussions which Comte condemned as "metaphysical". In the *Philosophy* he wrote that "Physical discoverers have differed from barren speculators, not by having no metaphysics in their heads, but by having good metaphysics ...".[34] This point had not been recognized by those writers "accustomed to talk with contempt of all past controversies, and to wonder at the blindness of those who did not at first take the view, which was established at last".[35] These metaphysical controversies brought various subjects "into a condition in which error is almost out of our reach ...". Nevertheless, progress – in the sense of a wider coverage of empirical facts by laws – remained a key mark of a successful science for Whewell; and like other writers of the time, he hoped that a study of the history of science would reveal the method by which further advances could be made.

METHOD

Whewell proposed that a survey of past discoveries, such as the *History* provided, "may not only remind us of what we have, but may teach us how to improve and increase our store, and afford us some indication of the most promising mode of directing our future efforts to add to its extent and completeness".[36] This work was conceived as a part of a neo-Baconian project for classifying the sciences, their development, and the means of future progress. But Whewell also pointed to the significant difference between his position and Bacon's: as a later observer, Whewell had the data of two centuries of successful scientific practice before him.[37]

There is a definite resemblance between this project and the Enlightenment notion of the history of science as a collection of lessons and clues for those seeking new discoveries. According to Joseph Priestley, for example, the various branches of science would be assisted by "an historical account of their rise, progress and present state". Priestley's rationale was that such an historical inquiry "cannot but animate us in our attempts to advance still further, and suggest methods and experiments to assist us in our further progress".[38] But it is arguable that this resemblance is superficial and that the conception of the nature of scientific progress differs from Priestley to Whewell – shifting from an account of discoveries as the consequence of the use of proper method, to an emphasis on individual qualities which transcend method.

The lessons of Priestley's historiography were egalitarian: a closer examination of great discoveries lessened their heroic quality and suggested that they were achieved by methods open to all. But in Whewell's *History,* significant discoveries were attributed to the uncommon characteristics, both intellectual and moral, of exceptional individuals: Copernicus, Kepler, Newton, Lavoisier, Fresnel, Faraday.[39] Whewell's *Inductive Epochs* – those periods in which the Inductive process 'has been exercised in a more energetic and powerful manner' – centred on key discoveries or generalizations of a 'fundamental conception' in the work of particular scientists.[40] And although the *History* aimed to learn something of the method by which progress in science occurred, Whewell arrived at the conclusion that truly great men of science, in their epoch-making contributions, had transcended any specifiable rules of method. While he was prepared to describe the necessary characteristics of a great scientific mind – "distinctness of intuition, tenacity and facility in tracing logical connection, fertility of invention, and a strong tendency to generalization" – this was hardly the basis of an egalitarian method of discovery.[41]

In the *Philosophy,* Whewell developed the theory of induction which supported his views on method. He used the term induction to refer to the general process by which laws and theories were attained; but he stressed that this was more than a mere generalization from the facts, because it involved the addition of a conception from the mind of the scientist. "For in Induction ... besides mere collection of particulars, there is always a *new conception,* a principle of connexion and unity, supplied by the mind, and superinduced upon the particulars".[42]

These ideal conceptions, such as that of the ellipse in Kepler's astronomy, derived from the Fundamental Idea – for example, Space, Number, Resemblance – appropriate to a particular branch of science. Whewell claimed that these, and other Ideas, "regulate the active operations of our mind", and were the grounds of the necessary truths which certain branches of science had so far established.[43] Furthermore, he stressed that the ability to derive from a Fundamental Idea the appropriate conception, capable of distilling a collection of facts into a law, required the "peculiar Sagacity which belongs to the genius of a Discoverer".[44] Significantly, in reviewing Whewell's work, John Herschel noted approvingly his break with eighteenth-century Enlightenment attitudes which had been carried on into the present by the utilitarians: "It is too much the present fashion to ascribe all progress – at least all modern progress – in inductive science ... to 'the Age', as if there were some magic in the word, and as if by its use it were possible to elude or abate down the acknowledgement of individual pre-eminence".[45] Although, like Priestley, Whewell strongly recommended the study of the history of science to students and the public, the message was not meant to encourage the reader with a vision of him or herself as a future discoverer. Whewell's historiography prescribed a clear division of scientific labour, and so his speeches to the members of the British Association for the Advancement of Science stressed the difference between humble cultivators and the theoretical elite; he warned against centralized conceptions of science policy, since "we cannot create, we cannot even direct, the powers of discovery".[46] While still seeking a general 'Philosophy of Discovery', Whewell's own view on method and his reading of the history of science, resisted any simple 'Art of Discovery'.

Whewell thus shared the belief that advances in science depended upon the proper method of inquiry; but he did not endorse the notion of an accessible scientific method which Priestley and, later, the members of the Geological Society and the utilitarian reformers promoted.[47] Whewell distinguished between the capacities involved in foundational discoveries and those required for scientific work once the intellectual framework of the discipline was established. In this respect his *History* did bear on the concerns of disciplinary histories because it included the concept of a *Sequel,* following each Inductive Epoch, during which the key discovery was diffused to "the wider throngs of secondary cultivators ... and traced to its distant consequences". "This", he said, "is always a work of time and labour, often of difficulty and conflict".[48] Herschel provided an interpretation of Whewell's meaning when he suggested that the period of the Sequel resembled "the occupation and settling of the country under the dominion of the conquerers, quelling the insurrectionary movements of ignorance and prejudice under the new regime, and partitioning out the land in provinces and domains".[49] In this sense, the history of science did not show how to make epochal discoveries, but it said something about the making of disciplines.[50]

REVOLUTION

Now it can be seen that Whewell's stress on the dependency of scientific advances on past contributions, including past metaphysical controversies, coexisted with his account of the sudden and dramatic character of major inductive discoveries. As noted in the treatment of his views on method, it was this facet of significant breakthroughs in science, involving imaginative leaps, that told against simple, mechanical rules of scientific method. This sustained a tension between images of continuity and discontinuity in his historical writing, and led to some ambivalence in his references to the term 'revolution'.

I.B. Cohen has recently pointed to the occurrence of the phrase "revolution in science" in Whewell's accounts of the work of Copernicus, Bacon, Harvey, Locke and Lavoisier.[51] But Whewell's use of the term 'revolution' in conjunction with references to profound changes in science requires some scrutiny, since the political connotations of the word in the 1830s were far from favourable. British politics of this decade were conducted against the background of the July 1830 overthrow of the Bourbon monarchy in France, and of course, against the terrible symbol of the French Revolution itself. Even the compromises surrounding the Reform Bill of 1832 met with strong conservative opposition, and Whewell's known Tory leanings suggest a less than enthusiastic attitude to hasty political and social change. In considering the possibility of arriving at socio-political knowledge, Whewell said that such a search must not begin with the notion that "the conditions of this generation" were "novel and unparalleled", but with the awareness that "there is a great deal of truth already in the world, and a great deal of it embodied in the framework of society".[52] When reflecting on his own work he remarked that "however much I may feel a craving for a new view of the truth, suited to our age, ... I have no hope of finding any view of the truth, which is really true, if it do not include and rest upon that which has been true up to the present time".[53] These comments are consistent with Whewell's earlier opposition to the view that advances in science involved the rejection of "what was known for something new". In discussing this view of novelty with his friend, the theologian H.J. Rose, he was anxious to stress that "the novelty, if the philosophy have been duly inductive, includes old truths and shews them from a new point of view".[54] Thus in both the political and intellectual spheres, although Whewell believed in progress, there is reason to suggest that he was sceptical about the notion of radical change involving drastic breaks with the past. In this sense he shared, to some degree, the dilemma of the Whig political historians: that of balancing an approval of change and reform with respect for values and traditions enshrined in the past.

In his study of Victorian historians, J.W. Burrow has restored complexity and nuance to the image of Whig history. While accepting the general definition proposed by Butterfield which emphasised the confidence and optimism of this outlook, Burrow draws attention to another dimension of Whig historians: their sensitivities towards the past. Burrow argues that even in the most ardent celebrations of present achievements, such as those in the writings of T.B. Macaulay, there was a tension between progress and continuity. Although Whig history was a success story, it was not one which ridiculed the past: the past was revered because it embodied the tradition by which progress was sustained.[55] In this respect, the Whig historians cannot easily be equated with the Enlightenment rationalists whom Forbes identified as the opponents of his Liberal Anglicans. Indeed, on Burrow's reading, the Whigs, like the Liberal Anglicans, may have imbibed the 'Burkean tradition', in so far as the notion of 'change in continuity' was central to their historical consciousness, in which the stress on progress was constrained by the idealization of continuity.[56]

This account opens the way to further readings of Whewell's *History*. That is, without denying the insights offered by the comparisions with Liberal Anglicanism, it is possible to take advantage of Burrow's portrait of Whig historiography. For example, the Whig acknowledgement and approval of progress – both intellectual and social – is closer to Whewell's outlook than the cyclical vision contemplated by the writers in Forbes' study.[57] At the same time, the problem of reconciling a support for institutional adaptations and political reforms with a commitment to the continuity of a particular tradition, appears to have an analogue in Whewell's conception of scientific change. In both cases, the notion of revolution becomes problematic. Thus in Lord Acton's view, Burke and Macaulay regarded the settlement after the 'Glorious Revolution' of 1688 as conservative rather than revolutionary: "it was little more than a rectification of recent error, and a return to ancient principles".[58] Burrow draws attention to the fact that it was the Tories and radicals who could entertain images of dramatic ruptures in English history; whereas the Whigs' need for continuity compelled them to see bridges between past and present.[59]

An awareness of these tensions in Whig historiography allows some perspective on the ambivalance in Whewell's references to the concept of revolution; for although he clearly invoked this term, his overall response is arguably less unequivocal than Cohen suggests. Indeed, when Whewell used the term he often qualified its possible implications. This is not to suggest that there is no sense of dramatic and significant change in Whewell's *History*. He regarded the first successful inductive generalization in a particular area as the commencement of a new epoch in a given science, the inductive epoch. These breakthroughs placed the relevant science on a new intellectual footing and set the scene for significant redirection, subsequently recorded by changes in terminology – a process Whewell compared with "the change of the current coin" accompanying "great political revolutions".[60] As noted in the previous section, Whewell stressed the genius of individual scientists – such as Copernicus, Newton, Lavoisier – who grasped the ideal conceptions necessary for these discoveries and, in this sense, acknowledged their often revolutionary departure from preceding scientific ideas.

It is important, however, to recognize that Whewell restrained the possible implications of this account of scientific change in ways that underlined the continuities between past and present. Thus, although the crucial discovery of an inductive epoch required a grasp of the relationship between facts and ideas not achieved by the previous generation, Whewell noted the extent to which the "principal discoverer" was preceded by "trials, seekings, and guesses, on the part of others".[61] He reinforced this point when considering the dispositions of major scientists: "Undoubtedly this deference for the great men of the past joined with the talent of seizing the spirit of their methods when the letter of their theories is no longer tenable, is the true mental constitution of discoverers".[62] Whewell also stressed that sciences were not formed by a single act but by a long series of progressive changes involving "apparently contradictory" principles. He went on to say that:

> The principles which constituted the triumph of the preceding stages of the science, may appear to be subverted and ejected by the later discoveries, but in fact they are, (so far as they were true), taken up into the subsequent doctrines and included in them. They continue to be an essential part of the science. The earlier truths are not expelled but absorbed, not contradicted but extended, and the history of each science, which may thus appear like a succession of revolutions, is, in reality, a series of developments.[63]

This perespective allowed the celebration of significant advances constituting an 'epoch' together with an acknowledgement of the dependence on past efforts. It is here that the idea of a conservation of past constitutional wisdom, apparent in the historical vision of Burke and Macaulay, has it analogue in Whewell's image of a "vast patrimony of science" in which real achievements live on: "Thus the final form of each science contains the substance of each of its preceding modifications; and all that was at any antecedent period discovered and established, ministers to the ultimate development of its proper branch of knowledge".[64] Like the Whig historians, therefore, Whewell sought a compromise between the notions of revolutionary change and gradual development, between progress and continuity.

This was necessary because he wanted to acknowledge the fact of major conceptual shifts in science while still stressing the significant links between new and older theories. But this tension could not be resolved because it derived from Whewell's insistence on the novel and imaginative element of major inductive discoveries.

UNITY OF SCIENCE

The comment by Sarton cited at the beginning of this article suggested that Whewell's *History* was an unintegrated collection of special histories. Yet while it was organized around separate branches of science, Whewell's approach to the history of science contrasted in significant ways with the histories of disciplines offered by contemporaries. Yet there is another sense in which Whewell's work gave a central place to the special character of the various disciplines, in spite of his fear that the "cultivators of special sciences" would look askance at one who pretended to philosophize about all of them. This point can be illustrated with reference to the work of Baden Powell who, like Whewell, was concerned with the intellectual and moral lessons to be derived from the history of science.[65]

In all his major works, Powell maintained that the inductive method depended upon the conviction of a "universal and permanent uniformity of nature": scientific explanation assumed the uniformity of presently known causes and eschewed references to any causes supposedly different in kind from those open to inductive investigation.[66] He argued that all branches of science should adhere to this principle and would therefore share a unity of method. But furthermore, Powell envisaged a substantive unity of the sciences. Thus his argument moved from the uniformity of nature, as the basis of induction, to "the great archetype of Unity", the union of all the sciences within an explanatory framework constituted by common, general laws of nature.[67] In his historical references, he claimed that this was the direction of scientific thought since the ancients: that is, away from isolated branches and a diversity of explanatory concepts, towards those manifesting "harmony, simplicity, and unity of character". The progress of science could thus be witnessed in discoveries – such as those of Faraday – which combined apparently distinct phenomena under single principles. Convinced about the inevitability of this common direction, Powell contended that boundaries between various scientific disciplines for the purpose of classification were only practical and provisional. In an obvious reference to Whewell's work, he claimed that:

> It is a reversal of the order of inductive advance to endeavour to isolate each department of science, and to place it on a separate base, by a theory which would assign to each branch certain real differences of principle and peculiar fundamental ideas essentially characterizing it. If such a distinction could be made out, it could be but a temporary and provisional ground of classification to a higher common principle.[68]

This passage attacks one of the central features of Whewell's project – a concern with the special metaphysical foundations of different disciplines. This emphasis was not explicit in the *History*, for although the term 'Fundamental Idea' was introduced in this work, the full implications of Whewell's idealist epistemology only appeared in the *Philosophy*. In the *History*, Whewell argued that successful inductive discovery required the combination of a set of observations with an appropriate idea; but in the *Philosophy* he claimed that such discoveries involved the grasp of a conception derived from a Fundamental Idea *peculiar* to the specific branch of inquiry.[69] This meant that, for Whewell, different scientific disciplines rested upon distinct conceptual foundations.

Admittedly, when Powell spoke of a progress "in one direction", that is, towards unity, he could well have been thinking of Whewell's own references to the consilience of inductions and the convergence of theories towards greater simplicity. This was the story Whewell portrayed in his inductive tables for the history of astronomy and optics.[70] But whereas Powell anticipated the convergence of mechanics and chemistry under some general law, Whewell emphasized their connection with distinct, appropriate conceptions (mechanics with causation and chemistry with substance). In Powell's opinion, this position implied some "mysterious connection" of mechanics, as such, with the idea of causation; he argued that this subject displayed the characteristics which could also be expected in other fields when phenomena were reduced to "simple and intelligible causes of force and motion".[71] These conflicting perspectives suggested different scenarios for the mode of scientific advance in new areas. While Powell urged the extension of previously successful conceptions, Whewell warned that new truths were usually achieved by "the application of new ideas, not by the modification of old ones"; researchers had to resist the tendency to enlist familiar causes in order to seek "distinctly new conceptions" for "newly-studied phenomena".[72] Whewell emphasized the special conceptual framework of different disciplines and opposed Powell's image of a substantive unity of the various branches of science under general laws. In one of his many political metaphors, Whewell announced the intellectual sovereignty of certain fields of inquiry: "The Mechanical, the Secondary Mechanical, the Chemical, the Classificatory, the Biological Sciences", together formed "so many great Provinces in the Kingdom of Knowledge, each in a great measure possessing its own peculiar fundamental principles". In his view there was indeed a unity in the nature of scientific knowledge and the manner in which it was achieved, but not in a unity of the laws of nature.[73] Whewell was therefore in conflict with an influential view of scientific progress, popularized not only by Powell and those attracted to Comtean Positivism, but also by writers like Charles Babbage and Robert Chambers.[74]

CONCLUSION

This article has discussed Whewell's views on the history of science in relation to other contemporary accounts of the scientific past: the histories of disciplines and the general survey of the physical sciences represented by Powell's work. Whewell's historical writing qualified some of the central themes of both, largely because it was informed by a different philosophy of science. We can now review these points of conflict and their implications.

Like the disciplinary histories, Whewell's work certainly affirmed progress, but in doing so, it did not support the image of scientific advances as complete breaks with an obscurantist past dominated by metaphysical thinking. The history of science was not entirely one of successive inductive generalizations leading gradually towards modern doctrines; rather, older, erroneous theories were replaced by new theories, often in sudden and dramatic fashion. But Whewell's position did not support the notion of a single, revolutionary liberation from the dark pre-history of a science. Firstly, as already noted, he stressed the important intellectual preparation provided in periods before the first crucial inductive discoveries of a particular science. Secondly, in his account of scientific change there was often more than one significant shift within the history of an individual discipline. Thus, even in formal (positional) astronomy there were revolutionary shifts from Hipparchus to Copernicus to Kepler. The implication here was that more recent disciplines, which were immature by comparison with astronomy, could not assume that their present theories were immutable. In this way, the triumphalism of the disciplinary histories was tempered by a comparison with other branches of science in Whewell's more general survey.

Although more comprehensive than Powell's works on the history of science, Whewell's *History* did share the image of progress in science as a movement towards more simple and general theories embracing a wide range of phenomena. Like Powell, Whewell also spoke of the unity of science, but by this he meant the common intellectual processes, including the dialectic between facts and ideas, by which knowledge was acquired, not the eventual incorporation of many sciences under one law. On this point, the epistemology set out in the *Philosophy,* with its emphasis on the distinct Fundamental Ideas underlying different disciplines, produced an historical message in conflict with the more general, synthetic histories which Powell and, later, Sarton, wished to advocate.

Whewell's writing on the history of science can also be approached from the perspective of contemporary political historiography. Geoffrey Cantor has argued that Whewell's *History* was influenced by Liberal Anglicanism and its qualification of Enlightenment assumptions. And by taking up J.W. Burrow's analysis of Whig political historiography, it can be suggested that Whewell's work had closer affinities with this tradition than it did with the triumphalist disciplinary histories which epitomized what has come to be seen as 'whiggish' history of science. Furthermore, by deploying motifs from general history – such as progress, revolution, and tradition – Whewell related the scientific enterprise to other political and cultural activities. In this sense, his work may be seen as legitimating not particular scientific disciplines, but science itself.

ACKNOWLEDGEMENTS

I would like to thank the editor and referees for helpful comments and suggestions on an earlier draft of this article.

Notes

1. See E.W. Strong, "W. Whewell and J.S. Mill: Their Controversy About Scientific Knowledge", *Journal of the History of Ideas* (1955), *16*:209-231; J.B. Schneewind, "Whewell's Ethics", *American Philosophical Quarterly* (1968), *1*:108-141; R.R. Yeo, "William Whewell, Natural Theology, and the Philosophy of Science in Mid-Nineteenth-Century Britain", *Annals of Science* (1979), *36*:493-512; H.W. Becher, "William Whewell and Cambridge Mathematics", *Historical Studies in the Physical Sciences* (1980), *11*:1-48; M. Garland, *Cambridge Before Darwin: The Ideal of Liberal Education 1800-1860* (Cambridge, Eng.: Cambridge University Press, 1980).
2. W. Whewell to Lord Northhampton, 5 October 1840, in I. Todhunter, *William Whewell, D.D. An Account of his Life and Writings* (2 vols.; London: Macmillan, 1876), vol. 2, p. 293.
3. G. Sarton, *The Study of the History of Science* (Cambridge, Mass.: Harvard University Press, 1936), p. 63.
4. W. Whewell, *The Philosophy of the Inductive Sciences* (2 vols., 2nd ed.; London: J.W. Parker, 1847), vol. 1, p. 1. Subsequent references to this work are also from this second edition.

38

5. For an analysis of Whewell's wider project, see Yeo, "William Whewell's Philosophy of Knowledge and Its Reception", in M. Fisch and S. Schaffer (eds.), *William Whewell: A Composite Portrait – Studies of His Life, Work and Influence* (Oxford: Oxford University Press, forthcoming).

6. See, for example, G.C. Seward, *Die Theoretische Philosophie William Whewells und der Kantische Einfluss* (Tubingen, 1936); and D.R. Oldroyd, *The Arch of Knowledge: An Introductory Study of the History of the Philosophy and Methodology of Science* (Kensington: University of New South Wales Press, 1986), pp. 157-164. While Whewell found Kant's work important when formulating his philosophy during the mid-1830s, he nevertheless rejected key aspects of Kantianism. See M. Fisch, "Necessary and Contingent Truth in William Whewell's Antithetical Theory of Knowledge", *Studies in History and Philosophy of Science* (1985), *16:*275-314 (p.279). For the theological side of his distance from Kant, see Yeo, *op.cit.* (ref. 1), pp.512-514.

7. P. Levine, *The Amateur and the Professional: Antiquarians, Historians, and Archaelogists in Victorian England, 1838-1886* (Cambridge, Eng.: Cambridge University Press, 1986).

8. A.N.L. Munby, *The History and Bibliography of Science in England: The First Phase, 1833-1845* (Berkeley: University of California School of Librarianship, 1968), p. 2.

9. See, for example, S.P. Rigaud, *Historical Essay on the First Publication of Sir Isaac Newton's Principia* (Oxford, 1838): *Correspondence of Scientific Men of the Seventeenth Century* (2 vols.; Oxford, 1851). I have dealt with the relevance of this kind of work to Newton's reputation in my paper, "Genius, Method and Morality: Images of Newton in Britain, 1760-1860", presented to the Newton Workshop, Tel Aviv and Jerusalem, April 1987. For perceptions of Bacon, see Yeo, "An Idol of the Marketplace: Baconianism in Nineteenth-Century Britain", *History of Science* (1985), *23:*251-298.

10. L. Graham, W. Lepenies, and P. Weingart (eds.), *Functions and Uses of Disciplinary Histories* (Dordrecht: Reidel, 1983).

11. T. Thomson, *History of the Royal Society from its Institution to the End of the Eighteenth Century* (London: R. Baldwin, 1812), preface.

12. Quoted in D. Outram, "Uncertain Legislator: Georges Cuvier's Laws of Nature in Their Intellectual Context", *Journal of the History of Biology* (1986), *19:*323-368 (p.352).

13. Thomson, *The History of Chemistry* (London: H. Colburn, 1830), p. 1.

14. R. Porter, *The Making of Geology: Earth Science in Britain 1660-1815* (Cambridge, Eng.: Cambridge University Press, 1977), pp. 138, 146; R. Laudan, "Ideas and Organization in British Geology", *Isis* (1977), *68:*527-538.

15. For these citations, see R. Porter, "Charles Lyell and the Principles of the History of Geology", *British Journal for the History of Science* (1976), *9:*91-103. For a less elaborate, but similar perspective, see J. Phillips, "Geology", *Encyclopaedia Metropolitana* (29 vols; London, 1845), vol. 6, pp. 530-534.

16. See Yeo, *op.cit.* (ref. 9, 1985).

17. See Laudan, *op.cit.* (ref. 14). For an analysis of these appeals to method, see D.P. Miller, "Method and the 'Micropolitics' of Science: The Early Years of the Geological and Astronomical Societies of London", in J.A. Schuster and R.R. Yeo (eds.), *The Politics and Rhetoric of Scientific Method: Historical Studies* (Dordrecht: Reidel, 1986), pp.227-258.

18. See Porter, *op.cit.* (ref. 15) and Laudan, "Reflections of a Discipline: Histories of Geology and Geological History", in Graham *et al.*, *op.cit.* (ref. 10), pp. 79-104.

19. Dionysius Lardner's collection also included volumes on geology, chemistry and natural history, all of which carried historical introductions. Powell's volume was published in 1834.

20. H. Butterfield, *The Whig Interpretation of History* (London: G. Bell, 1931); cf. Oldroyd, "Sir Archibald Geike and the Problem of Whig Historiography", *Annals of Science* (1980), *37:*441-462.

21. Whewell, *History of the Inductive Sciences* (3 vols., 3rd ed.; London: J.W. Parker, 1857), vol. 1, p. 4.

22. See G.N. Cantor, *Optics After Newton: Theories of Light in Britain and Ireland, 1704-1840* (Manchester: Manchester University Press, 1983), pp. 1-3.

23. Whewell, *op.cit.* (ref. 21), vol. 1, pp. 11-12.

24. Whewell, *op.cit.* (ref. 4), vol. 2, pp.647-668. The essay was originally published in the

Transactions of the Cambridge Philosophical Society for 1844. For an analysis of this concept in Whewell's work, see Fisch, *op.cit.* (ref. 6).

25. Whewell, *op.cit.* (ref. 21), vol. 1, pp. 5-7.
26. D. Forbes, *The Liberal Anglican Idea of History* (Cambridge, Eng.: Cambridge University Press, 1952); Cantor, "Between Rationalism and Romanticism: Whewell's Historiography of the Inductive Sciences", in Fisch and Schaffer (eds.), *op.cit.* (ref. 5).
27. B. Powell, *Essays on the Spirit of the Inductive Philosophy, the Unity of Worlds, and the Philosophy of Creation* (London: Longman, 1855), pp. 63-74.
28. *Ibid.*, p. 60.
29. Quoted in Y. Elkana (ed.), *William Whewell. Selected Writings on the History of Science* (Chicago and London: University of Chicago Press, 1984), p. xviii.
30. Whewell, *op.cit.* (ref. 21), vol. 1, pp. 215, 233.
31. *Ibid.*, pp. 319-320. It is worth noting that Whewell resisted the proposition of a simple connection between medieval alchemy and modern chemistry: in this case there was not a substantive intellectual link between a mystical doctrine and a scientific discipline; but alchemy was one factor in the revolt against scholastic authority, and in this way it assisted the rise of scientific thought. *Ibid.*, pp. 232-233, 246.
32. *Ibid.*, p. 204.
33. *Ibid.*, pp. 12-13.
34. Whewell, *op.cit.* (ref. 4), vol. 1, p. x.
35. *Ibid.*, vol. 2, pp. 377-378.
36. Whewell, *op.cit.* (ref. 21), vol. 1, p. 4.
37. For Whewell's attitude to Bacon, see Whewell, *op. cit.* (ref. 4), pp. 226-251 and Yeo, *op.cit.* (ref. 9, 1985), pp. 272-277.
38. J. Priestley, *The History and Present State of Electricity with Original Experiments* (2 vols., 3rd ed.; London: C. Bathurst, 1775), vol. 1, p. vi.
39. On attitudes to genius, see Yeo, *op.cit.* (ref. 9, 1987) and P. Theerman, "Unaccustomed Role: The Scientist as Historical Biographer – Two Nineteenth-Century Portrayals of Newton", *Biography* (1985), *8*:145-162.
40. Whewell, *op.cit.* (ref. 21), vol. 1. pp. 9-10.
41. *Ibid.*, vol. 2, p. 139. Whewell did note the unsuccessful guesses of discoverers, but not with the intention of endorsing democratic notions, such as those of Priestley, about the nature of discovery. *Ibid.*, vol. 1, pp. 317-331.
42. Whewell, *op.cit.* (ref. 4), vol. 2, p. 77; also pp. 49-50.
43. *Ibid.*, vol. 1, p. 66.
44. *Ibid.*, vol. 2, p. 40.
45. [J. Herschel], "Whewell on Inductive Sciences", *Quarterly Review* (1841), *68*:177-238 (p.187).
46. Whewell, *British Association Reports* (London, 1834), pp. xii, xxiv. On the various appeals to notions of method in debates concerning the British Association, see Yeo, "Scientific Method and the Image of Science, 1831-1890", in R.M. MacLeod and P. Collins (eds.), *The Parliament of Science: the British Association for the Advancement of Science, 1831-1981* (Northwood: Science Reviews Ltd, 1981), pp.65-88.
47. See Yeo, "Scientific Method and Rhetoric of Science in Britain, 1830-1917", in Schuster and Yeo (eds.), *op.cit.* (ref. 17), pp. 259-298 (pp. 264-273). Whewell acknowledged that there was more general participation in the early stages of a new science, but stressed the increasing gap between specialized disciplines and the public.
48. Whewell, *op.cit.* (ref. 21), vol. 1, p. 10.
49. Herschel, *op.cit.* (ref. 45), p. 186.
50. See S. Schaffer, "Scientific Discoveries and the End of Natural Philosophy", *Social Studies of Science* (1986), *16*:387-420.
51. I.B. Cohen, *Revolution in Science* (Cambridge, Mass.: Harvard University Press, 1985), pp. 529-532. Cohen sees Whewell's use of the term 'revolution' as part of a "fully developed theory" (p. 532); but this view requires some qualification because Whewell's emphasis depended on the opponents he addressed. Thus when replying to the cumulative, empiricist view of scientific development, he stressed qualitative shifts more than he did in parts of the *History*. See Whewell, "Mr. Macaulay's Praise of Superficial

40

Knowledge", *Fraser's Magazine* (1849), *40:*171-175 (p. 172).
52. Whewell to J.G. Marshall, 27 December 1842, in J.M. Stair-Douglas, *The Life and Selections from the Correspondence of William Whewell, D.D.* (2nd ed.; London: Kegan Paul, 1882), p. 281.
53. *Ibid.,* p. 282.
54. Whewell to H.J. Rose, 12 December 1826, in Todhunter, *op. cit.* (ref. 2), vol. 2, p. 79.
55. J.W. Burrow, *A Liberal Descent. Victorian Historians and the English Past* (Cambridge, Eng.: Cambridge University Press, 1981), pp.2-3.
56. *Ibid.,* pp. 22-23, 47-48, 52. For the influence of Burke on the Liberal Anglicans, see Forbes, *op. cit.* (ref. 26), pp. 1, 6.
57. See also Cantor, *op. cit.* (ref. 26), for a recognition of this tension.
58. Cohen, *op. cit.* (ref. 51), p. 71.
59. Burrow, *op. cit.* (ref. 55), pp. 16-17, 34.
60. Whewell, *op. cit.* (ref. 21), vol. 1, p. 9.
61 *Ibid.,* vol. 2, p. 99.
62 *Ibid.,* vol. 1, p. 286.
63. *Ibid.,* p. 8.
64. *Ibid.,* pp. 3, 8. For a refinement of this view, see "On the Transformation of Hypotheses in the History of Science", in Whewell, *On the Philosophy of Discovery* (London: J.W. Parker, 1860), p. 151.
65. See D.M. Knight, "Professor Baden Powell and the Inductive Philosophy", *Durham University Journal* (1967), *29:*81-87.
66. Powell, *On Necessary and Contingent Truth* (Oxford: T. Combe, 1849), p. 35.
67. *Ibid.* See also Powell, *op. cit.* (ref. 27), p. 98.
68. Powell, *op. cit.* (ref. 27), pp. 44-45; see also pp. 49-50 on the example of Faraday's work.
69. For an analysis of the tension in Whewell's work resulting from this, see Yeo, *op. cit.* (ref. 5).
70. Whewell spoke of a "Consilience of Inductions" as taking place when an induction based on one class of facts coincided with one derived from a different class of facts. He saw the discoveries of astronomy and optics as a "constant advance towards unity, consistency, and simplicity"; but this was *within* particular branches of science and not across disciplines. Whewell, *op. cit.,* (ref. 4), vol. 2, p. 78. The tables are inset after p. 118. In his unpublished notes, Whewell also attempted, apparently without satisfaction, similar inductive tables for the discoveries in 'thermotics' and chemistry. See Whewell Papers, R. 18. 10 no. 8 and Add Ms. a 217, no. 4. These are held at Trinity College, Cambridge. For access to these papers, I wish to thank the Master and Librarian of the College.
71. Powell, *op. cit.* (ref. 27), pp.45-46; and *op. cit.* (ref. 66), pp. 24-31.
72. Whewell, *op. cit.* (ref. 4), vol. 2, pp. 100-105.
73. *Ibid.,* p. 19; also vol. 1, p. 15. For further discussion of this see Yeo, *op. cit.* (ref. 5).
74. See C. Babbage, *Ninth Bridgewater Treatise: a Fragment* (2nd ed.; London: J. Murray, 1838), p.32; [R. Chambers], *Explanations: a Sequel to the Vestiges of the Natural History of Creation* (2nd ed.; London: J. Churchill, 1846), pp. 3-5, 147-149. Even close supporters such as James David Forbes felt constrained by Whewell's focus on the disciplinary field and, in reviewing the *History,* suggested that it be translated into a four-stage general history of science from the 1450 to 1850. J.D. Forbes, "The History of Science; and Some of its Lessons", *Fraser's Magazine* (1858), *57:*283-294 (p. 287).

Reading Encyclopedias

Science and the Organization of Knowledge in British Dictionaries of Arts and Sciences, 1730–1850*

> In everything that relates to *science,* I am a whole Ency-
> clopaedia behind the rest of the world.—CHARLES LAMB,
> "The Old and the New Schoolmaster" (1821).

> The names of the last authors he has consulted come
> back to my mind: Lamberty, Langlois . . . Lavergne. It
> is a revelation; I have understood the Autodidact's
> method: he is teaching himself in alphabetical order.—
> Antoine Rocquentin, in JEAN-PAUL SARTRE, *Nausea*
> (Penguin, 1978).

FOR AT LEAST THE LAST THOUSAND YEARS encyclopedias—arguably the most striking publishing enterprise of Western culture—have had
to confront an apparent absurdity: the combination of universal knowledge and
alphabetical order. In the ancient world such works were organized on various
philosophical and educational principles, but from the Renaissance on most
adopted the alphabetical arrangement. However, the idea that encyclopedias
embodied the unity of the "circle of knowledge" remained, coexisting uneasily
with the practice of placing topics in alphabetical order. Accordingly, from the
eighteenth century the prefaces of encyclopedias were devoted to explanations of
the ways in which the particular publication did in fact exemplify logical or systematic relationships between the various parts of knowledge that lay scattered
throughout its pages.[1]

Recent reflections on the role of encyclopedias sometimes seek to recover this
notion of unity. In 1974 the fifteenth edition of the Encyclopaedia Britannica
reconsidered this idea and sought to illustrate it in a *Propaedia,* the first volume
of the work. The next nine volumes of this edition are described as showing "the
knowledge of the universe we possess by means of the disciplines"—ranging
from physics to sociology. The *Propaedia* acknowledges the extreme specializa-

*I am grateful to John Gascoigne, Rachel Laudan, and Dorinda Outram for advice and encourage-
ment on earlier drafts of this article; to participants in seminars at Griffith University, University of
New South Wales, University of Hawaii, and Monash University, where versions of it were presented, for helpful comments; and to the Inter-Library Loan staff at Griffith University and the
rare-book librarians Margaret Dent of Australian National University and Richard Overell of Monash
University, for guidance and assistance. Research was supported by an Australian Research Council
grant.
[1] For a historical review see A. W. Read, "Encyclopaedias and Dictionaries," in *New Encyclopaedia Britannica,* 15th ed., ed. P. W. Goertz, 32 vols. (Chicago: Encyclopaedia Britannica, 1974), Vol.
XVIII, pp. 365–394.

tion of disciplines and discusses whether the knowledge they produce can be seen as complementary, as constituent parts of an integrated whole. Thus the editor of the *Propaedia,* Mortimer Adler, asks: "Do the various disciplines adhere together harmoniously?" Adler suggests that today there may indeed be a "circle of learning instead of a hierarchy of the branches of knowledge" found in traditional schemes of classification, and he proposes the conception of the encyclopedia as a totality, exemplifying the conviction that "the whole world of knowledge is a single universe of discourse."[2]

But this ideal was already insecure by the eighteenth century, precisely when the modern encyclopedic project was launched.[3] Two points need to be made here. First, the concept of systematic unity, in either a hierarchy or a circle of knowledge, was never effectively implemented in encyclopedias in the eighteenth century, although it did play an important role—variously conceived—in their legitimating rhetoric. In Ephraim Chambers's *Cyclopaedia* of 1728 and the *Encyclopédie* of 1751, edited by Denis Diderot and Jean d'Alembert, prefatory disquisitions on the unity of knowledge were accompanied by pronouncements about a society united by the power of science or reason. In Chambers this was couched in a form acceptable to the established regime: the dedication to the king recommended scientific knowledge (which for Chambers was the answer to superstition) as an appropriate symbol of royal prestige. Abraham Rees, a later editor of the *Cyclopedia,* put the message more directly, saying that cultivation of the arts and sciences offered a way of diffusing "a glory over this country, unattainable by conquest and dominion."[4] In the *Encyclopédie* the social vision was linked with an explicit challenge to the traditional order and the affirmation of a cosmopolitan ideal.[5] But by at least the mid-nineteenth century, encyclopedias were no longer associated with any political or social challenge, and the concept of a unity of knowledge was no longer central to their purpose.

Second, encyclopedias have recorded and reinforced the divisions within knowledge, especially science, as much as they have evinced connections and interactions. At the end of the nineteenth century the intellectual historian J. T. Merz noted the decline of the assumption that "the sum of human knowledge is an organic whole." He contended that the encyclopedic enterprise and its collection and classification of knowledge had led to the "disintegration, not the unification, of knowledge and thought."[6] It is arguable that rather than being a force against specialization, as is often suggested, encyclopedias have reflected and

---

[2] *New Encyclopaedia Britannica,* Vol. I: *Propaedia,* ed. Mortimer J. Adler, pp. 5–8, 475–477. For an earlier attempt to recover unity see Otto Neurath, Rudolf Carnap, and Charles Morris, eds., *International Encyclopaedia of Unified Sciences,* Vol. 1 (Chicago: Univ. Chicago Press, 1938).

[3] Robert McRae, *The Problem of the Unity of the Sciences: Bacon to Kant* (Toronto: Univ. Toronto Press, 1961), pp. 69–73, notes that Leibniz had projected the idea of a "demonstrative encyclopaedia" that would organize the key principles of various arts and sciences, but his conception was based on a deductive and hierarchical view of the sciences and was never actualized.

[4] Ephraim Chambers, *Cyclopaedia; or, An Universal Dictionary of Arts and Sciences,* ed. Abraham Rees, 5 vols. (London: C. Rivington, 1786–1788), Vol. I, dedication. On Chambers's position see Philip Shorr, *Science and Superstition in the Eighteenth Century: A Study of the Treatment of Science in Two Encyclopaedias of 1725–1750* (New York: Columbia Univ. Press, 1932).

[5] Thomas S. Schlereth, *The Cosmopolitan Ideal in Enlightened Thought* (Notre Dame, Ind.: Univ. Notre Dame Press, 1977); E. J. Hundert, "A Cognitive Ideal and Its Myth: Knowledge as Power in the Lexicon of the Enlightenment," *Social Research,* 1986, *55*:133–157.

[6] John T. Merz, *A History of European Thought in the Nineteenth Century,* 4 vols. (London/Edinburgh: Blackwoods, 1894–1914), Vol. I, p. 37.

facilitated the crystallization of natural knowledge into disciplines from the late eighteenth to the mid-nineteenth century.

Since the eighteenth century there has been a strong correlation between the multiplication of encyclopedias and the growth of scientific knowledge. Most of the encyclopedias published from this time gave an important place to science and technology—they were often called dictionaries of arts and sciences; subjects such as biography, history, geography, and literature were usually later additions. While they were a key vehicle for the diffusion of science and practical arts, it would be wrong to view encyclopedias as simply agents of popularization. In contrast to their successors, encyclopedias of the early nineteenth century carried articles that were effectively pieces of original research at the leading edge of various fields. In a pioneering survey of science in British encyclopedias from 1704 to 1874, Arthur Hughes noted the difficulties faced by editors in keeping new editions current with new scientific data and theories. The result of their efforts—the daunting kilometers of library space—has itself made detailed studies of encyclopedias an infrequent and heroic endeavor.[7]

One can, however, approach these massive texts by analyzing the rationales and assumptions underlying their organization of scientific material. This article attempts to examine debates over systematic classification and alphabetical arrangement from the time of Chambers's *Cyclopaedia* until the mid-nineteenth century. It argues that over this period systematic schemes were generally abandoned in favor of alphabetical organization. Special attention is given to the *Encyclopaedia Metropolitana*—the last significant attempt at a philosophical ordering of subjects—and to the various editions of the *Encyclopaedia Britannica*—the most famous of the alphabetical dictionaries of arts and sciences. In spite of their different formats, both these works highlighted, and possibly reinforced, the transition from eighteenth-century natural philosophy to the specialized scientific disciplines of the nineteenth century. A related aim is to show how the predominant alphabetical arrangement, rather than presenting an unordered array of information, contributed to this consolidation of disciplines.

CLASSIFICATION AND EIGHTEENTH-CENTURY ENCYCLOPEDIAS

In 1829 one publisher referred to the previous century as the age of encyclopedias: there has been "six principal encyclopaedias" from the time of "the great Father of the Encyclopaedic enterprize, Mr. Chambers."[8] The eighteenth century is also renowned as the age of classification, and the link between these two phenomena is an obvious one to make. Indeed, attempts to catalogue expanding

---

[7] Arthur Hughes, "Science in English Encyclopaedias, 1704–1874," *Annals of Science,* 1951, 7:340–370; 1952, *8*:323–367; 1953, *9*:233–264. Major historical and literary articles began to appear in the second edition of the *Encyclopaedia Britannica* (1778) and occupied a significant part of other nineteenth-century encyclopedias. Examining the distinction between arts and sciences in these publications is beyond the scope of this article. For a general account of the encyclopedic enterprise see Robert Collison, *Encyclopaedias: Their History through the Ages* (New York: Hafner, 1964). On specific publications see Herman Kogan, *The Great EB: The Story of the* Encyclopaedia Britannica (Chicago: Univ. Chicago Press, 1958); Paul Kruse, "The Story of the *Encyclopaedia Britannica,* 1768–1943" (Ph.D. diss., Univ. Chicago, 1958); Jacques Proust, *L'Encyclopédie* (Paris: Cohn, 1962); J. C. Guedon, "The Still Life of a Transition: Chemistry in the *Encyclopédie*" (Ph.D. diss., Univ. Wisconsin, 1974); John Lough, *The* Encyclopédie *in Eighteenth-Century England* (Newcastle: Oriel, 1970); and Robert Darnton, *The Business of Enlightenment: A Publishing History of the* Encyclopédie (Cambridge, Mass.: Belknap Press of Harvard Univ. Press, 1979).

[8] Thomas Curtis, *The London Encyclopaedia,* 22 vols. (London: T. Tegg, 1829), Vol. I, p. iii.

forms of knowledge could easily be seen as a reflection of the Enlightenment concern with ordering the world. Thus Bernard Groethuysen interpreted the *Encyclopédie* as an expression of the bourgeois desire to acquire, possess, and accumulate goods; it was an inventory of commodities, a tour of humanity's intellectual estate.[9] But it is important to recognize that while used by editors of encyclopedias—and given more public prominence by them—philosophical schemes of classification had a long tradition as an independent and self-critical genre.[10]

There is no doubt that early encyclopedias exhibited an interest in the classification of knowledge. In a review of works prior to Chambers's *Cyclopaedia*, Robert Shackleton lists several titles from the sixteenth century that used the term *encyclopaedia*. Some included intricate diagrams showing the relationships of faculties and subjects. This concern partly overlapped with the contemporary attempts of Locke, Descartes, Leibniz, and Kant to analyze the intellectual relationships between different branches of knowledge and arrange them according to some philosophical principle. The longevity of this tradition was charted by Robert Flint in 1904, when he surveyed the extensive array of classificatory schemes developed by thinkers from Aristotle to Bacon to Comte.[11] During the seventeenth century there were links between this activity and the taxonomic enterprises that governed research in natural history and sponsored interest in schemes of universal language.[12] But this enterprise was sustained by epistemological and pedagogic concerns that preceded the major eighteenth-century effort to record and disseminate "universal knowledge." From the publication of Chambers's work in 1728, systematic classification of the sciences was not the dominant impetus or the central organizing element of the encyclopedic project.

The major encyclopedias of the eighteenth century, such as Chambers's *Cyclopaedia* and the *Encyclopédie*, were primarily concerned with the collection of knowledge in an accessible form. They were partly motivated by the anxiety Leibniz expressed in 1680, when he spoke of that "horrible mass of books which keeps on growing," so that it would soon be a disgrace rather than an honor to be an author.[13] Chambers warned that "a reduction of the body of learning" in the form of an encyclopedia was "growing every day more necessary."

[9] Bernard Groethuysen, cited in Herbert Dieckmann, "The Concept of Knowledge in the *Encyclopédie*," in Dieckmann, Harry Levin, and H. Motekat, *Essays in Comparative Literature* (St. Louis, Mo.: Washington Univ. Studies, 1961), pp. 73–107, on p. 84. See also Michel Foucault, *The Order of Things* (London: Tavistock, 1972), Ch. 5, on the preoccupation with classifying as central to the classical *episteme*.

[10] McRae, *Problem of Unity* (cit. n. 3); R. G. A. Dolby, "Classification and the Sciences: The Nineteenth-Century Tradition," in *Classifications and Their Social Context*, ed. R. Ellen and D. Reason (London: Academic Press, 1979), pp. 167–193.

[11] Robert Shackleton, "The Encyclopaedic Spirit," in *Greene Centennial Studies: Essays Presented to Donald Greene*, ed. Paul J. Korshin and Robert R. Allen (Charlottesville: Univ. Virginia Press, 1984), pp. 377–390, on pp. 384–387 for the list; and Robert Flint, *Philosophy as Scientia Scientarum and a History of the Classifications of the Sciences* (Edinburgh: Blackwood, 1904). See also Dolby, "Classification and the Sciences" (cit. n. 10); P. Speziali, "Classification of the Sciences," in *The Dictionary of the History of Ideas*, ed. Philip Wiener, 5 vols. (New York: Scribners, 1968–1974), Vol. I, pp. 462–467; and Henry E. Bliss, *The Organization of Knowledge and the System of the Sciences* (New York: Holt, 1929).

[12] Comparisons could be drawn between Chambers's chart of knowledge (see second paragraph below) and the classifications of animals in the Aristotelian tradition. See M. M. Slaughter, *Universal Languages and Scientific Taxonomy in the Seventeenth Century* (Cambridge: Cambridge Univ. Press, 1982), pp. 30–31.

[13] G. W. Leibniz, "Precepts for Advancing the Sciences and Arts," in *Leibniz Selections*, ed. Philip Wiener (New York: Scribners, 1951), pp. 29–45, on pp. 29–30.

Without this he feared that the progress of science would be handicapped because those who should make discoveries would have to spend their "whole life learning what is already found out." This sense of urgency was also reinforced by the belief that certain "inventions and improvements" might "sink into oblivion." Chambers's language certainly lends support to Groethuysen's analogy between intellectual and material possessions: he relied heavily on geographic metaphors, writing in the preface to the *Cyclopaedia* that "in the wide field of intelligibles, appear some parts which have been more cultivated than the rest; chiefly on account of the richness of the soil, and its early tillage; but partly too, by reason of the skillful and industrious hands under which it has fallen. These spots, regularly laid out, and conveniently circumscribed, and fenced round, make what we call the *Arts and Sciences.*"[14]

Chambers offered a chart of knowledge but did not see it as central to his aims. He apologized for having to explain the principle of organization adopted in the *Cyclopaedia,* although admitting that this was the "great, but obscure hinge, on which the whole encyclopaedia turns." Yet he then said that the arts and sciences, the two largest parts of the map of knowledge, have been divided into "subordinate provinces" in a "wholly arbitrary" manner: "I do not know whether it might not be more for the general interest of learning, to have the partitions thrown down, and the whole laid in common again, under one undistinguished name."[15] Here the territorial metaphor is also political: Chambers was not enthusiastic about the enclosure of knowledge in various categories, preferring that its common field be left for free cultivation.

Chambers did not endorse current classifications or acknowledge that such divisions were useful. His main concern was the collection of knowledge in manageable form as a base point for further inquiry; the most powerful image of the *Cyclopaedia,* contributed by Abraham Rees, was one of a map that marked out the limits of the *"terra cognita"* for the direction of those who might extend "the boundaries of knowledge." Chambers betrayed no conviction that such scientific progress might depend on further sophisticated classification. While favoring the ideal of "polymathy," which sought the connections between subjects, he believed that new relationships were likely to be found because of the way in which the *Cyclopaedia* and its alphabetical order threw things together: "To do justice to a *collection,* I mean a general and promiscuous one, it has its advantages. Where numbers of things are thrown precariously together, we sometimes discover relations among them, which we should never have thought of looking for."[16] From this perspective, a tight scheme of categories could actually prevent new discoveries, many of which would be made by accident.

The *Encyclopédie,* on the other hand, at first glance seems to be preoccupied with classification, opening as it does with a preliminary discourse explaining the adoption of Francis Bacon's map of knowledge. Robert Darnton highlights the issue of classification in his analysis of this text, arguing that Diderot and D'Alembert recast traditional maps of knowledge in order to displace theology

[14] Ephraim Chambers, *Cyclopaedia; or, An Universal Dictionary of Arts and Sciences,* 2nd ed., 2 vols., Vol. I (London: Midwinter, 1738), pp. xxiv, ix. Subsequent references are to this edition, which will hereafter be cited as **Chambers,** *Cyclopaedia.*

[15] *Ibid.,* pp. vii, ix.

[16] Abraham Rees, preface to an enlarged version of the *Cyclopaedia,* 4 vols. (London: C. Rivington, 1786–1788), Vol. I, p. i; and Chambers, *Cyclopaedia,* Vol. I, p. xxv.

and privilege the natural sciences. But as Darnton notes, D'Alembert also indicated the arbitrary character of such arrangements: "One can create as many different systems of human knowledge as there are world maps having different projections." D'Alembert listed a range of possible classifications and then explained that Bacon's had been chosen for practical purposes and slightly modified; he stressed that the editors did not view this version as absolute: "We do not wish to resemble that multitude of naturalists . . . whose energies have been ceaselessly devoted to dividing the productions of Nature into genera and species, consuming an amount of time in this labor which would have been employed to much better purpose in the study of those productions themselves."[17]

In his article "The Encyclopaedia" Diderot seemed to move even further from the qualified classificatory aims of the prospectus and the preliminary discourse. He stressed that no one person could "understand all that is comprehensible" and repeated D'Alembert's warning about the relative character of intellectual maps. Robert McRae has argued that the notion of unity in the *Encyclopédie* is "not a unity in diversity, nor the unity of a whole in relation to its parts, but unity of continuity among parts."[18] Like Chambers, Diderot was less concerned with schemes of rational classification than with the imperative of collecting human knowledge in some manageable form. Given the current expansion of knowledge, he observed, one can "predict that a time will come when it will be almost as difficult to learn anything from books as from the direct study of the whole universe."[19]

THE *ENCYCLOPAEDIA BRITANNICA* AND THE DEMISE OF THE MAP OF KNOWLEDGE

The first edition of the *Britannica* displays a noticeable lack of interest in grand schemes of classification. No doubt this partly reflected the practical, business-like attitude of its editors and publishers: William Smellie, Colin Macfarquhuar, and James Tytler. These men were interested in science, but they were not philosophes concerned with reorganizing the intellectual world. Nevertheless, the preface to this first edition supported the idea that an encyclopedia should prevent any disintegration of the sciences, for it contained a critical reference to the manner in which both Chambers's *Cyclopaedia* and the *Encyclopédie* had dismembered the sciences by treating "various technical terms arranged in an alphabetical order." But the editors offered an alternative in terms of a coherence within sciences, not between them (see Figure 1). The unity they conceived of was not evinced through a relationship between different sciences on the map of knowledge; it was founded on deductive steps from self-evident or discovered principles that constituted "the very idea of science." Furthermore, the preface

[17] Robert Darnton, *The Great Cat Massacre and Other Essays in French Cultural History* (New York: Vintage Books, 1985), Ch. 5, on pp. 194–195; and Jean D'Alembert, *Preliminary Discourse to the* Encyclopédie, trans. R. N. Schwab (Indianapolis: Bobbs-Merrill, 1963), pp. 48, 50. See also Thomas L. Hankins, *Jean d'Alembert: Science and the Enlightenment* (Oxford: Clarendon Press of Oxford Univ. Press, 1977), pp. 107–108.

[18] Denis Diderot, "The Encyclopaedia," trans. in Jacques Barzun and Ralph Bowen, *Rameau's Nephew and Other Works* (New York: Doubleday, 1956), pp. 291–323, on pp. 292, 306; and McRae, *Problem of Unity* (cit. n. 3), p. 119. See also Hankins, *D'Alembert*, pp. 104–105.

[19] Diderot, "Encyclopaedia," p. 314. In the event of a catastrophe, he hoped that the *Encyclopédie* would "become a sanctuary, where the knowledge of man is protected from time and from revolutions"; see Diderot, "Prospectus," in D'Alembert, *Preliminary Discourse* (cit. n. 17), p. 121.

to the third edition, completed in 1797, expressed doubts about the integrity of general schemes of classification, such as those associated with the encyclopedias of Chambers and Diderot. The authority of Thomas Reid was invoked to censure such efforts to "contract the whole furniture of the human mind into the compass of a nutshell."[20]

The six-volume *Supplement* to the fourth, fifth, and sixth editions, edited by Macvey Napier in 1824, introduced extensive "preliminary dissertations" by leading authorities in science and philosophy. Dugald Stewart, the author of the first of this series, had convinced the new publisher, Archibald Constable, of the value of such treatises as the philosophical framework of the other entries.[21] But this did not signal a new attempt at classification. In the preface to his dissertation, Stewart made a stringent critique of D'Alembert's preliminary discourse, suggesting that such general schemes for the classification of knowledge were deeply flawed. Explaining that he had originally intended to adapt D'Alembert's use of Bacon's "intellectual map to the present advanced state of the sciences," he confessed that he found himself in disagreement with D'Alembert on so many points that he began to doubt the value of substituting "a new map of my own."[22] By investigating these areas of contention it is possible to understand some of the reasons why the *Encyclopaedia Britannica* did not attempt to displace the alphabetical arrangement with a systematic one.

Stewart noted D'Alembert's distinction between the genealogy of the sciences and the arrangement of subjects on an intellectual map or "Encyclopediacal Tree." This was meant to distinguish between the genesis of ideas crucial to particular sciences and the presentation of their mutual relations in a chart. Stewart accepted the need for this distinction because the order in which sciences were invented was not necessarily the best order for their "didactic communication." However, he found a continuing confusion in D'Alembert's treatment between "the history of the human species, and that of the civilized and inquisitive individual." D'Alembert had offered a psychological and epistemological account of the human mind, derived from Locke and Condillac, that was divorced from any serious investigation of its historical development. Stewart claimed that the latter should have been the prime concern of a discourse that sought to show the genealogy of sciences and how this compared with their classification in an encyclopedia. Instead, there was merely a "conjectural" history of science constructed on the basis of a theory of mental development; moreover, there was no connection between D'Alembert's epistemological reflections and his adoption of Bacon's scheme.[23]

[20] William Smellie, ed., *Encyclopaedia Britannica; or, A Dictionary of Arts and Sciences, Compiled upon a New Plan*, 1st ed., 3 vols. (Edinburgh: Bell & Macfarquhar, 1768–1771), Vol. 1, p. v; and Colin Macfarquhar and George Gleig, eds., *Encyclopaedia Britannica; or, A Dictionary of Arts, Sciences, and Miscellaneous Literature*, 3rd ed., 18 vols. (Edinburgh: Bell & Macfarquhar, 1788–1797), Vol. 1, p. vii.

[21] See Dugald Stewart to Archibald Constable, 15 Nov. 1812, in *Archibald Constable and His Literary Correspondents*, ed. Thomas Constable, 3 vols. (Edinburgh: Edmonston & Douglas, 1873), Vol. II, pp. 319–321; see also *Biographical Notice of Macvey Napier, Esq., Editor of the* Edinburgh Review (London: R. Griffin, 1847), p. 5.

[22] Dugald Stewart, *Preliminary Dissertation, Exhibiting a General View of Metaphysical, Ethical and Political Philosophy since the Revival of Letters in Europe*, in *The Collected Works of Dugald Stewart*, ed. William Hamilton, 11 vols. (Edinburgh: T. Constable, 1854–1877), Vol. I, pp. 1–2. This was published in two parts, the first in 1815, the second in 1821. They appeared together in Vol. 1 of the *Supplement* of 1824.

[23] *Ibid.*, pp. 2–4. See also Hankins, *D'Alembert* (cit. n. 17), pp. 83–84.

ENCYCLOPÆDIA BRITANNICA;

OR, A

DICTIONARY

OF

ARTS, SCIENCES,

AND

MISCELLANEOUS LITERATURE;

Conſtructed on a PLAN,

BY WHICH

THE DIFFERENT SCIENCES AND ARTS

Are digeſted into the FORM of Diſtinct

TREATISES OR SYSTEMS,

COMPREHENDING

The HISTORY, THEORY, and PRACTICE, of each,
according to the Lateſt Diſcoveries and Improvements;

AND FULL EXPLANATIONS GIVEN OF THE

VARIOUS DETACHED PARTS OF KNOWLEDGE,

WHETHER RELATING TO

NATURAL and ARTIFICIAL Objects, or to Matters ECCLESIASTICAL,
CIVIL, MILITARY, COMMERCIAL, &c.

Including ELUCIDATIONS of the moſt important Topics relative to RELIGION, MORALS,
MANNERS, and the OECONOMY of LIFE:

TOGETHER WITH

A DESCRIPTION of all the Countries, Cities, principal Mountains, Seas, Rivers, &c.
throughout the WORLD;

A General HISTORY, Ancient and Modern, of the different Empires, Kingdoms, and States;

AND

An Account of the LIVES of the moſt Eminent Perſons in every Nation,
from the earlieſt ages down to the preſent times.

*Compiled from the writings of the beſt Authors, in ſeveral languages; the moſt approved Dictionaries, as well of general ſcience as of its parti-
cular branches; the Tranſactions, Journals, and Memoirs, of Learned Societies, both at home and abroad; the MS. Lectures of
Eminent Profeſſors on different ſciences; and a variety of Original Materials, furniſhed by an Extenſive Correſpondence.*

THE THIRD EDITION, IN EIGHTEEN VOLUMES, GREATLY IMPROVED.

ILLUSTRATED WITH FIVE HUNDRED AND FORTY-TWO COPPERPLATES.

VOL. I.

INDOCTI DISCANT, ET AMENT MEMINISSE PERITI.

EDINBURGH.
PRINTED FOR A. BELL AND C. MACFARQUHAR.
MDCCXCVII.

Figure 1. The title page of the third edition of the Encyclopaedia
Britannica, emphasising the large treatises on the arts and sciences.

D'Alembert employed Bacon's division of knowledge into history, philosophy, and poetry and the corresponding division of faculties into memory, reason, and imagination. Stewart believed that the French philosopher had taken a dubious scheme and exaggerated its errors. The main problem was that in accepting the equation of certain mental faculties with particular sciences, D'Alembert was ignoring the fact that they were all "perpetually blended" in almost every branch of human knowledge. It was patently wrong to regard memory as something peculiarly linked with history rather than with philosophy or poetry; and, he contested, the faculty of abstraction was just as essential "to the Poet" as to the "Geometer and the Metaphysician." Another problem was the way in which this classification "confounded together the Sciences and the Arts under the same general titles." Stewart admitted that the criticism must also apply to Bacon, the originator of this scheme, but deflected it by suggesting that Bacon intended a sketch of the general domains of knowledge rather than "an accurate survey of the intellectual world." Not surprisingly, therefore, D'Alembert's attempt to apply it to the complex and changing state of modern knowledge was bound to fail. Indeed, Stewart predicted that "every future attempt of the same kind may be expected to be liable to similar objections," and the message of his preface was that the abandonment of such an exercise was a necessary prelude to any encyclopedia. Thus, after criticizing D'Alembert's speculations on the "limits by which contiguous departments of study are defined and separated," he remarked that the appropriate place for such considerations "would be as a branch of the article of Logic;—certainly *not* as an *exordium* to the Preliminary Discourse."[24]

As well as doubting the ability of such schemes to represent adequately the relationships among sciences, Stewart was convinced that the "present state of science" made it impossible to contemplate the overview that the two earlier editors of the *Encyclopédie* had attempted. Recent advances in mathematics and physics, in the mechanical arts, and most clearly in "the rapid succession of chemical discoveries" begun by Joseph Black and Antoine Lavoisier necessitated a division of labor in the *Britannica*. The significant recommendation was that the introductory dissertations should avoid "all innovations in language" and describe the different arts and sciences in the "prevailing and most intelligible phraseology." Here the division of labor came into play, for Stewart explained that the "task of defining [sciences], with a greater degree of precision, properly devolves upon those to whose province it belongs, in the progress of the work, to unfold in detail their elementary principles."[25] This marked a crucial shift of intellectual authority from the preface of the encyclopedia to the entries in its alphabetical volumes; it was no longer the editors or the writers of introductory dissertations who mapped out the relationships of the sciences. This enterprise was effectively abdicated. Insofar as it partially remained, it was in the form of comparative remarks about sciences in entries on particular disciplines by relevant experts.

[24] Stewart, *Preliminary Dissertation*, pp. 10–12, 15. By the mid-nineteenth century, classification was indeed covered under "Logic" in encyclopedias. It also featured in the problem of library cataloguing. On this see John W. Lubbock, *Remarks on the Classification of the Different Branches of Human Knowledge* (London: Charles Knight, 1838); and Augustus De Morgan, "Libraries and Catalogues," *Quarterly Review*, 1843, 72:1–25, on p. 15, for the view that libraries suffered from the same problem as encyclopedias: the lack of a "settled canon of classification."
[25] Stewart, *Preliminary Dissertation*, pp. 20–22.

The preliminary dissertations to the *Encyclopaedia Britannica* did not assume the task of classification, but they were seen as an important framework for the rest of the work. Both Stewart and Napier regarded them as a way of surveying the various departments of human knowledge without repeating the confusion of D'Alembert's discourse: namely, that between a historical or genealogical account of the sciences and their representation in some logically ordered system. The alternative approach was later summarized by James Macintosh, who described his dissertation on the progress of ethics as seeking to "make a development of ethical principles as they historically arose,—a new attempt in our language." In introducing these dissertations to readers, Napier claimed that "every Encyclopaedia must be so far incomplete, which does not furnish a connected view of the Progress of the Sciences, as well as the details of their Present State." The "Discourses on the History of the Sciences" in the *Britannica* were both novel and essential.[26]

This recourse to a historical framework was not peculiar to the *Britannica*. In fact, the idea of a chronological or historical presentation of knowledge has strong Baconian affiliations and was contrasted by seventeenth-century writers with the logically deductive mode of scholasticism.[27] Chambers saw the historical mode of exposition as the most suitable for a dictionary or encyclopedia: "The Dictionary . . . supposes the advances and discoveries made, and processed to explain or relate them. The Lexicographer, like an historian, comes after the affair; and gives a description of what passed." Charles Hutton's *Dictionary*, published in 1795 and again in 1815, although its focus was limited to the mathematical and physical sciences, attempted to provide "regular historical details . . . of the origin and progress of each of these Sciences."[28]

These precedents were reinforced by the assumptions of eighteenth-century Scottish moral philosophy and its effort to recover the historical development of both the human mind and forms of social organization. The "Scottish Discourses," as Napier hailed them, were part of this tradition, and this is why Stewart, when criticizing D'Alembert's historical survey as "conjectural," was careful not to reject the notion of conjectural history itself, since in the hands of Adam Smith or Adam Ferguson this was a method by which undocumented stages of human social history could be reconstructed. But the authors of the preliminary dissertations, focusing on the growth of the physical sciences, did not face a lack of sources; they believed that the history of human understanding of nature provided "some of the most instructive views of the advances and failings of human reason, and the progress of genuine science." Napier cited this historical perspective as one remedy for the defects of Chambers's *Cyclopaedia*, in which component parts of a science remained disconnected despite his attempts at cross-reference between alphabetically ordered terms.[29] Furthermore,

---

[26] James Macintosh to Macvey Napier, 3 Feb. 1829, in *Selections from the Correspondence of the Late Macvey Napier*, ed. M. Napier (London: Macmillan, 1879), p. 58; and Napier, ed., *Supplement to the Fourth, Fifth, and Sixth Editions of the Encyclopaedia Britannica, with Preliminary Dissertations on the History of the Sciences*, 6 vols. (Edinburgh: A. Constable, 1824), Vol. I, pp. 2, xxxi (hereafter **Napier, *Supplement***).

[27] Wilbur Howell, *Logic and Rhetoric in England, 1500–1700* (Princeton, N.J.: Princeton Univ. Press, 1956).

[28] Chambers, *Cyclopaedia*, p. xvii; and Charles Hutton, *A Mathematical and Philosophical Dictionary*, new ed., 2 vols. (London: Rivington, 1815), Vol. I, p. vi.

[29] Napier, *Supplement*, Vol. I, p. xxxi. For the Scottish conjectural method see Gladys Bryson,

Napier saw the preliminary dissertations as providing a perfect vehicle for the recognition of heroic discoverers, those "great lights of the world by whom the torch of science has been successively seized and transmitted." This focus on individuals had been absent in the *Encyclopédie* and in the early editions of the *Britannica,* so the combination of historical exposition and the identification of major discoveries was a significant innovation.[30]

In summary, the preliminary dissertations to the *Encyclopaedia Britannica* were not unaltered successors of D'Alembert's famous discourse. They sanctioned Stewart's rejection of the search for a systematic map or tree of knowledge, offering in its place a historical survey of the physical and moral sciences as the intellectual framework for the encyclopedia, the focus of which was now the major entries in the alphabetical volumes.

COLERIDGE, CLASSIFICATION, AND THE *ENCYCLOPAEDIA METROPOLITANA*

The case of the *Britannica* should not lead us to conclude that all attempts at logical or systematic organization had ended. While Napier, Stewart, and the publisher Constable were casting the new form of the *Britannica,* Samuel Taylor Coleridge was commissioned to edit the *Encyclopaedia Metropolitana,* a projected work of twenty-five volumes on a philosophical plan devised by the great Romantic poet. The prospectus for the *Metropolitana,* written by Coleridge, was circulated in April 1817, but the final volume did not appear until 1848. Early in the project there was a major disagreement between Coleridge and the original publisher, Rest Fenner, over the implementation of the classification of knowledge embodied in the prospectus and in his "General Introduction; or, Preliminary Treatise on Method," also first published in 1817. Coleridge was annoyed because Sir John Stoddart, who had revised the prospectus for publication, had reduced the eight divisions of knowledge to four. He also claimed that Thomas Curtis, one of the editors, had abandoned the full systematic plan, so that the version of the preliminary treatise that appeared in January 1818 in the first part of Volume I had been "so bedeviled, so interpolated and topsy turvied," that he was "equally ashamed of it as a man of letters and as a man of common honesty." Only five serial parts of the encyclopedia were issued by the original owners; it was taken over by B. Fellowes and others in 1819 and placed under the editorship of Edward Smedley, Hugh J. Rose, and Henry J. Rose. It finally appeared between 1829 and 1845.[31]

In the prospectus Coleridge was clear about the novelty of his project: he intended to return to the ideal of "*rational* arrangement," the central feature of the encyclopedia when it was "really *Instruction in a cycle*" of arts and sciences that constituted the education of the higher classes. This had been lost since

---

Man and Society: The Scottish Inquiry in the Eighteenth Century (Princeton, N.J.: Princeton Univ. Press, 1945). Napier also criticized the manner in which the *Cyclopaedia* gave the conclusions of sciences with little reference to the "experimental details" of their formulation. This point can be related to the Scottish interest in historical and inductive presentation of science. See Wilbur S. Howell, *Eighteenth-Century British Logic and Rhetoric* (Princeton, N.J.: Princeton Univ. Press, 1971), Ch. 5.

[30] Napier, *Supplement,* Vol. I, p. xxxi; see also Hutton, *Dictionary* (cit. n. 28), Vol. I, pp. vi–vii.

[31] *Collected Letters of Samuel Taylor Coleridge,* ed. Earl L. Griggs, 6 vols. (Oxford: Oxford Univ. Press, 1956–1971), Vol. IV, pp. 723–725; see also Robert Collison, "Samuel Taylor Coleridge and the *Encyclopaedia Metropolitana,*" *Journal of World History,* 1966, 9:751–767.

modern encyclopedias abandoned the application of a strict *"scientific* method" capable of producing a philosophical plan for the relationship between subjects in the work. Now they were characterized by "more or less complete disorganization of the Sciences and Systematic Arts." This situation was particularly regrettable because the recent "scientific and moral revolutions," no less profound than the political changes, required a philosophical response. The new discoveries and fresh facts flowing from the different branches of experimental philosophy could not simply be added as a postscript to the existing corpus of knowledge, because they affected the "whole theory and consequent arrangement of the Art or Science to which they belong." Coleridge claimed that there was a "manifest tendency" in the arts and sciences, from the purely intellectual to the practical, to "lose their former insulated character, and organize themselves into one harmonious body of knowledge."[32] The distinction of the *Metropolitana* would be to provide a "repository" for this new constellation.

Trevor Levere and others have shown that Coleridge drew upon German *Naturphilosophie* as an antidote to the Newtonian picture of a mechanical world.[33] It is also worth contemplating the sense in which Coleridge regarded his encyclopedia as a response to the materialist philosophy allegedly advanced by the French philosophes in the *Encyclopédie*. In the "Preliminary Treatise on Method" he referred to their successors, such as Condillac and Condorcet, as men whose names it would be "absolutely ridiculous to mention, in a history of science," if it were not for the number of their followers. He mentioned their "exploded doctrine of *perfectibility,*" but acknowledged that there was hope of a "permanent revolution in the moral world" provided that this was guided by "the historic sense," one that appreciated the organic connection of past and present. Similarly, he believed that the intellectual revolution had to be approached "methodically."[34] The *Metropolitana*, with its emphasis on the systematic relationships between sciences, provided the framework in which such developments could be accommodated.

Coleridge sought to explain how the *Metropolitana* could be called "a METHODICAL *Compendium of Human Knowledge.*" In claiming that other encyclopedias had forsaken the attempt to "methodize" knowledge, he meant that they did not demonstrate the principles of "unity" and "progression." Coleridge argued that no science could be categorized until the principle by which the mind gave it unity was recognized. Some of these principles involved *"relations of things"* perceived by the mind as necessary, and these formed the *pure sciences;* others, although still "originating in the mind," constituted what were commonly called laws of nature and formed the groundwork of the "mixed sciences," such

[32] Samuel Taylor Coleridge, *Prospectus: Coleridge on the Science of Method* (London: J. J. Griffin, 1849), pp. 2, 4. This twelve-page version was prefixed to at least some of the separate volumes of the Cabinet edition of the *Encyclopaedia Metropolitana*, 40 vols. (London: J. J. Griffin, 1849–1858). Alice Snyder, *S. T. Coleridge's Treatise on Method, as Published in the* Encyclopaedia Metropolitana (London: Constable, 1934), pp. 71–72, believes that the first thirteen paragraphs of this 1849 publication are from the lost original.

[33] Trevor Levere, *Poetry Realized in Nature: Samuel Taylor Coleridge and Early Nineteenth-Century Science* (Cambridge: Cambridge Univ. Press, 1981), Ch. 3.

[34] Samuel Taylor Coleridge, "General Introduction; or, Preliminary Treatise on Method," in *Encyclopaedia Metropolitana; or, Universal Dictionary of Knowledge, on an Original Plan: Comprising the Twofold Advantage of a Philosophical and Alphabetical Arrangement*, ed. Edward Smedley, Hugh J. Rose, and Henry J. Rose, 29 vols. (London: Fellowes & Rivington, 1829–1845), Vol. I, pp. 1–43, on pp. 32, 43 (hereafter **Smedley et al., *Metropolitana*).

as mechanics and astronomy. Coleridge stressed that in the perception of the relationships that constituted the "guiding idea" of a science, the mind was active, not passive; although there were differences between mathematical and physical studies, both involved "a previous act and conception of the mind," or what he called an "initiative."[35]

The second part of the "Science of Method" revealed a progression from one key idea to another in a way that placed "one or more *particular* things or notions, in subordination, either to a pre-conceived *universal* idea, or to some lower form of the latter." Coleridge could therefore speak about sciences or studies as being arranged in terms of class, order, genus, or species, each of which derived its "scientific worth, from being an ascending step towards the universal." This was the philosophical basis of his division of the sciences into pure sciences (both formal and real), mixed sciences, and applied sciences. In his original formulation the fine arts were to form another division, linking the mixed and the applied sciences, but this was one of the casualties of the publisher's alterations.[36]

The category of pure sciences represented "pure acts of mind," such as those involved in the "formal" sciences of grammar, logic, and mathematics, which related to the "forms of our mental conceptions." Second, it included the subjects of metaphysics, morals, and theology, which dealt with "real" principles or conditions of the universe and human nature. In explaining the categories of mixed and applied sciences, Coleridge said that these depended on theory and not solely on the perception of a necessary law, as in the pure sciences. The initiative of the mind was still essential, but in the case of theory "it may be an image or conception received through the senses, and originating from without." Mixed sciences were formed when "certain ideas of the mind, are applied to the general properties of bodies, solid, fluid, and aerial; to the power of vision, and to the arrangement of the universe."[37] Thus mechanics, hydrostatics, pneumatics, optics, and astronomy were all mixed sciences—that is, they combined ideas and observation and so could not be certain sciences in the same sense as the pure sciences.

The applied sciences were a step further from the pure sciences than the mixed sciences. In contrast to the latter, they did not apply ideas to "general and permanent properties of all bodies," but more particularly to the changes in these properties. Magnetism, electricity, galvanism, and chemistry were thus applied sciences; and Coleridge believed that, at present, all of these suffered from a lack of clarity in their "first idea." In any case, he also indicated that the theories of such sciences were usually imperfect because they were "necessarily progressive": it was always possible that some new discovery might alter the theory. But a progressive theory was better than no theory at all, and Coleridge argued that useful arts were "capable of being reduced to Method," and hence to the theories of the mixed and applied sciences. Political economy, agriculture, commerce, and manufactures, despite some resistance from their traditional practitioners, were now "considered scientifically."[38]

[35] Coleridge, "Preliminary Treatise," pp. 7–10.
[36] *Ibid.*, p. 34 (quotations); and Coleridge, *Collected Letters*, ed. Griggs (cit. n. 31), Vol. IV, pp. 801, 816.
[37] Coleridge, "Preliminary Treatise," pp. 35–38.
[38] *Ibid.*, pp. 39–40.

This scheme placed Coleridge's efforts within the long tradition of the classification of the sciences, an enterprise not seriously pursued by the *Encyclopédie* or the *Britannica*. Furthermore, Coleridge sought to forge a nexus between the hierarchy of the sciences in his classification and the order of their treatment in the *Metropolitana*. His prospectus therefore announced that it aimed to convey scientific instruction and information "not in a confused mass, but in the natural sequence of the sciences." This sequence was derived from the "philosophical system" of the "Preliminary Treatise on Method" and was intended to bring those sciences "capable of mutual dependency" within one volume so that they could be read in "their natural sequence."[39] The assumption here was that the philosophical order was also a pedagogic one and that a student could, without the aid of a tutor, begin with the first of the pure sciences in Volume I and work through to the mixed and applied sciences. With this in mind, the references in the *Metropolitana* were designed to be "retrospective," never looking to a forthcoming volume for the "explanation of a term, the establishment of a principle, or the demonstration of a proposition." Thus, Coleridge explained, the theory of balance was given in the treatise on mechanics; but information on particular constructions of balances could not be introduced into a scientific treatise "without destroying the symmetry of its parts by a suspension of the logical order," and hence such information was located in the alphabetical volumes of the work. John Herschel was enacting this principle in his article on physical astronomy when he proceeded on the understanding that the reader "will be presumed acquainted with the general *facts* of astronomy, with the principles of mechanical philosophy, and so much of analysis, of the differential and integral calculus . . . as shall render it unnecessary to interrupt the general train of our reasoning."[40]

The idea of a rational progression through the sciences as the pedagogic order for their study distinguished the *Metropolitana* from its major counterparts. The editors of the *Encyclopédie* and the *Britannica* did not see their works as courses of orderly instruction. They did stress the connections between different branches of science, but the cross-references were not intended to enforce a linear progress through the volumes. Outside encyclopedias, however, this interest in pedagogic order was not unusual. It was found, for example, in introductory surveys of science such as Neil Arnott's popular *Elements of Physics* of 1827. Arnott's scope was almost encyclopedic, and the subjects were displayed in a "table of Science," commencing with the study of "physics" and passing through chemistry, life, mind, and quantity. "From the mutual dependency of the different sciences . . . [he said,] it follows, that 'the Table' exhibits the order in which they should be methodically studied, so as to prevent repetitions and anticipations." Arnott spoke of a "pyramid of Science, with Physics for the base," and his inference from this indicates the strong association between the encyclopedic movement and the autodidactic tradition from the eighteenth century. The truths of physics, he explained, were now reduced to a few, and these "are referred to by the words *atom, attraction, repulsion* and *inertia*. It gives an astonishing, but true idea of the nature and importance of methodical *Science*, to

---

[39] Coleridge, *Prospectus* (cit. n. 32), p. 5 (quotations). H. J. Rose endorsed this aim in his preface to the work; see Smedley *et al., Metropolitana*, Vol. I, p. ix.

[40] Coleridge, "Preliminary Treatise" (cit. n. 34), p. 42 (Vols. XIV–XXIX covered miscellaneous topics in alphabetical order); and John Herschel, "Physical Astronomy," in Smedley *et al., Metropolitana*, Vol. III, pp. 647–734, on p. 648.

be told that a man, who understands these words . . . understands the greater part of the phenomena of nature." At a more sophisticated level, the tree of knowledge that Jeremy Bentham proposed in 1817 was associated with an educational program that stressed the order of study.[41]

Nevertheless, the preface of the *Metropolitana* almost predicted its demise. Citing the fashion for publishing "small and superficial works," the editors asserted that such a large exercise was not likely to be repeated.[42] Their belief was confirmed by the encyclopedia's commercial failure and by the presentation of the second Cabinet edition, published by J. Griffin from 1849. Although the prospectus to this edition claimed that there would be a "more rigid adherence" to the *"principles of Method,"* the order of publishing effectively dismantled the system it purported to embody. The advertisement drew the attention of subscribers to the "difference that exists between the *order* in which the SUBJECTS *occur* in the general system of this Encyclopaedia and the *order* in which it may be *advisable to publish* the TREATISES on those subjects." This decision was taken with the presumed reading habits of the audience in mind: it was not thought agreeable to "the great body of the SUBSCRIBERS to so comprehensive a work" to receive weekly parts, for months at a time, devoted solely to mathematics or geography "or indeed to *any* Department, in its order—*all other subjects being, for the time, systematically excluded.*" Even the numbers on the bound volumes now referred to their order of publication, correctly described as "indeterminate," since it alternated between history, science, and art, sometimes separating two parts of a single treatise. The place of each treatise in the philosophical system was still indicated on its title page, but this system was no longer a functional order prescribing a pedagogic sequence. After the first four volumes, which were taken from Volume I of the original edition, the subjects were presented in apparently random order: treatises on miracles, political economy, Greek literature, Roman antiquities, and botany appeared within the next ten volumes.[43] Instead of an organized succession of volumes proceeding through the sciences, the Cabinet edition was a collection of introductory manuals or textbooks on particular subjects, any of which could be read and studied independent of the others. In this sense it repeated the plan of the *Cabinet Cyclopaedia,* produced by Dionysius Lardner between 1830 and 1840 and eventually comprising 133 volumes. These were not prefaced by a reference to any explicit scheme of classification, although major areas of science were preceded by a preliminary discourse that outlined their scope and method.[44]

[41] Neil Arnott, *Elements of Physics; or, Natural Philosophy, General and Medical, Explained Independently of Technical Mathematics,* 2nd ed. (London: Underwood, 1827), pp. xi, xix; and Jeremy Bentham, "Essay on the Nomenclature and Classification of Arts and Sciences," in *Chrestomathia,* ed. M. J. Smith and W. H. Burston (Oxford: Clarendon Press of Oxford Univ. Press, 1983), p. xxviii. In this case the order was the reverse of Coleridge's: natural history came first, grammar last. Bentham agreed with Stewart's critique of D'Alembert's map of knowledge, but he offered one of his own (p. 159).

[42] Smedley *et al., Metropolitana,* Vol. I, p. xxiv.

[43] *Prospectus,* Cabinet edition of the *Encyclopaedia Metropolitana* (London: J. J. Griffin, 1849), Vol. I, pp. 10–12. These pages appear after the reprint of Coleridge's *Prospectus.* The original plan of the encyclopaedia is given on p. 13. See also Collison, "Coleridge and the *Metropolitana*" (cit. n. 31), p. 764.

[44] E.g.: John Herschel, *A Preliminary Discourse on the Study of Natural Philosophy* (London: Longman, 1830); and William Swainson, *A Preliminary Discourse on the Study of Natural History* (London: Longman, 1834).

The original *Metropolitana,* under Coleridge's plan, deliberately opposed this arrangement. Yet although this encyclopedia was intended as a systematic work displaying the logical order of subjects, its treatises also highlighted major disciplines. There were two main reasons for this. First, in spite of the stress on the circle of sciences in Coleridge's classification, his epistemology reinforced distinctions between different sciences. As explained earlier, he postulated specific ideas as the epistemological basis for different disciplines and categorized disciplines with reference to the relationship of these ideas to the external world. Second, the treatises on scientific subjects were written by experts who sought to define the intellectual basis of their discipline and its boundaries. This arguably facilitated the move to the Cabinet edition, where the deference to a systematic approach existed only in a transcendental, not in a functional, sense. Now the major organizing element was the treatise on a single subject or discipline, as was the case in most of the alphabetical encyclopedias to which the *Metropolitana* was to have been a glorious alternative.

THE ALPHABET AND THE CONSOLIDATION OF DISCIPLINES

The alphabetical arrangement of terms and subjects has been common in encyclopedias for at least the last five hundred years. Coleridge realized this when he sought to revive the systematic form. Writing in 1803 to Robert Southey, who had just been involved in an abortive project for a *Bibliotheca Britannica,* he remarked: "What a strange abuse has been made of the word encyclopaedia! It signifies, properly, grammar, logic, rhetoric, and ethics and metaphysics, which . . . formed a circle of knowledge. . . . To call a huge unconnected miscellany of the *omne scibile,* in an arrangement determined by the accident of initial letters, an encyclopaedia, is the impudent ignorance of your Presbyterian bookmakers."[45] Since Coleridge promoted a systematic encyclopedia, it is not surprising that he attacked the alphabetical arrangement. But the latter's virtues and vices had been discussed since the time of Chambers, by editors who had abandoned systematic maps of knowledge as the organizing principles of their encyclopedias. An analysis of this discussion suggests that the effect of alphabetical ordering was not entirely a random presentation of material. Rather, the manner in which it developed from the eighteenth century had the effect of privileging and consolidating the identity of scientific disciplines.

In the prospectus to the *Metropolitana* Coleridge listed some of the problems of an alphabetical encyclopedia. First, scientific terms with several synonyms were dispersed throughout the whole work. Second, the reader might have to search through still more volumes to find the terms that had to "be *previously* understood" in order to comprehend the term first sought. Coleridge explained that alphabetical position itself was subject to the "caprice of convenience of the compiler," who had to decide whether subordinate scientific and technical terms were treated in one or several entries. There were also more practical difficulties. For example, Charles Hutton noted the tendency for entries late in the alphabet to be condensed in order to meet printing deadlines or financial limits.[46] In the

---

[45] Samuel Taylor Coleridge to Robert Southey, July 1803, in Coleridge, *Collected Letters,* ed. Griggs (cit. n. 31), Vol. II, p. 956. See also, more generally, Read, "Encyclopaedias and Dictionaries" (cit. n. 1), p. 366.

[46] Coleridge, *Prospectus* (cit. n. 32), pp. 2–3; and Hutton, *Dictionary* (cit. n. 28), Vol. I, p. viii.

original *Britannica,* the letters *A* and *B* occupied 697 pages, with the remainder squeezed into 2,000. The preface to the third edition apologized for some omissions in the last six volumes caused by the imposition of a paper tax. The usual mode of publication in successive parts also entailed certain problems.[47] One of these, as an appreciative reviewer of the *Metropolitana* observed, was that in alphabetical encyclopedias "a particular branch of science presents itself long before the fundamental principles on which it depends have been considered, and some years may elapse before they can find their place in the alphabet." Thus "air" and "air pump" preceded "pneumatics"; "aberration" appeared before an account of "optics."[48]

Yet while problems such as these were recognized, they were not regarded as incompatible with the aim of presenting a coherent account of scientific knowledge. Chambers believed that the systematic and alphabetical orders could be usefully combined. He explained that he had provided a chart showing how the "several branches of knowledge" were placed in terms of a division between "natural" and "artificial" knowledge. However, it was not this grand order that most concerned Chambers, but rather the integrity of particular subjects or areas. He therefore included a long set of notes that gave the constituent terms and subjects belonging to each major science. Thus meteorology was broken down into separate entries on air, ether, fire, vapor, and so on. Chambers hoped that this would assist the reader "both in reinstating the scattered articles in the book, and in connecting them together." Furthermore, this guide would indicate "the order they are most advantageously read in." However, this scheme was attacked in the first edition of the *Britannica,* its editors claiming that readers of Chambers and the *Encyclópedie* would have discovered the "folly of attempting to communicate science under the various technical terms arranged in an alphabetical order." This led to "the dismembering [of] the Sciences" and the loss of the "connected series of conclusions" deduced from principles that constituted the "very idea of science."[49]

The third edition continued this criticism, making it clear that the integrity of individual sciences was endangered by Chambers's failure to provide sufficiently detailed accounts in one part of the work. Given the multitude of tiny articles into which a science had been dispersed by the alphabet, it was ridiculous to expect that they could be reconstituted by the reader, since it would require the amalgamation of twenty-four entries to restore meteorology, forty-eight to restore metaphysics. It is difficult to see how such a feat could be accomplished, even by what a contemporary publication called "the *Systematic Reader.*" Nevertheless, a reviewer in 1809 still thought it worthwhile that editors supply a "sketch of the order" in which articles might be read by those who were using them for more than occasional reference.[50]

[47] The seventh edition of the *Britannica* was sold in monthly parts, with six forming a volume: each part contained 130 pages and sold for six shillings. One volume of 800 pages cost 36 shillings, and this was said to be "lower than the cost of any similar publication of the day": Napier Papers, British Library, ms 34,600, fol. 68.

[48] Review of *Encyclopaedia Metropolitana, Monthly Review,* 1829, *88*:187–191, on p. 191. Another writer referred to "two different plans," the "alphabetical and scientific": "Analytical Notices," *Blackwood's Magazine,* 1817, *1*:180–188, on p. 186.

[49] Chambers, *Cyclopaedia,* Vol. I, pp. iii, v; and Smellie, ed., *Britannica,* 1st ed., Vol. I, p. v.

[50] Macfarquhar and Gleig, eds., *Britannica,* 3rd ed., Vol. I, pp. viii–ix; G. S. Howard, *The New Royal Encyclopaedia Londonensis,* 3 vols. (London: A. Hogg, 1796–1802), Vol. I, p. vi (quotation); and "Dictionaries of Arts and Sciences," *Eclectic Review,* 1809, *5*:541–553, on p. 542.

This vestige of Chambers's method was not adopted in the *Britannica*. Its solution was to combine "distinct treatises" on each science with shorter entries on particular terms. Both large and small entries were alphabetically ordered. This format, although still alphabetical, did distinguish the *Britannica* from previous dictionaries and encyclopedias.[51] The combination of large treatises or "systems" and short entries on technical terms was a powerful one, and the balance between them could effectively alter the form of an encyclopedia and the manner in which it might be read. A more strict alphabetical coverage of terms would have failed to highlight particular subjects. This was the case, as Napier indicated, in John Harris's *Lexicon Technicum* of 1704 and Chambers's *Cyclopaedia*, where the major subjects usually had no more space devoted to them than specific terms. Napier argued that the new plan was an improvement on Chambers, who had sought to achieve the desirable goal of "continued discourse" by references from "general to particular heads" and vice versa. This had failed because the sciences could not be taught without being "viewed continuously, in their natural state of unity and coherency." In short, the *Britannica* had effectively answered the objection about "breaking the Sciences into fragments, scattered fortuitously among the Letters of the Alphabet."[52]

The crucial decision in this plan concerned the manner in which words and terms were selected, first, as candidates for separate entries and, second, for inclusion in the cross-references. Chambers raised this problem when he attempted to define "terms" as assemblages of ideas. He realized that something more specific was called for to prevent the *Cyclopaedia* from becoming an unmanageable collection of words, indistinguishable from a dictionary. He thus distinguished between "terms of knowledge" and terms of science. The former, he explained, were one degree simpler than the latter; in contrast, terms of science and art "were before members of the commonwealth of knowledge: but they are now incorporated into some certain province, or city thereof; where they become of further significance and consideration than before." He gave the example of the simple word *force* becoming a term of science, as in "central force" or "centripital force." Chambers's intention was then to "give a course of references from generals to particulars . . . from more to less complex; and from less to more; a communication might be opened between the several parts of the work." But he confessed that he had not been thorough in this and had often left the reader "without his clue."[53] When pursued more conscientiously, however, this practice had the effect of asserting the primacy of major subjects, making particular terms and words the satellites of larger disciplinary entries. This was not the case in most eighteenth-century encyclopedias, because although there was cross-referencing, the system was not imperial: that is, terms were not exclusively owned by particular subjects; rather, the emphasis—theoretically in Chambers and effectively in the *Encyclopédie*—was on multiple cross-connections.[54]

---

[51] Smellie, ed., *Britannica*, 1st ed., Vol. I, p. v (quotation); Macvey Napier, ed., *Encyclopaedia Britannica; or, Dictionary of Arts, Sciences, and General Literature*, 7th ed., 21 vols. (Edinburgh: A. and C. Black, 1830–1842), Vol. I, pp. viii–ix, xiii; and "Dictionaries" (cit. n. 50), p. 547. In the original *Britannica* there were forty-four large treatises; besides the major sciences, it included topics such as bookkeeping, gardening, and tanning.

[52] Napier, ed., *Britannica*, 7th ed., Vol. I, p. ix; and Napier, *Supplement*, Vol. I, p. xxxvi.

[53] Chambers, *Cyclopaedia*, Vol. I, pp. xiv, ii, xxiii.

[54] McRae, *Problem of Unity* (cit. n. 3), pp. 118–120; see also Hugh M. Davidson, "The Problem of

The combination of the larger systems with the practice of cross-referencing in the *Britannica* significantly contributed to the profile of major subjects or disciplines. Indeed, the third edition had drawn the distinction between these systems and the "various detached parts of knowledge," which it also promised to explain. The preface said that in addition to the "compendious" treatises, there were three kinds of "subordinate articles": those independent of particular systems and thus able to be treated separately, those requiring some discussion under the system to which they belonged, and those belonging to "systems of which all the parts must be elucidated together." This had implications for cross-referencing, in that terms in the first category required none, those in the second were partially explained in subordinate articles but completed in the relevant treatise, and those in the third were "wholly referred to the system of which they are constituents."[55] This scheme was not consistently applied, but it was reinforced in the *Supplement* produced by Napier in 1824, which updated and extended the major treatises. By the seventh edition, large entries on scientific disciplines were effectively the central element of the encyclopedia. Now the cross-references highlighted the major articles, and shorter entries on terms pointed back to the subjects in which they had a precise technical or theoretical meaning. This system was still not always enforced, and in any case it was partly undermined by the exigencies of publishing already mentioned: some references pointed to volumes or parts that had not yet appeared. On the other hand, another factor served to affirm the importance of the major entries on disciplines. The *Britannica* did not have an index until Robert Cox compiled one in 1842 for the seventh edition. Napier admitted that without an index information on particular topics was often hidden "where it might not be expected or readily found"; consequently, it was not possible to find material relating to part of a large entry except by looking for a shorter entry on a term covered by it.[56] This meant that disciplines were presented as wholes to be read as such.

There were some exceptions to this format. For example, Abraham Rees's *Cyclopaedia*—the successor to Chambers's and the *Britannica's* major competitor in the first quarter of the nineteenth century—argued against comprehensive treatises. The preface to the first volume of 1819 referred to the "distinct treatises" of the *Britannica* and to the use of "separate dictionaries appropriate to each particular science" in the *Encyclopédie Méthodique*, but rejected both plans. Rees claimed that "the advantage of separate treatises under each head of science, such as the limits of a dictionary will allow, seems to be more imaginary than real." He did not question the idea that encyclopedias should preserve the coherence of scientific subjects and admitted that without separate treatises it might seem that the sciences would be "mutilated and mangled." But he believed that cross-references would "lead the reader to subordinate articles, that form, by their mutual connection and dependence, an aggregate or whole, superseding in all common cases the necessity of a distinct treatise."[57] This was a continua-

Scientific Order versus Alphabetical Order in the *Encyclopédie*," *Studies in Eighteenth Century Culture*, 1972, 2:33–49.

[55] Gleig, ed., *Britannica*, 3rd ed., Vol. I, title page, p. ix. The first edition referred to "technical terms" that were supposed to be closely connected to relevant sciences treated in large entries; see Smellie, ed., *Britannica*, 1st ed., Vol. I, p. vi.

[56] Napier, ed., *Britannica*, 7th ed., Vol. I, p. xl.

[57] Abraham Rees, ed., *The Cyclopaedia; or, Universal Dictionary of Arts, Sciences and Literature*, 39 vols., Vol. I (London: Longman, 1819), pp. v–vi. See Darnton, *Business of Enlightenment* (cit. n.

tion of Chambers's hope that the reader might reconstitute a science from its separate terms, except that no chart or map of knowledge was provided. Similarly, the *London Encyclopaedia* of 1829 no longer saw its role as one of providing "distinct and full cultivation of any one science"; rather, it would present an "ample abridgement of the entire mass of human knowledge."[58]

Cheaper publications of this sort reflected a judgment that comprehensive articles were not sought by "the general class of readers," and this began to mark a difference between encyclopedias and what Rees called "a Scientific Dictionary." Napier too acknowledged the existence of two audiences. Writing in 1827 to A. and C. Black, who had bought the *Encyclopaedia Britannica* after the collapse of Constable, he urged that the seventh edition be extended to twenty-four volumes so that it could serve both general readers and "men of science." He sought to include "miscellaneous matter more particularly adapted to the wants and taste of ordinary readers," without treating significant subjects "in a way too curt and superficial to satisfy those of a higher class."[59]

EXPERTS, EDITORS, AND DISCIPLINES

From the early nineteenth century the notable feature of major British encyclopedias—the *Britannica,* the *Metropolitana,* and the *Edinburgh Encyclopaedia*—irrespective of their avowed organizational principles, was the presentation of detailed entries on scientific disciplines by expert contributors. Even Rees's *Cyclopaedia,* which did not have these large treatises, included separate entries on subjects such as comparative anatomy, botany, chemistry, electricity, magnetism, and mineralogy, and it advertised the collaboration of authorities such as William Lawrence, John Dalton, and Humphry Davy. This emphasis on discrete areas of expertise was apparent as early as the *Encyclopédie Méthodique,* the successor to the work of Diderot and D'Alembert, which commenced in 1782. In discussing its contents, Darnton has remarked that they announce "the modern world, in which certified experts rule over carefully demarcated territories." It is significant that this form of presentation occurred in the period that is now seen as the point of transition from eighteenth-century natural philosophy to the emergence of modern scientific disciplines, perhaps marking a second scientific revolution.[60]

There is certainly a temptation to regard the profiling of major disciplines as a

---

7), pp. 419–420, for a description of the separate alphabetical dictionaries on twenty-six sciences in the *Encyclopédie Méthodique.*

[58] Curtis, *London Encyclopaedia* (cit. n. 8), Vol. I, preface. See also A. Whitelaw, ed., *The Popular Encyclopaedia; or, "Conversations Lexicon,"* 7 vols. (Glasgow: Blackie, 1841–1867), Vol. I, p. 12; and G. Long, ed., *The Penny Cyclopaedia,* 27 vols., Vol. I (London: C. Knight, 1833–1858), p. 1. While this latter work eschewed large articles, it had contributions from leading men of science, notably Augustus De Morgan.

[59] Rees, ed., *Cyclopaedia* (cit. n. 57), Vol. I, p. vi; and Napier, *Correspondence* (cit. n. 26), p. 53. See David Layton, "Diction and Dictionaries in the Diffusion of Scientific Knowledge: An Aspect of the History of Popularization in Great Britain," *The British Journal for the History of Science,* 1965, 2:221–234.

[60] Darnton, *Business of Enlightenment* (cit. n. 7), p. 451. There is no detailed treatment of this notion of a "second scientific revolution," but see P. Steven Turner, "The Great Transition and the Social Patterns of German Science," *Minerva,* 1987, 25:56–76; Simon Schaffer, "Scientific Discoveries and the End of Natural Philosophy," *Social Studies of Science,* 1986, 16:387–420; and Andrew Cunningham, "Getting the Game Right: Some Plain Words on the Identity and Invention of Science," *Studies in History and Philosophy of Science,* 1988, 19:365–389.

reflection of this shift. In the case of the *Britannica,* there was an explicit concern with the integrity of disciplines treated in large treatises. There is evidence that these were seen as replacing the older categories, such as natural history and natural philosophy, that had figured in the first edition. In the seventh edition the entry on natural history advised that this category was best divided into five great branches: meteorology, hydrography, mineralogy, botany, and zoology. Significantly, this entry referred to the "chain of being" as the most famous concept of eighteenth-century natural history and reported that it was now shown to be specious by the *specialist* evidence of anatomy and chemistry. Similarly, natural philosophy was no longer a distinct subject but only a general term indicating "a cluster of sciences," including mechanics, hydrostatics, optics, astronomy, magnetism, and electricity.[61] Readers were referred to the separate treatises on these different sciences and also to John Playfair's dissertation on the physical sciences for their historical background.

In the *Supplement* of 1824 and the seventh edition of 1842 these disciplinary treatises took a fairly standard form and were characterized by four interrelated elements that had not appeared together in any earlier editions: a definite historical introduction, a statement of the methodological principles of the discipline, claims about the specific domain of the subject and a defense of its boundaries, and celebration of the heroes of the subject and their part in making it a modern science.[62] It is not possible to discuss these in detail, but a brief notice of the third theme reveals the way in which disciplines were profiled.

Many of the articles in the *Britannica* commented on the recent dramatic changes in certain sciences, particularly those that Thomas S. Kuhn has called the Baconian sciences. As early as the *Supplement* to the third edition, which aimed to update previous entries on major subjects, contributors mentioned the need for extensive alterations. The entry on chemistry said that progress had been so rapid that, although the existing article was only ten years old, the "language and reasoning" of the subject had improved considerably, making it necessary to trace "over again the very elements of the science." The author of "Magnetism" announced a "decision to write a new treatise."[63] This theme was more pronounced in the *Supplement* edited by Napier, where even the apparently stable science of botany was seen as requiring a new treatment, and William Brande's dissertation spoke of a "new epoch" and a "chemical revolution" associated with Lavoisier. In 1842 Napier looked back to the third edition and remarked that the changes in chemistry and natural history at that time were such that nothing "of any stable value could be added to the work."[64]

This emphasis on the recent developments in several sciences was linked with a strong concern about the boundaries between them. Especially in the treatises

[61] Napier, ed., *Britannica,* 7th ed., Vol. XV, pp. 738, 740–741.

[62] These elements can also be found in the major articles in the *Metropolitana* and the *Edinburgh Encyclopaedia.* In the latter, most of the entries on scientific disciplines began with a section headed "History."

[63] Thomas Thomson, s.v. "chemistry"; and John Robinson, s.v. "magnetism"; both in *Supplement to the Third Edition of the Encyclopaedia Britannica,* ed. George Gleig, 2 vols. (Edinburgh: T. Bonar, 1801), Vol. I, p. 210; and Vol. II, p. 112. See also Thomas S. Kuhn, "Mathematical versus Experimental Traditions in the Development of Physical Science," in *The Essential Tension: Selected Studies in Scientific Tradition and Change* (Chicago: Univ. Chicago Press, 1977), pp. 31–65.

[64] Napier, ed., *Supplement,* Vol. II, p. 376; William T. Brande, "Dissertation Third, Exhibiting a General View of the Progress of Chemical Philosophy," *ibid.,* Vol. III, pp. 1–79, on pp. 50, 69; and Napier, ed., *Britannica,* 7th ed., Vol. I, p. xvii. For a recent view of these changes see Arthur Donovan, ed., *The Chemical Revolution: Essays in Reinterpretation,"* *Osiris,* N.S., 1988, *4.*

devoted to the new sciences of geology and physiology, but also in those dealing with phenomena whose disciplinary framework was unclear, such as electricity and heat, conscious attention was paid to this issue. The articles acknowledged the mutual relationships between sciences but also stressed the integrity of each subject as a discipline with its own intellectual domain. The author of "Geology," for example, identified the problem of defining it as a separate subject even though it could "have no existence before the phenomena of chemistry, zoology, and mechanics" were established. But while admitting that geology relied on the "general progress of the natural sciences," the article referred to the danger of "listening to the reasonings of the mathematician, the astronomer, the chemist, or the zoologist, when applied to the history of the globe."[65]

Writers of entries on chemistry were also preoccupied with drawing boundaries. First, they sought to mark off the modern subject from its alchemical past and link it with the Newtonian program, using an appropriate historical introduction.[66] Second, they were confronted with the emergence of new fields of experimental research: heat, magnetism, and electricity. These eventually became classified as part of the new discipline of physics, but in the early nineteenth-century encyclopedias their membership in a larger subject or discipline was not clearly defined.[67] The subject of heat was often treated in a separate entry, but cross-references indicated its connection with chemistry. In 1835 the *British Cyclopaedia* noted that the phenomenon of heat was one that upset the editor's attempt to distinguish between chemistry and natural philosophy, "to which the term physics is applied." The seventh edition of the *Britannica* had a separate entry on heat, but the references it gave for further reading were to chemistry texts. On the other hand, Francis Lunn, who wrote the articles on heat and chemistry for the *Metropolitana*, made a bold bid to exclude heat, together with light, electricity, and magnetism, as *"objects* of investigation" in chemistry, while admitting that they may be involved as *"agents."* He acknowledged the sensitivity of this issue, explaining in the article on heat that "as the connection with Chemistry is in some points most intimate, we have taken some care that by our line of demarcation the definitions including both sciences are not violated."[68]

There is, however, a need for caution in taking the demarcation of subjects in encyclopedias as an unproblematic indicator of agreed divisions within the sciences. To begin with, the presence of expert contributors may have increased the level of specialization and sharpened the boundaries between subjects in encyclopedias to a degree not typical of scientific transactions, journals, or meetings. Moreover, the recruitment of these writers introduced a new set of negotiations with the editor; if this complication is ignored, features attributable to editorial style might be regarded as reflections of definite perceptions in the scientific community.

[65] Napier, ed., *Britannica*, 7th ed., Vol. XV, pp. 173, 231. "Geology" was written by John Phillips and formed part of the entry on mineralogy. The editor explained that the rapidly progressive state of geology did not allow sufficient time to prepare the entry for inclusion under *G* in the alphabet: *ibid.*, Vol. X, p. 421.

[66] Brande, "Chemical Philosophy" (cit. n. 64), p. 18.

[67] Keith A. Nier, "The Emergence of Physics in Nineteenth-Century Britain as a Socially Organized Category of Knowledge" (Ph.D. diss., Harvard Univ., 1975).

[68] Charles F. Partington, *The British Cyclopaedia of the Arts and Sciences,* 10 vols. (London: Orr & Smith, 1835–1838), Vol. I, p. ii; and Francis Lunn, "Chemistry," in Smedley *et al., Metropolitana,* Vol. IV, pp. 587–746, on p. 587; and Lunn, "Heat," *ibid.*, pp. 225–340, on p. 225.

What, briefly, can be said about these two factors? By the early nineteenth century encyclopedias began to advertise expert authorship. Some reference to the assistance of learned writers did occur in previous works, but it was common for editors to assemble most of the material themselves and to assume responsibility for the encyclopedia as whole. In the original *Britannica* Smellie himself wrote the entries for "fifteen capital sciences" and confessed that he made a "Dictionary of Arts and Sciences with a *pair of scissors,* clipping out from various books a *quantum sufficit* of matter for the printer." But from the third edition and its *Supplement,* the major scientific contributions of John Robison and Thomas Thomson were acknowledged, and the prestige of the *Britannica* was now seen to reside in its ability to offer articles attributable to respected writers in a particular field. George Gleig distinguished between material associated with a particular writer and accounts that "are floating everywhere on the surface of science, and are the property, therefore, of no living author."[69] Napier explained that the *Supplement* of 1824 would contain a "key" at the end of each volume by which "the purchasers of the Work will be regularly informed" of the identity of authors of major articles. "This plan," he said, "seems eminently calculated to increase the confidence of the public in such undertakings." Later, in the seventh edition, he drew a contrast between the anonymity of reviews in the leading quarterlies and the policy of the encyclopedia: in the latter entries were signed "in order to give each article the authority due to its author's name." Some textbooks, such as Thomson's on chemistry, began as large articles in encyclopedias.[70]

This involvement of experts in the production of large articles in particular fields had significant consequences. It meant that some articles, rather than summarizing a particular body of facts or the consensus on a subject, represented the leading edge of a field: for example, "Polarization of Light," by François Arago, which Napier solicited. David Brewster remarked that "it is rarely elsewhere than in the encyclopaedias of the day that we can expect new and original treatises containing all the recent discoveries which have been made in the exact sciences." He suggested that this was due to the lack of demand for accessible books on the highest branches of mathematical and physical sciences. Thus, as well as being a vehicle for the dissemination of science to the general public, these articles were also a medium, in addition to that of the publications of scientific societies, for the presentation of original research. In introducing his own *Edinburgh Encyclopaedia* Brewster noted that it covered "new sciences"— "POLARIZATION of LIGHT and ELECTROMAGNETISM"—which were not even known at the commencement of the work."[71]

This new reliance on experts did not necessarily reduce the control of editors. Darnton has said that Charles-Joseph Panckoucke, the editor of the *Encyclopédie*

---

[69] Robert Kerr, *Memories of the Life, Writings, and Correspondence of William Smellie,* 2 vols. (Edinburgh: J. Anderson, 1811), Vol. I, pp. 362–363; and Gleig, ed., *Supplement to Third Edition,* Vol. I, p. v.

[70] Napier, *Supplement,* Vol. I, pp. 3–4; and Napier, ed., *Britannica,* 7th ed., Vol. I, p. xli. Napier referred to the success of the "separate publication of the articles in octavo form . . . introduced as Class Books in several of the Universities of the country": Napier Papers, MS 34,630, fol. 78. Thomas Thomson, *A System of Chemistry,* 4 vols. (Edinburgh: J. Brown, 1802), reached a seventh edition in 1831.

[71] David Brewster, "The *Encyclopaedia Britannica,*" *Quarterly Rev.,* 1842, 70:44–72, on p. 53; and Brewster, ed., *The Edinburgh Encyclopaedia,* 18 vols., Vol. I (Edinburgh: W. Blackwood, 1830), p. vii. Brewster began this work in 1809.

Méthodique, got his authors to agree on the boundaries between the sciences. We do not have similar detailed studies of the main British encyclopedias, and this article does not aim to supply such information; but it is clear that editors did see their role as strongly managerial and, at times, interventionist. The large number of contributors—150 in the case of the *Edinburgh Encyclopaedia*—made this imperative. Napier outlined "the peculiar labours of an Editor," including that of "judging what subjects or sciences had received an imperfect or unphilosophical examination." His aim was to make the *Britannica* a dignified and efficient medium "for the diffusion of *matured* knowledge." But given that there appears to have been no definite agreement among Napier's authors on what constituted an appropriate scientific entry, it is possible that the extremely uniform presentation of disciplines noted above was the result of editorial policy.[72] In short, the practical need to parcel out topics and set standard guidelines for authors may have produced artificially clean demarcations.

Apart from this interaction between experts and editors, there may have been commercial pressures behind some of the distinctive features of the nineteenth-century encyclopedias. They existed in a competitive market for information that had created a plethora of dictionaries, manuals, and abridgements, all offering access to a variety of topics. They had to compete against other encyclopedias and, in a real sense, against their own previous editions. Thus the prospectuses for new editions underlined the continuous progress of science, promising that the latest advances were captured and that older issues were obsolete.[73] Furthermore, within each encyclopedia commercial logic reinforced the lines between various articles so as to ensure that each specific entry, or volume, was a unique product and did not cover topics found elsewhere in the work. In this way publishers tried to ensure that the whole encyclopedia was bought and not just some of its volumes. It is quite likely, for example, that certain volumes of the unsuccessful *Metropolitana* were purchased separately, to the detriment of the whole, since it was possible to acquire all the "mixed sciences" in four volumes.[74]

In his examination of English encyclopedias Hughes emphasized the process whereby older material was carried over in new editions, often for practical reasons, thus giving encyclopedias a conservative character. While not doubting that outdated content was carried across various editions of the *Britannica,* my analysis draws attention to the significant shifts in the presentation of science, particularly from the *Supplement* of 1801. Hughes saw the famous and still usable ninth edition of 1875 as the major intellectual watershed, in that "the amount of material which survives from the past" is less than in any other edition.[75] However, it may be that this concentration on content prevents a recognition of the important shifts in format and rhetoric that characterized some of the earlier editions as they registered or interpreted the emergence of modern disciplines.

---

[72] Darnton, *Business of Enlightenment* (cit. n. 7), p. 422; and Napier, ed., *Britannica,* 7th ed., Vol. I, p. xix (emphasis added). For the queries of William Wollaston, Thomas Malthus, and Thomas Young about the nature of a scientific article see Napier, *Correspondence* (cit. n. 26), pp. 12, 29–31.
[73] Rees, ed., *Cyclopaedia* (cit. n. 57), Vol. I, p. iv. In 1834 the prospectus to the seventh edition of the *Encyclopaedia Britannica* boasted new articles "to supersede those of former editions rendered obsolete by the progress of discovery": Oxford, Bodleian Library, G[eneral] Pamphlets 2921, pp. 123–124. Some reviewers seemed convinced by this: see W. E. Donne, "The *Encyclopaedia Britannica,*" *Westminster Review,* 1862, 22:394–433, on p. 405.
[74] Smedley *et al., Metropolitana,* Vol. I, p. xii, Vols. III–VI covered the mixed sciences.
[75] Hughes, "English Encyclopaedias" (cit. no. 7), 1951, p. 349; see also p. 340.

CONCLUSION

By the second quarter of the nineteenth century, the departure from the ideal of the unity of knowledge was twofold. First, in spite of the presence of the *Metropolitana* and some continued appeals to the older rhetoric, the organization of most encyclopedias recognized and perhaps reinforced the primacy of specialist disciplines. This occurred at a time when it had just become possible to satirize the prospect of the omniscient individual. The satirical novelist Thomas Love Peacock did this in 1816 when he described Mr. Panscope (probably Coleridge) as "the chemical, botanical, geological, astronomical, mathematical, metaphysical, meteorological, anatomical, physiological, galvanistical, musical, pictorial, bibliographical, critical philosopher, who had run through the whole circle of the sciences, and understood them all equally well." The need for discourse between the increasingly specialized branches of science was commonly urged, but encyclopedias were not usually seen as the answer. It is significant that the editor of the *British Cyclopaedia* referred to Mary Somerville's warning on the hazards of "investigating one science independent of another" while explaining that his own publication separated chemistry from natural philosophy.[76]

One of the most direct critiques of the ideal of unity was made in 1824 by the Scottish philosopher Alexander Blair. Seriously questioning the assumptions that he regarded as fundamental to the encyclopedic project, Blair proclaimed that "all attempts at bringing knowledge into *encyclopaedic* forms seem to include an essential fallacy. Knowledge is advanced by individual minds wholly devoting themselves to their own part of inquiry. But this is a process of separation, not of combination." He proceeded to argue that the attempts to "reduce knowledge into encyclopaedic forms," to present the "CIRCLE of the Sciences," were based on the misconception that this unity could be grasped by individual minds. Furthermore, Blair claimed that the progress of knowledge was not impelled by a search for unity but by the activities of specialized research; the implication, which he accepted, was that "the division of minds from one another becomes more and more the principle, or condition, . . . of farther advancement to the separate Sciences."[77] When David Masson, another Scottish literary figure, reflected on this issue in 1862, he acknowledged that "while we speak of the universality of knowledge the thing is impossible." Instead, he observed, there was "the learning of the Chemist, the learning of the Physiologist, the learning of the Natural Historian . . . and so on."[78] As we have seen, all these special domains were represented as such in encyclopedias.

The demand for easy access to these new sciences also threatened the ideal of unity: by valuing information higher than knowledge it fostered a disjunctive over an integrative approach. The long list of scientific subjects in Peacock's

[76] Thomas Love Peacock, *Headlong Hall,* ed. R. Garnett (London: J. M. Dent, 1893), p. 68; and Partington, *British Cyclopaedia* (cit. n. 68), Vol. I, p. ii. Mary Somerville, *On the Connexion of the Sciences* (London: J. Murray, 1834), sought to illustrate the unity of the laws and processes of the physical sciences.

[77] Alexander Blair, "Thoughts on Some Errors of Opinion in Respect to the Advancement and Diffusion of Knowledge," *Blackwood's Magazine,* 1824, 6:26–33, on pp. 26, 32. See J. P. Klancher, *The Making of English Reading Audiences, 1790–1832* (Madison: Univ. Wisconsin Press, 1987), Ch. 2, for the place of this in a more general *Blackwood's* strategy. But see James Wylde, ed., *The Circle of the Sciences,* 2 vols. (London, 1862–1867), for the continuation of this notion at a popular level.

[78] David Masson, "Universal Information and the 'English Cyclopaedia,' " *MacMillan's Magazine,* 1862, 5:357–370, on pp. 362–363.

parody constituted a form of knowledge that demanded attention. Writing in the *Edinburgh Review* in 1810, Francis Jeffrey indicated its implications for contemporary ideas about the appropriate body of "general knowledge": "Now-a-days . . . a man can scarcely pass current in the informed circles of society, without knowing something of political economy, chemistry, mineralogy, geology and etymology." He warned that there was a real danger that profound knowledge could be overwhelmed by the attraction of superficial information on a variety of topics, a mere "Encyclopaedical trifling." "Various and superficial knowledge is now not only so common, that the want of it is felt as a disgrace; but the facilities of acquiring it are so great that it is scarcely possible to defend ourselves against its intrusion."[79] Over forty years later, Masson found that Jeffrey's analysis was largely vindicated, so much so that there were now two kinds of encyclopedias: those with large disciplinary entries beyond the capacity or needs of the general reader, and "Dictionaries of Universal Information" giving short entries on various terms.[80]

By this stage the salient contrast was no longer between systematic and alphabetical encyclopedias, but between those providing extensive, specialist entries and those offering comparatively disconnected pieces of information. To some extent it is fair to describe this transition, as Jeffrey did, in terms of the fragmentation of profound knowledge into superficial information. But such a process should not be confused with the shift from systematic to alphabetical arrangement. This article has sought to show that the movement away from systematic classification was not in fact one from order to miscellany, but rather one that allowed, and arguably effected, the consolidation of scientific knowledge into separate disciplines. If readers had consulted the scientific articles in the *Britannica* or the *Metropolitana* in the early nineteenth century, they would have received not disconnected information, but disciplinary knowledge.

[79] Francis Jeffrey, "Stewart's Philosophical Essays," *Edinburgh Review*, 1810, *17*:167–211, on pp. 168, 169.
[80] Masson, "Universal Information," p. 366.

X

Alphabetical lives: scientific biography in historical dictionaries and encyclopaedias

> Theophilus Hopkins was a moderately famous man. You can look him up in the 1860 *Britannica*. There are three full columns about his corals and his corallines, his anemones and starfish. It does not have anything very useful to say about the man. It does not tell you what he was like. You can read it three times over and never guess that he had any particular attitude to Christmas pudding.
>
> Peter Carey, *Oscar and Lucinda*, 1988, p.7

Today we conventionally consult encyclopaedias for biographical information. Probably at least as many readers seek this in the *Encyclopaedia Britannica* as in the *Dictionary of National Biography*, or *Who's Who*, or other specialist biographical dictionaries. But in the famous encyclopaedias of the Enlightenment such biographical information was absent. Neither Ephraim Chambers's *Cyclopaedia* of 1728 nor the more renowned *Encyclopédie* (1751–80) edited by Denis Diderot and Jean D'Alembert, contained any biographical entries: when personal names appear these usually represent some larger intellectual doctrine, such as Baconianism, Cartesianism, or Newtonianism.[1] From the early nineteenth century encyclopaedias began to include the biographical entries we now expect. However in its 1985 revision, the *Britannica* decided to relegate all biographical articles from its Macropaedia to the Micropaedia, except for those on '100 people who profoundly affected

---

[1] See for example 'Baconisme', 'Cartesianisme' and 'Newtonianisme' in Diderot and D'Alembert (1751–80), vol. 1, 8–10, vol. 2, 716–26 and vol. 11, 122–5.

world history'. This was a departure from its previous twentieth-century editions, but still a long way from the rigorous exclusion practised by its eighteenth-century predecessors.[2]

The absence of biography in the eighteenth-century encyclopaedias was a conscious choice, in response partly to the separate genre – the 'historical' dictionary – which included biography and geography. Influential works of this kind by Louis Moreri and Pierre Bayle in France, and their English successors such as the *Biographia Britannica*, edited by William Oldys (and later by Andrew Kippis), carried entries on the lives of aristocracy, political, religious and military leaders, and also on philosophers and men of science, although usually not in any comprehensive manner.[3] We often tend to lump these encyclopaedias and dictionaries together indiscriminately as reference works, thus missing their distinctive rationales. In this essay I discuss some of the assumptions governing the initial exclusion of biography from encyclopaedias and, secondly, the implications of its eventual inclusion by the end of the eighteenth century. This requires some attention to the historical dictionaries deriving from the late seventeenth-century French publications just mentioned. I refer to important English translations (and extensions) of these but I do not attempt a systematic study of the place of scientific biography in these or in the encyclopaedias that followed them. Such an investigation is beyond the scope of this discussion. However, I believe that the point raised here about the conventions regarding biography in these works needs to be considered in any larger treatment of the representation of scientific lives in the major reference works of the period.

Where is the biography? Historical dictionaries vs encyclopaedias

The omission of biographical entries from encyclopaedias was more than a convenient division of labour. Encyclopaedias did not include biography because they were seeking to record knowledge, not lives. By the eighteenth century most of these publications were alphabetical in arrangement, but their contents continued to be influenced by some of the principles that guided the older, systematic encyclopaedic compendia, such as Johann Heinrich Alsted's *Encyclopaedia* of 1630 which recorded the major branches of

---

[2] Whiteley (1992), 85.
[3] Lipking (1970), 67 says Oldy's *Biographia Britannica* (1747–66) 'is an initial great landmark in that trail of systematic lives which would lead eventually to the *Dictionary of National Biography*'. See Retat (1991), 505 on the importance of historical dictionaries in the republic of letters in the eighteenth century.

X

scientia: namely, systematic knowledge embodied in the seven liberal arts of the medieval university curriculum.[4] From the 1690s a new kind of scientific and technical dictionary appeared: for example, Antoine Furetiere's *Dictionnaire Universel* (2 vols, 1690), John Harris's *Lexicon Technicum* (2 vols, 1702; 2nd edition 1710), and Chambers' *Cyclopaedia* (2 vols, 1728). These were subtitled 'dictionaries of arts and sciences', and they sought to record the major terms and concepts of these categories or fields of knowledge. However, by this time there had been shifts in the definition of *scientia*: the new experimental sciences, although not strictly fitting the axiomatic character expected in the traditional definition, were now included, together with the mechanical arts.[5] These works went through several editions and were aimed both at scholars and educated readers. For example, Chambers' *Cyclopaedia* was published by subscription in 1728 at a cost of four guineas for the two folio volumes. The first volume carried the names of 375 initial subscribers and its immediate success encouraged Chambers and his printers to begin a second edition; this was issued in 1738 and again in 1741. The range of subjects covered was broader than in Harris' *Lexicon*, but both works disseminated scientific terminology, including that of Newtonian natural philosophy, beyond scientific circles. Editions of these two English works usually ran to about 2000 copies; both were still being published and supplemented in the 1750s and hence prepared an audience and a market for the major encyclopaedias, such as the *Encyclopédie* and the *Encyclopaedia Britannica*. But the readers of these scientific dictionaries could not learn anything about individual natural philosophers. There was still no rationale for a coverage of lives, even the lives of the persons closely involved in the various arts and sciences.

The first edition of the *Encyclopaedia Britannica*, advertised to appear in 100 weekly parts from January 1768, was published in three volumes in 1771. Although differing markedly from Chambers' *Cyclopaedia* in the way it organised scientific knowledge – in large systems, rather than small entries on terms[6] – it continued the exclusion of biography. It did carry a short entry on 'Biography' which praised the moral, didactic value of the genre and remarked that there were not enough suitable lives being written for the edification of the young. But the editors felt no compulsion to remedy this

[4] See Alsted (1630) for the full title.
[5] Chambers (1728), vol. 1, ii for his chart of knowledge.
[6] See *Proposals for printing by subscription a work, intitled Encyclopaedia Britannica* (Edinburgh, 1768). On its plan and organisation, see Yeo (1991). For a study of the first edition, see Kafker (1994).

in their own publication. When it defined the two leading systems of natural philosophy – 'Cartesianism' and 'Newtonian Philosophy' – the *Britannica*, like Chambers', allowed no comment on the lives or character of the two individuals who founded these doctrines, although the 'Cartesians' were said to embrace a 'romantic system'. The entry on 'Newtonian philosophy' refers the reader to the one on 'Optics', where Newton is mentioned only in a description of the reflecting telescope.[7] Francis Bacon's method is praised in 'Aether', where Newton's speculation about elastic fluids is censured; Boyle is mentioned in 'Chemistry'. Even major religious leaders such as Luther, Calvin and Knox do not receive special treatment, although the first two are mentioned in entries on their doctrines. Thus eighteenth-century readers seeking a biographical account of men of science needed to consult one of the 'historical' or biographical dictionaries. How did these dictionaries treat the lives and discoveries of men of science?

Louis Moreri's *Grand Dictionnaire Historique, ou mélange curieux de l'histoire sacrée et profane*, first published in Lyon in 1674 and then issued in an expanded second edition, is usually regarded as the first reference work (other than bibliographies) to summarise a range of subjects in strictly alphabetical order.[8] Although the word 'biography' was not mentioned in the title it was the first 'modern' work to include biographical entries. It set a pattern for the 'historical dictionaries' of the eighteenth century, professing to include history, geography and genealogy as well as the lives of famous people. This biographical content was intelligible given the contemporary definition of biography as the history of a life. Although history and biography were classified as distinct genres by Francis Bacon and John Dryden, they were rarely distinguished in practice until the mid eighteenth century.[9]

---

[7] [Smellie] (1768–71), vol. 1, 555 on 'Biography'; vol. 2, 39–40; on Cartesians'; vol 3, 399 on 'Newtonian Philosophy'; vol. 3, 424 on the reflecting telescope. The absence of biography is also apparent in Scott (1765). Here there is an entry for 'Copernicus' but this describes the 'astronomical instrument invented by Mr Whiston'. I have not done a careful check of the German encyclopaedia edited by Johann Heinrich Zedler, but Collison 1964, 105 says it was unusual in having biographical entries of living persons. In the volume published in 1740 there is a one column entry on Newton which describes him as the greatest philosopher and mathematician of his time, and gives some account of his early studies. Newton's epitaph is printed under his name and the reader is referred to the larger entry on *'Newtonische Philosophie'*. See Zedler (1731–50), vol. 24, 411–12 and 413–16.

[8] The second edition appeared in two volumes in 1681. I use the English translation, from the 1692 Utrecht edition, published in two volumes in 1694. The French version reached a 24th edition in 1759. For further information, see Miller (1981); Retat (1987), 505.

[9] Bacon (1974), 72–5; Stauffer (1941), 517. In 1728 Chambers's *Cyclopaedia*, which did not include biographies, nevertheless gave the accepted definition of the art: a 'Biographer' was 'an author who writes the history, or the life of one or more persons'. Chambers (1741), vol. 1. Also Hall 1797, vol. 1: 'Biography: a species of history which records the lives and characters of remarkable persons.'

Thus in describing the 'historical' contents of his work Moreri promised that:

> besides the Lives of the Ancient and Modern Emperors, Kings and
> Princes, which are to be found in other Authors, here are the Histor-
> ies of all who have ever been Famous for Arts, Arms, or any Thing
> else; which are either not to be met with, or but very slightly
> touch'd upon in the Chronicles of Nations, etc. Here are also Lists
> of Learned Men in all Faculties, with Catalogues of their most
> Remarkable Writings.[10]

More than half of Moreri's content was biographical but many of his entries gave the names of aristocratic figures with very little account of their lives.[11] It is significant that as a compendium that broke with the systematic (non-alphabetical) organisation of encyclopaedic works, Moreri's dictionary also introduced biography – a subject dealing with contingent matters that clashed with the axiomatic and deductive character of *scientia*. Biography, together with geography and the mechanical arts, was a subject outside formal university studies; to the best of my knowledge, Moreri's was the first alphabetical reference work to include it.[12]

An English translation was published in 1694 as *The Great Historical, Geographical and Poetical Dictionary*. The editors said that Moreri 'thought himself oblig'd to gratify the Nobles and Gentry of that Kingdom [France], with an Account of their Illustrious Families, and the Famous Exploits of their Ancestors'. They followed his charter by 'paying the same regard to the Nobles and Gentry of England, Scotland and Ireland', thus expanding the content of the work. The lives of 'Philosophers' were also included but these entries were not extensive: Aristotle got two columns, Roger Bacon ten lines, Copernicus half a column, Kepler eleven lines, Francis Bacon half a column. Newton was not included because he was still alive, but Robert Boyle's name

---

[10] Tytler (1778–84), vol. 2, 1156: 'a species of history which records the lives and characters of remarkable person'. On the clash between the historical and biographical material of Moreri's dictionary and the ideal of a didactic and encyclopaedic order pursued by Diderot and D'Alembert, see Retat (1991), 508. For a deliberate distinction between 'biographical and historical matter', see Aiken and Enfield (1799–1815), vol. 1, 3 and the 'Discours Préliminaire' to the *Biographie Universelle*, Paris: M. Freres, 1811, vol. 1, vii–xviii.

[10] Moreri (1694), preface, no pagination. The copy I used was bound in one volume.

[11] R. Christie (1884), 194. When starting the *Dictionary of National Biography* (1895), Leslie Stephen still felt it necessary to say: 'I exclude names which are only names, because otherwise I should have to publish (amongst other things) all the parish registers. A biographical dictionary should surely consist of biographies, however brief '. R. Christie (1884), '6.

[12] Moreri's, however, was not the first compendium to include subjects outside the scholastic curriculum. Sebastian Munster's *Cosmographie* (1544) contained an even wider range of miscellaneous information, arranged under various 'heads'. See Strauss (1966).

is mentioned under 'Boyle, Richard of Burlington', from where the reader is sent to the letter 'R' to discover one of the longest entries (two and a half columns) under *'Robert Boyle'*. This entry contained hardly any description of Boyle's scientific work – apart from a list of his books – but it did present him as a pious man whose natural philosophy was compatible with 'the truth of the Christian Religion in General'. He was civil to strangers, offering charity to 'those in want', and unlike some men of learning, he was 'decently Cheerfull'. This was an early statement of the ideal Christian Philosopher or Virtuoso, following Gilbert Burnet's sermon at Boyle's funeral in 1692. But unlike later biographical accounts of natural philosophers, none of these entries in Moreri provided much discussion of the controversies in which the individuals had been involved. Brief portraits of character took precedence over intellectual details, and a list of works was appended at the end of each entry, curiously detached from the history of the life, which was largely presented in terms of personal appearance, behaviour and moral character.[13]

Despite the great success of Moreri's work, which reached a twenty-fourth edition in 1759, its reputation today derives partly from its influence on Pierre Bayle, whose famous two-volume *Dictionnaire Historique et Critique* of 1697 began as an attempt to remedy the errors of his compatriot. However, when Bayle wrote his *Dictionnaire* (an enlarged second edition appeared in 1702) he concerned himself only with the biographical, not the geographical or historical, articles.[14] This focus made it a stimulus and model for the more strictly biographical dictionaries published during the eighteenth century. But this genealogy should not be oversimplified, because while Bayle's *Dictionnaire* was explicitly biographical in content – that is, it listed names of people in alphabetical order – there was no clear emphasis on the 'life history' of the person; this was lost in the mass of intervening footnotes (the hallmark of Bayle's 'critical' style) which digressed into philosophical commentary. This method displayed his belief that intellectual claims needed constant reassessment. Bayle distanced himself from contemporary publications – particularly Moreri's – underlining the difference

---

[13] Moreri (1694), see under 'Copernicus', 'Bacon', 'Kepler', 'DesCartes'. Another English work based on Moreri was produced by Jeremy Collier in 1721 and extended in 1727. See Collier (1721). On Boyle's self-image and the assessments of his contemporaries, see Hunter (1990) and his chapter in this volume.

[14] The first edition of Bayle was published in Rotterdam to avoid French censorship; it reached a ninth edition in 1741. Bayle omitted names adequately covered in Moreri. For details, see Burrell (1981). For selections in English from the French edition, see Bayle (1965).

between 'other Historical Dictionaries and mine. I am not contented, according to the custom of those dictionaries, to give a general account of a man's life'.[15] Rather, he explained, his work offered comment and criticism on certain aspects of the lives and teachings of the persons covered in the *Dictionnaire*. These lives were selected to allow scope for the discussion of the topics nearest to Bayle's heart: Protestantism, religious doctrine and other philosophical issues concerning evidence and authority. There were no articles on Copernicus, Galileo, Boyle, Harvey, Leibniz or Newton. Thus although there was an entry on Kepler, and support of the heliocentric theory, the new science did not feature strongly in Bayle's work.[16]

Bayle's *Dictionnaire* appeared in English in two translations of 1709 and 1710 and these were followed by two further competing editions of 1734–38 and 1734–41. The most significant of these was a new and greatly enlarged one undertaken by Thomas Birch and a number of collaborators. This translation, A *General Dictionary, Historical and Critical*, published in ten volumes between 1734 and 1741, was in some ways a distinct work that added many new lives, especially English ones.[17] The editors explained that Bayle had chosen 'such Articles only as best suited his views, or for which he had materials already prepared, [and] he omitted a great many persons illustrious for their rank and dignity, as Emperors, Kings, Princes, etc or conspicuous for their knowledge in the Arts, the Sciences and Polite Literature.' The editors aimed to add entries on 'the most famous personages throughout the Dictionary of Mr BAYLE, and have enlarged and compleated his Articles, wherever we apprehended them to be defective'. Recognising the danger of being 'voluminous' they excluded 'whatever relates to Geography, as being foreign to a Work of this kind'.[18] This restraint did not stop the edition running to ten volumes, a very large number for that time.

The result was a more strictly biographical dictionary than Moreri's and a less critical one than Bayle's. While offering only lives, and not history or geography, it reinstated the illustrious names that Bayle had omitted when he decided on a more extensive discussion of significant figures. This also

[15] Bayle (1826), vol. 4, 164–5.
[16] Burrell (1981), 93–4. D'Alembert indicated that Bayle's work was not a '*dictionnaire historique*' but a philosophic and critical *dictionnaire*, where the text is only the pretext of the notes. Retat (1987), 510.
[17] See Bayle (1710); for the ten volume work edited by Birch, see, Bayle (1734–41). The rival five volume translation was Bayle (1734–38).
[18] Bayle (1734–41), vol. 1, preface. For Birch's major role in this, see Osborn (1938).

meant the addition of British names, because when Bayle had heard that the translation of Moreri was being greatly expanded he decided 'not to treat of the illustrious men of Great Britain.'[19] The English editors agreed that Bayle 'draws the characters of such persons, relates the particulars of their lives, discovers the several springs of their actions, and examines the judgement that has been, or may be formed of them'. But they suggested that the text seemed to be written merely for the sake of the notes. Nevertheless this English edition did not entirely escape the influence of Bayle's format – footnotes dominate many pages – yet in expanding the original two volumes into ten, it greatly extended the coverage of names.[20]

It is interesting that this coverage included a significant proportion of men of science from the seventeenth century. Thomas Birch contributed at least twenty new entries on both major and minor figures. Indeed, under his direction, the first volume was dedicated to Sir Hans Sloane, President of the Royal Society, and addressed its Fellows in these terms:

> Gentlemen, although you are so assiduously engaged in the most
> rational and most sublime Pursuits, those of the Mathematicks and
> of Nature, we yet presume to interrupt them a few moments . . . as
> the subject of it is the Lives of eminent Men, many of whom bear so
> near a resemblance to Your Selves.[21]

The editors also professed that 'the bare Names of NEWTON and of BOYLE raise the most exalted Ideas, and image to us something more than human'. But while the lives of natural philosophers were regarded as worthy of record, very few details of their work were given. The entry on Newton – partly derived from Fontenelle's *eloge* – is an exception in that his theory of gravity is summarised, with some of the mathematical reasoning given in the footnotes; but his work in optics is only noted in passing. Hooke's skill in experimentation is praised, but there is no description of this work or its significance.[22] Compared with Moreri's, there was slightly more integration of scientific material into the narrative of the life: publications were cited in footnotes rather than merely listed at the end of the article. The editors also published, for the first time,

---

[19] This remark occurs in the reprint of Bayle's preface to the original French edition. See Bayle (1734–8), vol. 1, 3.

[20] Bayle (1734–41), preface. Osborn (1938), 33 says 889 new lives were added, of which Birch wrote 618.

[21] Bayle (1734–41), vol. 1, dedication page; Osborn (1938), 33–4 for list of Birch's entries.

[22] Ibid, vol. 7, 776–802; vol 6, 219.

letters of Newton and John Wallis, and the entry on Henry Oldenburg drew heavily on his correspondence with Boyle. But there is no evidence of a concern with how the person's scientific ideas developed, and certainly no indication of the possibility of situating an individual's work in relation to that of contemporaries, as was suggested in William Wotton's planned, but uncompleted, biography of Boyle.[23] There was no clear acknowledgment of science as a special vocation, distinct from other kinds of scholarship.[24]

Biographical dictionaries: the Biographia Britannica

The major English biographical dictionary of the eighteenth century was the *Biographia Britannica*, published between 1747 and 1766, and again in a second edition between 1778 and 1793. Edward Gibbon paid it a somewhat ambiguous compliment, saying that: 'The author of an important and successful work may hope without presumption that he is not totally indifferent to his numerous readers: my name may hereafter be placed among the thousand of articles of a *Biographia Britannica*.' This was 'the first book in any language having the title of *Biographical Dictionary*'.[25] It excluded the geographical and historical material covered by Moreri and its founding editor, William Oldys, aimed to give 'a more methodical Collection of *Personal History*', 'in the manner of Bayle'.[26] Both editions defended the inclusion of men whose lives revolved around ideas, rather than military and political action. The first edition noted that 'we have very few memorials of PHYSICIANS, though scarcely any nation has produced better', and it advertised the inclusion of the 'most eminent Scholars, with a clear and rational account of their works'. It mentioned 'Men of Letters' as an important category, and 'Philosophers, Physicians, Mathematicians, Chemists etc' were said to be neglected by previous dictionaries of this kind. One practical rationale was supplied: namely, that when men of genius realised that their contributions would 'not be buried in oblivion' they would be 'more eager in pursuit of knowledge and virtue'. Nevertheless, the coverage of men of science in the *Biographia Britannica* looks weak when compared with that of a more

---

[23] Osborn (1938), 26 on the use of letters; Hall (1949) on Wotton; also Hunter, this volume. See Stauffer (1941), 252 on the lack of critical sense in Birch's expansion of Bayle.

[24] See Shapin (1991)

[25] Christie (1884), 202; Stauffer (1941), 249 on the others. For Gibbon's remark, see Reese (1970), 3.

[26] Preface to first edition by Oldys reprinted in Kippis (1778–93), vol. 1, xiv.

specialised work such as Benjamin Martin's, *Biographia Philosophica* of 1764.[27]

The *Biographia Britannica* also gave explicit attention to the scope and role of biography. At least in the preface by Oldys there was an interest in the *development* of a reputation: detailed personal histories of famous individuals, it averred, allow us to trace 'the beginnings of their greatness, and learn the steps by which they rose'. Kippis claimed that 'Biographical knowledge' like 'Natural Philosophy' could move from individual cases to 'general truths and principles'.[28] But this did not translate into a serious analysis of their intellectual activities. Paul Korshin has remarked that 'one can read dozens of lives in the *Biographia Britannica* without uncovering a single notion as to what somebody thought, how he composed, or whom he knew'.[29] While this may be a little extreme in the case of some of the entries on philosophers and men of science, it is generally true that scientific publications were treated as marks of *action* – the category operating for the more numerous entries on political and military figures. The entries on Boyle and Newton are quite detailed; the one on Newton occupies thirty-four pages and includes some discussions of his work, together with a liberal use of letters. This solid article here no doubt reflects the amount of material available for summary, some of it written by Newton's disciples such as Henry Pemberton and Colin Maclaurin.[30] However, most entries on scientific figures – such as those on Barrow, Bentley, Derham, Maclaurin, Ray – mention their scientific works but give no extensive summary of these, of developments in their thought, or of any debates in which they were involved. Instead, the evaluative emphasis centres on their character, with most being seen as examples

---

[27] Preface to first edition reprinted in Kippis (1778–93), vol. 1, xiv–xvi. Stauffer 1941, 257 says Kippis was more receptive to literary figures than Oldys. See Stanley (1701), preface, for the view that the Ancients gave greater attention to men of thought and contemplation. Martin (1764) has entries on Roger Bacon, Francis Bacon, Barrow, Bernoulli, Boyle, Brahe, Copernicus, Roger Cotes, Descartes, Flamsteed, Galileo, David Gregory, James Gregory, Halley, Hobbes, Hooke, Horrox, Huygens, Kepler, Leibniz, Locke, Maclaurin, Newton, Oldenburg, Pascal, Rohault, Torricelli, John Wallice [sic], Whiston, Wilkins, and Wren. Some of these give summaries of the theories associated with each person – especially those of Copernicus, Kepler, Galileo, Descartes, Newton – and their place in the history of science. Martin acknowledged the accounts by Colin Maclaurin and Henry Pemberton. See Pemberton (1728) and Maclaurin (1748); and Stewart, L. (1992), for the context of this popularisation of Newton.

[28] Oldys in Kippis (1778–93), vol. 1, xv. Kippis (1778–93), preface to 2nd edition, vol. 1, xxi. For comments on this work as an example of early literary biography, see Stauffer (1941), 249–56 and Lipking (1970), 79–81.

[29] Korshin (1974), 516.

[30] Oldys (1747–66), vol. 2, 913–34 on Boyle; vol. 5, 3210–44 on Newton. See Pemberton (1728); Maclaurin (1748).

of the 'Christian Philosopher'.[31] Significantly though, in the biographical entries this particular concept does seem to be most closely linked with *natural* philosophy – a nexus which, as Steven Shapin has noted recently, was not so clear in Burnet's original use of it in his sermon at Boyle's funeral.[32]

Nevertheless, the articles in this dictionary were longer and more detailed than those in Moreri or in Birch's ten-volume English edition of Bayle. There was also more sense of the contemporary context in which an individual operated, and Oldys recognised the possibility of 'a succinct account of any disputes or controversies in which they were engaged'.[33] But in Kippis's second edition, in particular, these accounts ran out of control, lost in irrelevant facts pursued under the compulsion for what Boswell called 'authentick information'.[34] A nineteenth-century critic said that Kippis did not rewrite the first edition of the dictionary 'by methodising those lives which were injudiciously or incorrectly given in the first edition', but rather gave the article verbatim and then added his additions and corrections. This gave 'the whole the air of a tedious controversy between himself and the preceding editors.' One consequence was that the second edition only reached volume five, ending with the letter F.[35]

Biography in encyclopaedias

By the late eighteenth century some lines of demarcation between historical dictionaries and encyclopaedias had collapsed. Biography now appeared on both sides of the border. The second edition of the *Encyclopaedia Britannica* departed from the original edition by including biographical articles along with those on the major 'systems' of the arts and sciences. This was sufficient to cause the resignation of William Smellie, the compiler of the first

---

[31] Ibid, vol. 3, 1649; vol 5, 3047, 3494; also Kippis (1778–93), vol. 5, 116 on William Derham. The concept of 'character' did not necessarily prescind such intellectual analysis. For its role in David Hume's historical works, appearing during the period of the first edition of the *Biographia Britannica*, see Wertz (1993).

[32] Shapin (1993), 338–9 and (1994), 170–2. This point would need to be confirmed by a more thorough check of a variety of entries in both Birch's edition of Bayle and the *Biographia Britannica*.

[33] Oldys (1447–66), vol. 1, xiv.

[34] Kippis (1778–93), vol. 1, xix aimed for biographies based on 'the most original information, to render them peculiarly authentic'. On Boswell, see Stauffer (1941), 402–55 and Dowling (1978); Boswell (1953), 22–3, 694, 770.

[35] Chalmers (1812–17), vol. 17, 382–86 at p. 384. The warning signs appeared in volume two, where Kippis had to explain why the letter 'B' had so many names. Kippis (1778–93), vol. 2, vii. Lipking (1970), 85 says that English works of this kind 'never resolved the mixture of antiquarian curiosity and critical ambition on which they were predicated'.

edition – or at least it was the reason he gave for severing his ties with the publication he had joined at the invitation of two Edinburgh printers, Andrew Bell and Colin Macfarquhar. One reporter says that when his partners insisted upon the introduction of 'a system of general biography' in the second edition, Smellie withdrew with the objection that this was 'by no means consistent with the title *Arts* and *Sciences*.'[36]

This incident highlights the assumption about the role of encyclopaedias as compendia of rational systems, contrasting with other dictionaries, such as the historical ones. Thus when biography *was* included in encyclopaedias from the end of the eighteenth century, it marked the introduction of foreign material, information belonging to a previously separate genre of reference publication – namely, the historical dictionary. Biographical articles were a new kind of entry – neither technical terms nor large systems – and thus fitted awkwardly into the dictionaries of arts and sciences that appeared from the mid eighteenth century, either as successors of Harris' *Lexicon* and Chambers' *Cyclopaedia*, or as smaller versions of the *Encyclopédie*.

In the second edition of the *Encyclopaedia Britannica*, edited by James Tytler in ten volumes between 1778 and 1784, the preface announced the appearance of biography. In doing so the editor stressed that it was 'a new department', not found in 'any collection of the same kind'.[37] He gave an enthusiastic endorsement of the genre, noting its sheer human interest and its moral value. The longest biographical entry (five and a half columns) on a scientific figure was, unsurprisingly, the one on Isaac Newton. This gave the standard account of the main moments in his life, referring to his work on astronomy and Biblical chronology, but not the work on optics or alchemy. However, like the earlier dictionaries, it still made 'character' the governing concept: Newton's avoidance of scientific controversy and his modesty about his intellectual achievements were treated under this rubric. Any detailed account of his scientific theories was still reserved, as in the first

---

[36] Kerr (1811), vol. 1, 363. Smellie did however write biographies of contemporaries such as David Hume. These were published posthumously in Smellie 1800. In a letter of 1797 about his own work on Adam Smith, Dugald Stewart said: 'I hate biography'. Stewart (1980), 265. For another criticism of anecdotal biography by Stewart, see Stauffer (1941), 538–9.

[37] Tytler (1778–84), vol. 1, vii. In fact, 'biography' is listed on the title page of a work which appeared in 1774. See Proctor and Castieau (1774). I have not been able to check the extent to which this advertising is confirmed by the contents. Tytler's edition of the *Britannica* does not use this word, but the last part of the extended content description on the title page promised 'an Account of the *Lives* of the most eminent Persons in every Nation, from the earliest ages down to the present times.' As far as I can tell, the word 'biography' never appeared on the title page of any edition of the *Britannica*.

edition, for the entry on 'Newtonian Philosophy' which followed immediately after the biographical one.[38] This separation of the biography of a scientific figure from an account of his scientific work (although less extreme in the case of Newton) set the pattern for the biographical entries in the *Encyclopaedia Britannica* between the 1780s and the 1820s.[39]

The inclusion of biography was justified by a link with the main content of the encyclopaedia – the large 'systems' on the various arts and sciences. Thus the preface suggested that 'after surveying any particular science, it will be found equally useful and entertaining to acquire some notion of the private history of such eminent persons as have either invented, cultivated or improved, the particular art or science'.[40] In giving the policy on the selection implied here, the editors of the third edition (Colin Macfarquhar, and later George Gleig) said that the biographical entries would treat persons who have 'distinguished themselves either in the theatre of action or in the recess of contemplation'. They accepted that this rule might lead readers to complain about the absence of 'their favourite philosopher, hero, or statesman', or conversely, about the inclusion of some obscure names 'who were no proper objects of such public regard'. But they insisted that selection was determined by the presence of some link 'with recent discoveries and public affairs'.[41]

Thus the second and third editions of the *Britannica* – appearing between 1778 and 1797 – did set limits on the scope of biography. The preface to the two volume *Supplement* of the third edition, undertaken by the Episcopal clergyman, George Gleig in 1801 (a second edition was issued in 1802), is quite revealing. Explaining the omission of 'articles (chiefly biographical)' which were advertised in the prospectus, Gleig admitted that he had to deviate from the original plan of supplementing the third edition with sketches of men whose lives did not meet the existing test for inclusion – that is, having some connection with 'science, art, or literature.' He now admitted

---

[38] Ibid., vol. 7, 5385–87 on Newton; 5388–99 on Newtonian philosophy.

[39] The entry on Roger Boscovich in the third edition seems more integrated than most, but recent research suggests that it was written in two parts. Michael Barfoot of Edinburgh University Library has referred me to Goldie (1991) who claims that Patrick Geddes wrote the article. However, Barfoot (private communication) argues that there is a case that Geddes wrote only the biographical part, and that John Robison wrote the pages on the scientific theories. For the Boscovich entry, see Macfarquhar and Gleig (1788–97), vol. 2, 92–9 on his life and character; then 99–107 begins: 'It now remains that one give an account of his Theory of Natural Philosophy'.

[40] Tytler (1778–83) vol. 1, vii. This sentence was repeated in the prospectus to the third edition, issued in 1788. See British Library: Mic. B. 896, reel 1505, no. 24.

[41] Macfarquhar and Gleig (1788–97), vol. 1, x–xi. This third edition was issued in 300 parts from 1788 and was completed as eighteen volumes in 1797.

that his subsequent experience confirmed the appropriateness of the stricter, original criterion:

> So many applications were made to me to insert accounts of per-
> sons who, whatever may have been their private virtues, were
> never heard of in the republic of letters, that I was under the necess-
> ity of excluding from the second volume [of the Supplement] the
> lives of *all* such as had not either been themselves eminent in litera-
> ture, or in some liberal art or science.[42]

This could be taken as a vindication of Smellie's stand against biography. It certainly showed the danger of a subject that touched human emotions and left editors to deal with the pleas of readers to perpetuate the memory and reputation of their dead friends and relatives. Nevertheless, James Millar, the editor of the fourth edition, which appeared from between 1801 and 1810, seemed to loosen the previous policy. Like his predecessors, he justified the inclusion of some 'obscure' names on the grounds that 'there has rarely passed a life of which a faithful record would not be useful'. But by combin-ing this statement with the announcement that the encyclopaedia now aimed for a 'more perfect biographical register than any which has hitherto been offered to the public', he threatened the existing rationale – the need for some link with a major discovery (in the case of men of science) as the condition for the inclusion of a biographical entry.[43]

It is possible that the introduction of biography allowed the *Britannica* to meet the challenge of its smaller rivals. Apart from Abraham Rees' editions of Chambers' *Cyclopaedia* (which appeared in various editions of four or five folio volumes between 1778 and 1788), it was undoubtedly the major English language encyclopaedia of the late eighteenth century; but it began to face competition from a variety of publications that experimented with size and price in order to attract sections of the market. For example, *The New Royal Encyclopaedia Londonensis*, appearing between 1796 and 1802, claimed to fit the best of the *Britannica* into three volumes – by *excluding* biography. It asserted that 'very expensive Works (stimulated by private Interest instead of public Utility) have thus loaded their Performances, by absurdly . . . running into a tedious Prolixity of Systems of BIOGRAPHY, GEOGRAPHY, ENGLISH, HISTORY etc' which take up three-quarters of the

---

[42] Gleig (1801), vol. 1, vi.
[43] Millar (1801–10), vol. 1, xiii. This was a reprint of the third edition, plus a two volume sup-plement.

work and are not 'proper to form a Part of an Encyclopaedia of Arts and Sciences'. A similar charge was brought by a larger rival – *The English Encyclopaedia: a collection of treatises and a dictionary of terms, illustrative of the Arts and Sciences*, issued in ten volumes in 1802. In its view, the *Britannica* had 'been swelled with such a variety of uninteresting biography; with tedious geographical descriptions of obscure towns and villages; with minute histories of fabulous heroes and divinities.' On the other hand, the *Encyclopaedia Perthensis, or Universal Dictionary of Knowledge* [1815] included biography because arts and sciences 'are far from comprehending every necessary subject of inquiry'.[44] But if they were to remain smaller and cheaper, most of these competitors could not afford to include biography while also covering the expanding material in the arts and sciences. The *Britannica* was prepared to do both; indeed, by 1802 it had no choice because Rees began his own projected 44 quarto volume *New Cyclopaedia*, thus dramatically breaking the tie with Chambers *and* including biographical articles.

Biography and history of science

Whatever the commercial strategy behind the decision of the *Britannica* to embrace biography, there were significant intellectual consequences. The justification given for biography led to a related case for the historical and geographical contexts of science. Thus the second edition predicted that most readers, having learnt something of the persons involved in a 'branch of human knowledge', will naturally wish to know 'something of the places where those transactions have passed'.[45] The preface of the fourth edition (repeating a sentence from the prospectus to the third edition) asserted that there was a 'natural and necessary connexion' between major achievements and the 'scenes where they were performed'.[46] This association between person, place and intellectual activity was also offered in Richard Pulteney's *Historical and Biographical Sketches of the Progress of Botany in England*, published in 1790. His preface noted that:

> In tracing the progress of human knowledge through its several gradations of improvement, it is scarcely possible for an inquisitive and liberal mind, of congenial taste, not to feel an ardent wish of

[44] [Howard] [1796–1802], vol. 1, vi; anon. (1802), vol. 1, vii; anon. [1815], vol. 1, ii.
[45] Tytler (1778–84), vol. 1, vii.
[46] Millar (1801–10), vol. 1, xiii; see also McKillop (1965) on the notion of local attachment in the eighteenth century.

information relating to those persons by whom such improvements
have severally been given; and hence arises that interesting sympathy which almost inseparably connects biography with the history of each respective branch of knowledge.[47]

This assumption meant that biography could function as a Trojan horse for
more historical treatments of science in encyclopaedias.

But did this happen? What form did this treatment of biography take? In
what follows I offer some remarks on the place of biographies in encyclopaedias – against the background of the earlier historical and biographical dictionaries. As already noted, the *Encyclopaedia Britannica* began to carry biographical entries after 1778 and these included some on men of science. I have not
been able to carry out a comprehensive survey of the proportion of scientific
lives in these editions, but I can offer some observations. It is clear that no
living figures were included. This followed the accepted view of biographical
dictionaries which, as the English edition of Bayle said, were 'a kind of general monument, to the memory of deserving persons'.[48] Biographical entries
were final accountings of a life; they were memorials having both a testi-
monial and moral function. The editors of the third edition put one of the
consequences rather starkly, explaining that some important figures were
missing because the letters of their names had passed before 'we had intelligence of their deaths'. Similarly, when David Brewster and the Reverend
John Lee were compiling the *Edinburgh Encyclopaedia*, Lee offered to write
an entry on the Scottish historian and philosopher, Adam Ferguson – on the
expectation that he would die before they reached the letter 'F'.[49] Another
implication was that during the last two decades of the eighteenth century
the *Britannica* carried no separate biographical entries on the leading contemporary figures of science, some of the heroes of what is now regarded as a
second scientific revolution.[50]

However, in the large treatises or 'systems' on particular sciences, the
contributions of key contemporary figures *were* mentioned, although still not

---

[47] Pulteney (1790), vol. 1, cited in Stauffer (1941), 505–6.
[48] Bayle (1734–8), vol. 1, dedication; Kippis (1778), vol 1, xviii. But presumably editors had to consider the 1721 House of Lords decision that it was a breach of privilege to print the 'life of a peer
without the consent of his heirs or executors.' Plant (1965), 120.
[49] Macfarquhar and Gleig (1788–97), vol. 1, x. Lee to Brewster, 6 December 1809, MS 3432, ff 231–
2, National Library of Scotland.
[50] See Kuhn (1961), 190 Cohen (1985), 97 on the idea of a second scientific revolution. Writing in
1845 Henry Brougham claimed that scientists of this period surpassed those of seventeenth century. Brougham (1855–61), vol. 1, v.

as part of a biographical treatment. For example, the article on 'Chemistry' in the third edition of the *Britannica* opens with an historical introduction organised around the contributions of those, beginning with Boyle, who freed the subject from its alchemical shackles.[51] There is a separate biographical entry on Boyle, but more recent participants, such as Joseph Priestley, Joseph Black, William Cullen, Henry Cavendish and Antoine Lavoisier, are not covered in biographical entries, although their theories and disputes are mentioned in 'Chemistry' and in some of the other articles, such as 'Aerology'.[52] With the appearance of the *Supplement* to this third edition in 1801, a biographical notice of Lavoisier (who died on the guillotine in May 1794) was included, and more space was given to his and other recent work in the large article on chemistry.[53]

When he began to compile the *Supplement* to the fourth, fifth and sixth editions of the *Encyclopaedia Britannica* in 1815, Macvey Napier introduced new biographical articles, 'mostly of recent lives'. Although some additions filled what he called 'palpable omissions' in the previous editions, Napier focused mainly on 'persons who have died during the last thirty years', and in doing so brought the number of biographies to 165 within a total of 600 articles in the six volumes of the *Supplement* completed in 1824. Of these biographical entries, about a quarter were the new ones Napier added: that is, on people who had recently died. He explained that 'the subjects of them [biographical articles] have been selected, for the most part, on account of their eminence in Science or Literature'.[54] In fact, of the 165 lives 58 (35%) were men of science. In contrast, only 5 (3%) of the biographical articles were on religious figures.[55] This seems to indicate Napier's preferences, although generalisations have to be tempered by the fact that 55% of all biographical articles in the *Supplement* were drawn from the first three letters of the alphabet. Another contingency is that forty-six of all the biographies

[51] Macfarquhar and Gleig (1788–97), vol. 4, 374–635; see 374–7 for mention of Boyle.

[52] Ibid., vol. 4, 394; vol. 1, 144–97. In the article on chemistry, Black, Cullen, Cavendish, Kirwan, Boerhaave are mentioned in both the index and the text.

[53] Gleig (1801), vol. 1, 210–403; this article takes up half of part one of volume 1. See also vol. 2, 70–2 for a very positive article on Lavoisier. There was no biography here of Priestley, Black or James Watt.

[54] Napier (1824), vol. 1, xxix.

[55] Gooding (1929), 66–7. In the case of the *Edinburgh Encyclopaedia*, started in 1809 but not completed until 1830, Brewster's coeditor, John Lee, drew up an alphabetical notebook (dated 1805) with forty-seven possible entries, mainly biographical ones, finishing at the letter T. These seem to be suggestions for lives not adequately covered in other encyclopaedias, but only two, Barrow and Robison, are natural philosophers. MS 3455, National Library of Scotland. By the time it appeared Brewster had remedied this to some extent.

were written by Thomas Young, the natural philosopher and Egyptologist. But there is little doubt that Napier was concerned to enlarge the representation of scientific lives in the encyclopaedia.[56]

Charles Babbage referred in 1830 to the 'diminutive' world of science.[57] This observation holds, even allowing for the polemical intent of commentators such as Babbage who wrote about the alleged decline of science in England in comparison with France. In the early nineteenth century, the number of individuals who could be counted as men of science was small when compared with other vocational categories. It follows that Napier must have been well disposed towards scientific lives for the biographical entries: otherwise the figure of 35% could not have been achieved. He also went beyond England and Scotland. Thus it is worth noting that the *Encyclopaedia Britannica*, unlike the *Biographia Britannica*, was not a strictly British reference work; indeed, its coverage of leading French and German figures in the biographical entries was quite extensive. This was complemented by the fact that Napier recruited foreign experts as contributors to some of the major articles on the sciences: Francois Arago on 'Polarisation', Jean Baptiste Biot on 'Pendulum', 'Electricity' and 'Galvanism'. On both counts the work presented itself as continuing the ideal of a cosmopolitan republic of learning at a time when biographical dictionaries, for both practical and political reasons, had become decidedly national.[58]

What was the nature of these biographical entries in Napier's *Supplement*? How do they compare with those in the 'historical' and biographical dictionaries? Any answers must first recognise that although Napier made a point of adding to the existing biographical articles of the *Britannica*, these were not his highest priority, even if they may have played a part in a marketing strategy. Like most early nineteenth-century encyclopaedia editors, he devoted most of his energy to the task of recruiting appropriate contributors for the major treatises on scientific and other subjects. Once engaged, some of these authors volunteered biographical pieces on men of science associated with the field they were covering: for example, James

---

[56] Napier (1824), vol. 1, xxix for mention of Young as responsible for 'a large proportion of such as relate to men of science'. See Peacock (1855), vol. 2, 446–623 for a selection of Young's biographical contributions.

[57] Babbage (1830), x–xi.

[58] Of the 165 biographical articles in Napier's *Supplement* at least 94 were on Continental European figures. For the earlier cosmopolitan ideal, see Daston (1991). The *Encyclopaedia Metropolitana*, projected by S. T. Coleridge in 1817, at first intended to arrange its historical sections in chronological order, starting with the ancient world; but the editors soon shifted to national histories. See Smedley *et al.* (1845), vol. 1, xviii.

Smith did some of those related to botany. Napier said that some accounts of distinguished figures were offered by friends who had 'long personal knowledge of them'. One effect of this pragmatic approach was that the biographical entries of the previous editions were republished, usually without revision, whereas alterations and additions to those on the various arts and sciences were a selling point of successive editions.[59] This meant that Napier did not recommission the entries on the major scientific personalities of the seventeenth century who were included in earlier editions. Thus the entry on Newton remained unaffected by new archival evidence regarding his alchemical views and heterodox theology until Brewster wrote for the seventh edition of the *Britannica*, completed in 1842.[60] Given these factors, it is unlikely that a concerted approach to biography can be discerned, but there are features that distinguish encyclopaedias from the earlier historical and biographical dictionaries.

The *Britannica* never extended its initial short notice of the word 'biography' into a separate article until the eleventh edition of 1910. Napier did contemplate this when preparing his *Supplement*, and asked William Hazlitt: 'Is there anything *good* and *striking* to be said upon *Biography*, its kinds, rules and cases?'[61] But this article never materialised. Nevertheless, the entry for James Boswell served as a discourse on the positive features of the genre. 'We commemorate him as an author, and particularly as a writer of Biography. Here he is almost an inventor'.[62] But the 'minuteness' of information, mentioned as one of the key features of this modern biographical approach, and attempted by Kippis in the *Biographia Britannica*, was rarely sought in the biographical entries of the *Supplement*. The extensive footnotes and quotations from letters were also omitted. One practical reason here was the limited space available in the *Supplement*: Napier cited this when asking Hazlitt to 'curtail' the length of a biographical entry.[63] Yet within the space allotted, choices of emphasis were made. The biographies of scientific figures ranged in length from under a page to five pages (making the latter ones

[59] Napier (1824), vol. 1, xxx; Gooding (1929); Yeo (1991), 47.
[60] See Brewster (1842). These issues were more fully confronted in his biography of Newton. See Brewster (1855) and Christie (1984). Augustus De Morgan also wrote an essay on Newton for a collection called *Old England's Worthies*. This was informed by his own extensive knowledge of the available archives. The material was seen as controversial, so De Morgan signed his name to it. De Morgan (1846). For the contemporary debate about Newton's character, see Yeo (1988) and (1993), ch. 5.
[61] Napier to Hazlitt, 24 August 1816, MS 674, f 69, National Library of Scotland.
[62] Napier (1824), vol. 2, 372–75, p. 373.
[63] Napier to Hazlitt, 24 August 1816, MS 674, f. 69, National Library of Scotland.

substantial entries of about 3000 words), but the focus was always on their scientific work rather than on their personal lives.

This suggests a contrast with the historical and biographical dictionaries in which the character of the person was usually more prominent than the details of their scientific or scholarly work. We should note, however, that the eighteenth-century notion of 'character' was closer to our idea of reputation and was not synonymous with later concepts of personality as the expression of inner psychological dynamics, often developing from childhood. Further consideration of this issue is beyond my scope here; but it is possible to say that character was often regarded as something directly displayed.[64] In many of these eighteenth-century publications, biographical articles on philosophers invariably referred to the physical body of the person: that is, to stature, health, diet, habits. Thus according to Moreri, Aristotle ate little 'and slept less'; Bacon's 'Port was Stately, his Speech flowing and grave'; Boyle 'was of a weak infirm Body, which renders it the more astonishing how he could write, meditate, read, and try Experiments as he did'.[65] The English edition of Bayle continued this concern: Galileo was 'of little stature but of venerable aspect and vigorous constitution. His conversation was affable, and free, and full of pleasantry.' On the other hand, Robert Hooke's person 'was but despicable, being very crooked, and always pale and meagre'.[66] Similarly, moral character was deduced from behaviour. These entries also allowed space for character assessments by contemporaries, including accounts of the last days of the person and their deathbed demeanour. In Kippis, this emphasis on character, both physical and moral, continued, but with greater attention to the doctrines and controversies associated with the person, often illustrated with long extracts from their letters. By the third edition of the *Encyclopaedia Britannica* (from 1788), this emphasis on character had not altered, although it offered much shorter biographical articles than the major eighteenth-century historical and biographical dictionaries.

---

[64] For some comments on the concept of 'character' as used by Hume, see Wertz (1993). D'Alembert's definition in the *Encyclopédie* points towards the modern sense: 'In a moral sense it signifies an habitual disposition of the soul, that inclines to do one thing preferably to another. Thus a man who *seldom* or *never* pardons an injury, is of a revengeful character. Observe, we say seldom or never, because a character results not from a disposition being rigorously constant at all times, but from its being generally habitual, and that by which the soul is the most frequently swayed.' D'Alembert (1772), 301. On nineteenth-century notions, see Collini (1991), ch. 3.

[65] Moreri (1694), no pagination.

[66] Bayle (1734–41), vol. 5, 372; vol. 6, 218. This portrait of Hooke was repeated in Oldys (1741), vol 4, 2659.

By the time of Napier's *Supplement* (1815–24) however, even the intention of discussing character seems to have gone. Thus the entry on Charles Coulomb states that: 'Mr Coulomb's moral character is said to have been as correct as his mathematical investigations', but there was no attempt to discuss it; indeed, this statement followed five pages summarising nineteen of his scientific memoirs and papers.[67] A passage from the entry on the eighteenth-century Russian natural philosopher Francis Aepinus confirms the priorities at work here:

> We regret that our means of information do not enable us to communicate any particulars in regard to his personal history; but we shall give some account of his contributions to science, and these, after all, form the most interesting memorials of a philosopher's life.[68]

In Napier's *Supplement*, the emphasis was on the contribution to the story of scientific progress; men of science were considered as *discoverers*, not as individuals whose whole lives were being weighed. The concern with details about the body and demeanour of the person was now usually absent; so too were the long extracts from their letters or character assessments by friends. As Peter Carey noticed in the extract from his novel quoted at the top of this chapter, there was rarely any room for the idiosyncrasies of the particular man of science. The articles in the *Supplement* to the *Britannica* began to distance 'the merely personal' – as Albert Einstein later called it – from its account of scientific lives. How far this set the direction for the treatment of scientific biography in other encyclopaedias, and indeed, later editions of the *Britannica*, remains a question for investigation. For example, some comparative analysis of Abraham Rees *New Cyclopaedia*, which included biographical entries, would be useful. Charles Hutton's *A Philosophical and Mathematical Dictionary* (1815) also advertised 'Memoirs of the Lives and Writings of the Most Eminent Authors, both ancient and modern, who by their discoveries and improvements have contributed to the advancement of them'.[69]

One reason for the limited detail about personal lives in the case of

---

[67] Napier (1824), vol. 3, 414–19.

[68] Ibid, vol. 1, 63.

[69] See Bernstein (1985), 296. For a comment on the evacuation of self from nineteenth-century scientific biography, see Gagnier (1991), 258. For the late eighteenth-century contrast, see Outram (1978). As might be expected, biographies in the Positivist *Calendar* of Auguste Comte evaluated the individual's contribution to the progress of science and gave little space to their views on other matters. See Harrison (1892).

Napier's work is the way he severely restricted the theological glosses that accompanied character portraits of natural philosophers in earlier publications such as the English edition of Bayle and the *Biographia Britannica*. (I have already cited the low percentage of religious leaders and theologians.) In his *Supplement*, biographical accounts of men of science very rarely include references to the natural theological framework that support the praise of their behaviour as Christian philosophers. This contrasted with the practice of the earlier English biographical dictionaries and with the evangelical emphasis in contemporary early nineteenth-century British society. Indeed, at some points Napier's *Supplement* seemed closer to the anti-clerical outlook of the French *Encyclopédie* — for example, the negative effects of fanatical religion were mentioned in some articles. Thus in noting that the Italian mathematician and philosopher, Maria Agnesi, retreated to a convent towards the end of her life, the writer cited this as 'another melancholy instance' of the 'darkening power of superstition over the brightest minds'.[70]

But the main reason for this reduced interest in personality was the new function of these short biographical entries as satellites to the larger articles on the various sciences.[71] Under this arrangement, biography was subservient to the history of science. As mentioned above, a large part of each entry under the name of a natural philosopher recorded his or her publications; it also ranked the person against other individuals using the more detailed account of the progress of science given in the larger articles on particular disciplines. Thus Coulomb was evaluated 'in the particular department of science which he cultivated', and it was concluded that 'he may be fairly ranked in the same class with Franklin, Aepinus, and Cavendish'.[72] The grounds for this assessment were to be found in the article on 'Electricity'. In retrospect, we can see a hint of this division of territory between biographical and scientific articles in the third edition of the *Britannica*. For example, the biographical entry on Copernicus pointed to the one on Astronomy for a summary of his theory, but there was no assessment, or even discussion, of the scientific work of a person in the short biographical notices.[73] In Napier's *Supplement* however, many of these entries awarded their subject, at least in a shorthand way, a position in the history of the relevant field of science — as

---

[70] Napier (1824), vol. 1, 113. See Carew (1831) for a response to Stewart's comments on rational religion, and Hilton (1988) for evangelical attitudes.

[71] For an analysis of the organisation of encyclopaedias, see Yeo (1991).

[72] Napier (1824), vol. 3, 419.

[73] Macfarquhar and Gleig (1788–97), vol. 5, 432–33 referring to vol 2, 421. This is only partly qualified in the case of the entry on Newton in the second edition referred to earlier.

this was presented in one of the larger articles, or in one of the three Preliminary Dissertations.[74]

These large 'Discourses on the History of the Sciences', as Napier called
them, aimed to provide a 'connected view of the Progress of the Sciences',
allowing contrasts and comparisons to be made across some of the disciplines that lay separated by different alphabetical letters in the body of the
work. They were not intended to overlook the role of individuals – indeed
Napier thought they could highlight the heroic discoverers who received no
special place in Chambers' *Cyclopaedia*, those 'great lights of the world by
whom the torch of science has been successively seized and transmitted'.[75]
But this approach still reinforced the emphasis on the progress of the particular science, not the intellectual or emotional or social biographies of the men
of science. Thus in William Brande's dissertation on the history of chemistry
'well known names' appeared alongside the major discoveries he judged as
crucial to the progress of the subject; others who failed to grasp, or resisted,
this direction had 'sunk into oblivion' and their names were not to be rescued. This approach not only excluded personal details but also various applications of science, if they were not central to a key discovery, as Brande
explained:

> I have diligently endeavoured to record every important event in
> the general history of the science. Of many who have attained
> deserved eminence in the exclusive pursuit of its distinct branches,
> no mention has been made: I have looked with attention into their
> works, and am well aware of their individual merits; but I should
> have swerved from the principal object of this Dissertation, *that of
> recording discoveries*, had I attempted even the superficial enumer-
> ation of their infinitely varied applications.[76]

Conclusion

When biography entered the *Encyclopaedia Britannica*, in the second edition
from 1778, this was recognized as a departure from the prevailing convention that allocated this subject, as well as more general history and geography, to the 'historical' dictionaries, such as those following Moreri's. There

---

[74] On these see Yeo (1991), 33–4 and (1993), 150–1.
[75] Napier (1824), vol. 1, xxxi.
[76] Brande (1818), vol. 3, 50, 79; my emphasis.

was also the suggestion that readers would appreciate information on the historical and geographical settings in which major personalities acted.[77] However, this did not take the form of a life history as found in the earlier historical or biographical dictionaries, and by the time of Napier's *Supplement* (1824) the biographical articles largely abandoned any attention to the character or personality of the individuals they noted. The focus was on their contributions to the march of discoveries surveyed in the preliminary dissertations and explained in the 'treatises' on the various sciences.

It would be pointless to assess this approach to biography in terms of modern conceptions of the subject. However, we can try to view the encyclopaedic treatment of biography in the context of early nineteenth-century notions about what an account of scientific lives might contain. In 1845 Henry Brougham gave the following prescription in his *Lives of Men of Letters and Science, who flourished in the time of George III*:

> The history of a philosopher's life, that is, of his labours, the tracing of those steps by which he advanced beyond his predecessors, the comparison of the state of the science as he found it, with that in which he left it, tends mightily to interest the reader, to draw him towards the same inquiries.[78]

These requirements were to some extent met by the biographies in Napier's *Supplement* to the *Encylopaedia Britannica*. Especially when taken in tandem with the larger articles on the sciences and the preliminary dissertations, Brougham's concern with ranking an individual discoverer is apparent in Napier's work. Although such accounts are now seen as overly 'presentist' or Whiggish, they were attuned to the 'state of science' at different historical periods, and hence fulfilled a need recognized by some of the leading members of the scientific community. Commenting on discussions at the British Association over the history of chemistry in 1845, George Peacock stressed the importance of explaining why a theory [phlogiston] now known to be false 'was so long considered to be true'; this involved, he said, an awareness of the precise conditions of opinion prevailing in specific periods of scientific history.[79] On a related issue, William Vernon Harcourt, the first secretary of the British Association, criticised Brougham's handling of the controversy over the roles of Black, Watt and Cavendish in determining the chemical

---

[77] See note 43 above and Macfarquhar and Gleig (1788–97), vol. 1, x–xi.
[78] Brougham (1855–61), vol. 1, ix–x.
[79] Peacock (1845–6), 108, 132.

composition of water. Writing to Brougham he suggested that such issues were complex and that if venturing on 'such dangerous ground, you should at least learn how to choose your authorities'.[80] The articles in the 1824 *Supplement* to the *Encyclopaedia Britannica* did, to some extent, provide the starting point for such historical inquiries relevant to what Harcourt called 'arbitration of the rights of discovery'.

From this perspective biographies did not need to ponder the character of the man of science, his social or religious views, nor even his reasons for such an unusual vocational choice. The 'lives' in encyclopaedias were small sections of the history of science in which the contributions of an individual were mapped against a larger story. There is a paradox here in so far as this historical genealogy (or even 'rational reconstruction') of discoveries coexisted with an increasing interest in the heroic genius of great discoverers.[81] Thus in early Victorian biographical writing outside encyclopaedias there was concern with the question of intellectual style and its connection with personal and private life. This became apparent when separate books (as distinct from dictionary entries or obituaries) began to be written on men of science. In 1831 Humphry Davy's biographer, John Paris, argued that an understanding of scientific figures would not be found in a collection of anecdotes, but 'in an analysis of human genius, and in the development of those elements of the mind, to whose varied combinations, and nicely adjusted proportions, the mental habits, and intellectual peculiarities of distinguished men may be readily referred'.[82] Another example from the middle of the century suggests that the celebration of the intellectual qualities of major scientific individuals still left an image of the man of science as remote from life, absorbed in strange journals and language, 'as one who refuses to conform to the conventionalities of society, rejects its enticements'. As an antidote to this perception, a writer for the *Eclectic Review* recommended Francois Arago's biographies of six French scientists (Baily, Fourier, Carnot, Malus, Fresnel, Laplace) who, 'while they pursued the most occult subjects of scientific research, were, for good or evil, foremost in the political movements of their age'.[83] A study of the biographical essays on major scientific

---

[80] Harcourt to Brougham, in Harcourt (1846). For a critical review on more general grounds, see Lockhart (1845) who said the 'public might reasonably expect more care' from Brougham about the history of science.

[81] Schaffer (1986).

[82] Paris (1831), vol. 1, iv, vii; also Knight (1992) and his chapter in this volume. On the emergence of popular scientific biographies, see Sheets-Pyenson 1990, 399.

[83] Anon. (1857), 201–2.

figures in the Victorian periodicals, and in scientific journals such as the *Philosophical Magazine*, might provide instructive comparisons with those in encyclopaedias.

These examples illustrate the variety of opinion on the appropriate form of scientific biography. Even as he announced his own view, Brougham's conception of this genre as a way of allotting the individual a place in the history of discovery did not satisfy all readers or authors. They had to look beyond encyclopaedias for a different approach, but also beyond the earlier 'historical' and biographical dictionaries which focused on moral character in the absence of any detailed account of the scientific work of the person. Similarly, as historians or biographers, we must choose our eighteenth- and nineteenth-century evidence carefully. Perhaps one of the lessons from this study is the need to appreciate the rationale of dictionaries and encyclopaedias before using them as primary sources for biographical material, or as evidence of contemporary images of scientists.

Acknowledgements

This study was supported by an Australian Research Council grant, the Ian Potter Foundation, and by the Faculty of Humanities, Griffith University. For research assistance I thank Judith Deppeler-Hagan and Diana Solano; for research involving French translation I thank Alice Addison. I am also grateful to John Gascoigne, Dorinda Outram, Michael Shortland and Bill Zachs for their comments.

Bibliography

Note: Articles in nineteenth-century journals appeared anonymously and have been identified in Houghton (1966–88).

Aiken, J. and Enfield, W. (1799–1815) *General Biography; or, Lives, Critical and Historical*, London: G. and J. Robinson.

Alsted, J. (1989) [1630] *Encyclopaedia . . . Serie Praeceptorum, Regularum, et Commentarium Perpetua*, ed. with foreword by W. Schmidt-Beggemann. Stuttgart–Bad Cannstatt.

Anon. (1802) *The English Encyclopaedia: being a Collection of Treatises and a Dictionary of Terms, Illustrative of the Arts and Sciences*, 10 vols. London: G. Kearsley.

Anon. [1815] *Encyclopaedia Perthensis; or, Universal Dictionary of Knowledge, collected from every source*, 23 vols. London: C. Mitchell.

Anon. (1857) *Biographies of distinguished scientific men*, *Eclectic Review*, 2, 201–25.

Arago, F. (1857) *Biographies of Distinguished Scientific Men*, trans. W.H. Smyth, B. Powell and Robert Grant. London: Longman.

Babbage, C. (1830), *Reflections on the Decline of Science in England*. London: B. Fellowes.

Bacon, F. (1974) [1605] *The Advancement of Learning and the New Atlantis*, ed. A. Johnson. Oxford: Clarendon Press.

Bayle, P. (1710) *An Historical and Critical Dictionary, by Monsieur Bayle. Translated into English with many Additions and Corrections, made by the Author himself, that are not in the French Editions*, 4 vols. London: C. Harper and others.

Bayle, P. (1720) [1697] *Dictionnaire Historique et Critique*, 3rd edition, 4 vols. Rotterdam: Chez M. Bohm.

Bayle, P (1734–8) *The Dictionary Historical and Critical of Mr Peter Bayle. The Second Edition*, trans. P. Des Maizeaux, 5 vols. London: D. Midwinter.

Bayle, P. (1734–41) *A General Dictionary, Historical and Critical*, ed. J.P. Bernard, T. Birch, J. Lockman, and other hands, 10 vols, London: J. Bettenham for G. Strahan.

Bayle, P. (1826) *An Historical and Critical Dictionary, selected and abridged from the Work of Peter Bayle: with a Life of Bayle*, 4 vols. London: Hunt and Clark.

Bayle, P (1965) *Historical and Critical Dictionary: selections*, trans. with introduction by R. Popkin, New York: Library of Liberal Arts.

Bernstein, J. (1985) The merely personal. *American Scholar* 54, 295–302.

Boswell, J. (1953) *Life of Johnson*. Oxford: Oxford University Press.

Bradshaw, L.E. (1981) Ephraim Chambers' Cyclopaedia. In Kafker (1981), pp. 123–40.

Brande, W.T. (1818) The preliminary dissertation on the progress of chemical philosophy from the early ages to the end of the eighteenth century. In Napier, vol. 3, pp. 1–79.

[Brewster, D.] (1842) Newton. In *Encyclopaedia Britannica*, 7th edition, 21 vols. Edinburgh: A. and C. Black, vol. 16, pp. 175–81.

Brewster, D. (1855) *Memoirs of the Life, Writings and Discoveries of Sir Isaac Newton*, 2 vols, Edinburgh: Constable.

Brougham, H. (1845–6) *Lives of Men of Letters and Science, who Flourished in the Time of George III*, 2 vols. London: C. Knight.

Brougham, H. (1855–61) *Works*, 11 vols. London and Glasgow: R. Griffin.

Burrell, P. (1981) Pierre Bayle's *Dictionnaire Historique et Critique*. In Kafker (1981), pp. 83–103.

Carew, Rev. P.J. (1831) *Remarks, Analytical and Historical, on the Connexion of Revealed Religion with Literary and Civil Liberty; being a Reply to Certain Statements in the First Preliminary Dissertation of the Encyclopaedia Britannica*. Dublin: Richard Coyne.

Carey, P. (1988) *Oscar and Lucinda*, Brisbane: University of Queensland Press.

Chalmers, A. (1812–17) *The General Biographical Dictionary, Containing an Historical and Critical Account of the Lives and Writings of the Most Eminent Persons in every Nation*, 32 vols. London: J. Nichol and Son *et al.*

Chambers, E. (1728) *Cyclopaedia, or An Universal Dictionary of Arts and Sciences*, 2 vols. London: J. Knapton *et al.*

Christie, R. (1902) *Selected Essays and Papers*. London and New York: Longman, Green and Co. Reprinted from Biographical Dictionaries. *Quarterly Review*, 157 (1884), 187–230.

Christie, J.J.R. (1984) Sir David Brewster as an historian of science. In A.D. Morrison-Low and J.J.R. Christie (eds) *Martyr of Science: Sir David Brewster 1781–1868*. Edinburgh: Royal Scottish Museum, pp. 53–6.

Cohen, I.B. (1985) *Revolution in Science*. Cambridge, MA: Harvard University Press.

Collier, J. (1721) *The Great Historical, Geographical, Genealogical and Poetical Dictionary; being a Curious Miscellany of Sacred and Profane History*, 3rd edition. London: H. Rhodes.

Collini, S. (1991) *Public Moralists: Political Thought and Intellectual Life in Britain, 1850–1930*, Oxford: Clarendon Press.

[D'Alembert, J.] (1772) Analysis of the word character. In *Select Essays from the Encyclopedy, being the most curious, entertaining, and instructive parts of that very extensive work, written by Mallet, Diderot, D'Alembert, and others, the most celebrated writers of the age*. London: Samuel Leacroft.

Daston. L. (1991) The ideal and reality of the Republic of Letters in the Enlightenment. *Science in Context*, 4: 367–86.

De Morgan, A. (1846) Newton. In *Old England's Worthies*, London: C. Knight, vol. 11, pp. 78–117. [I have used a proof copy in the De Morgan papers, Senate Libary, London University.]

Diderot, D. and D'Alembert, J. (1751–80) *Encyclopédie, ou Dictionnairé Raisonné des Sciences des Arts et des Métiers*, 35 vols, Stuttgart–Bad Cannstatt.

Dowling, W.C. (1978) *Boswell and the problem of biography*. In D. Aaron (ed.) *Studies in Biography*. Cambridge, MA: Harvard University Press, pp. 73–93.

Elwin, W. (1855) Arago and Brougham on men of science. *Quarterly Review*, 97, 473–513.

Gagnier, R. (1991) *Subjectivities: a History of Self-Representation in Britain 1832–1920*. Oxford: Oxford University Press.

Galloway, T. (1844) The martyrs of science. *Edinburgh Review*, 80, 164–98.

Gleig, G. (ed.) (1801) *Supplement to the Third Edition of Encyclopaedia Britannica*, 2 vols. Edinburgh: T. Bonar.

Goldie, M. (1991) The Scottish Catholic Enlightenment. *Journal of British Studies*, 30, 20–62.

Gooding, L.M. (1929) The *Encyclopaedia Britannica*: a critical and historical study of the three Constable editions. Unpublished Master of Science thesis, Toronto University.

Harcourt, W.V. (1846) *Letter to Henry Brougham FRS containing remarks on certain statements in his Lives of Black, Watt and Cavendish*, London: R. Taylor. Copy in William Whewell Papers, 103. c. 80.1 no. 2, Trinity College, Cambridge.

Hall, A.R. (1949) William Wotton and the history of science. *Archives Internationales d'Histoire des Sciences*, 9, 1047–62.

Hall, W. (1797) *The New [Royal] Encyclopaedia; or Modern Universal Dictionary of Arts and Sciences, on a New Improved Plan*, 3 vols, 3rd edition, revised by T.A. Lloyd. London: C. Cooke.

Harrison, F. (ed.) (1892) *The New Calendar of Great Men. Biographies of the 558 Worthies of all Ages and Nations in the Positivist Calendar of Auguste Comte*. London: Macmillan.

Hilton, B. (1988) *The Age of Atonement: the Influence of Evangelicalism on Social and Economic Thought 1785–1865*. Oxford: Clarendon Press.

Houghton, W. (1966–88) *The Wellesley Index to Victorian Periodicals*, 5 vols. Toronto: Toronto University Press.

[Howard, G.S.] [1796–1802], *The New Royal Encyclopaedia Londinensis; or, a Complete Modern and Universal Dictionary of Arts and Sciences*, 3 vols. London: printed for A. Hogg.

Hunter, M. (1990) Alchemy, magic and moralism in the thought of Robert Boyle. *British Journal for the History of Science*, **23**, 387–410.

Hutton, C. (1815) *A Philosophical and Mathematical Dictionary*, 2nd edition, 2 vols. London.

Jeffrey, F. (1848) The discoverer of the composition of water: Watt or Cavendish?. *Edinburgh Review*, **87**, 67–137.

Kafker, F. (ed.) (1981) *Notable Encyclopedias of the Seventeenth and Eighteenth Centuries: Nine Predecessors of the Encyclopédie*. Oxford: the Voltaire Foundation, *Studies on Voltaire*, vol. 194.

Kafker, F. (1994) William Smellie's edition of the *Encyclopaedia Britannica*. In F. Kafker (ed.) *Notable Encyclopedias of the Late Eighteenth Century: Eleven Successors of the Encyclopédie*. Oxford: the Voltaire Foundation, *Studies on Voltaire*, vol. 315, pp. 145–82.

Kerr, R. (1811) *Memoirs of the Life, Writings and Correspondence of William Smellie*, 2 vols. Edinburgh: J. Anderson.

Kippis, A. (1778–93) *Biographia Britannica; or, the Lives of the Most Eminent Persons who have flourished in Great Britain and Ireland*, 2nd edition, 5 vols. London: W. and A. Strahan.

Knight, D.M. (1992) *Humphry Davy: Science and Power*. Oxford: Blackwell.

Korshin, P.J. (1974) The development of intellectual biography in the eighteenth century. *Journal of English and Germanic Philology*, **73**, 513–23.

Kuhn, T.S. (1961) The function of measurement in modern physical science. *Isis*, **52**, 161–93.

Laertius, D. (1954) *Lives of Eminent Philosophers*, with a translation by R.D. Hicks. Cambridge, MA: Heinemann.

Lipking, L. (1970) *The Ordering of the Arts in Eighteenth-Century England*. New Jersey: Princeton University Press.

Lockhart, J.G. (1845) Lord Brougham's lives of men of letters. *Quarterly Review*, **76**, 62–98.

Macfarquhar, C. and Gleig, G. (1788–97) *Encyclopaedia Britannica; or, A Dictionary of the Arts and Sciences, and Miscellaneous Literature*, 3rd edition. 18 vols. Edinburgh: Bell and Macfarquhar.

Maclaurin, C. (1748) *An Account of Sir Isaac Newton's Philosophical Discoveries, in four books*. London: printed for the author's children by A. Millar.

Martin, B. (1764) *Biographia Philosophica, being an Account of the Lives, Writings, and Inventions of the Most Eminent Philosphers and Mathematicians who have flourished from the Earliest Ages of the World to the Present Time*, London: W. Owen.

Millar, J. (1801–10), *Encyclopaedia Britannica; or, a Dictionary of Arts, Sciences, and Miscellaneous Literature, enlarged and improved*, 4th edition, 20 vols. Edinburgh: A. Bell for A. Constable and Company.

Miller, A. (1981) Louis Moreri's *Grand Dictionnaire Historique*. In Kafker (1981), pp. 13–52.

McKillop, I.D. (1965) Local attachment and cosmopolitanism: the eighteenth-century pattern. In F.W. Hilles and H. Bloom (eds) *From Sensibility to Romanticism: Essays Presented to F.A. Pottle*. New York: Oxford University Press, pp. 191–218.

Moreri, L. (1694) *The Great Historical, Geographical and Poetical Dictionary; being A Curious Miscellany of Sacred and Profane History. Containing, in short, the Lives and most*

Remarkable Actions . . . of ancient and modern authors; of Philosophers, Inventors of Arts, and all those who have recommended themselves to the world, by their Valour, Virtue, Learning, or some Notable Circumstances of their Lives, 2 vols. London: printed for Henry Rhodes.

Napier, M. (ed.) (1815–24) *Supplement to the Fourth, Fifth, and Sixth Editions of the Encyclopaedia Britannica*, 6 vols. Edinburgh: A. Constable.

Oldys, W. (1747–66) *Biographia Britannica; or, the Lives of the Most Eminent Persons who have flourished in Great Britain and Ireland*, 6 vols. London: W. Innys.

Osborn, J.M. (1938) Thomas Birch and the *General Dictionary* (1734–41). Modern Philology, 36, 25–46.

Outram, D. (1978) The language of natural power; the *éloges* of Georges Cuvier and the public language of nineteenth-century science. *History of Science*, 16, 153–78.

Paris, J.A. (1831) *The Life of Sir Humphry Davy*, 2 vols. London: Colburn and Bentley.

Peacock, G. (ed.) (1855) *Miscellaneous Works of Thomas Young, D.D., F.R.S.*, 2 vols. London: John Murray.

Peacock, G. (1845–6) Arago and Brougham on Black, Cavendish, Priestley and Watt. *Quarterly Review*, 77, 105–39.

Pemberton, H. (1728) *View of Sir Isaac Newton's Philosophy*. London: S. Palmer.

Plant, M. (1965) *The English Book Trade: an Economic History of the Making and Sale of Books*, 2nd edition. London: George Allen and Unwin.

Priestley, J. (1765) *A Description of a Chart of Biography; with a catalogue of names inserted in it, and the dates annexed to them*. Warrington.

Proctor, P. and Castieau, W. (1774) *The Modern Dictionary of Arts and Sciences; or, Complete System of Literature*, London.

Pulteney, R. (1790) *Historical and Biographical Sketches of the Progress of Botany in England from its Origin to the Introduction of the Linnaen System*, 2 vols. London.

Reed, J.W. (1966) *English Biography in the Early Nineteenth Century 1801–1838*. New Haven: Yale University Press.

Rees, A. (1802–20) *The New Cyclopaedia; or, Universal Dictionary of Arts and Sciences*, 44 vols. London: Longman.

Reese, M.M. (1970) *Gibbon's Autobiography*. London: Routledge.

Retat, P. (1991) Encyclopédies et dictionnaires historiques au XVIII siècle. In A. Becq (ed.) *L'Encyclopédisme: Actes du Colloque de Caen, 12–16 Janvier 1987*. Paris: Editions Aux Amateurs de Livres, pp. 505–11.

Rose, H.J. (ed.) (1840–8) *A New General Biographical Dictionary*. London.

Schaffer, S. (1986) Scientific discoveries and the end of natural philosophy. *Social Studies of Science* 16, 387–420.

Scott, J., Green, C. and Meader, J. (1765) *A General Dictionary of Arts and Sciences; or a Complete System of Literature*. London: S. Crowder.

Shapin, S. (1991) 'A Scholar and a Gentleman': the problematic identity of the scientific practitioner in early modern England. *History of Science*, 29, 279–327.

Shapin, S. (1993) Personal development and intellectual biography: the case of Robert Boyle. *British Journal for the History of Science*, 26, 335–45.

Shapin, S. (1994) *A Social History of Truth: Civility and Science in Seventeenth-Century England*. Chicago: University of Chicago Press.

Sheets-Pyenson, S. (1990) New directions for scientific biography: the case of Sir Willian Dawson. *History of Science*, 27, 399–410.

Smedley, E., Rose, Hugh. J. and Rose, Henry J. (eds) (1829–45) *Encyclopaedia Metropolitana; or, Universal Dictionary of Knowledge, on an Original Plan*, 29 vols. London: Fellows and Rivington.

[Smellie, W.] (1768–71) *Encyclopaedia Britannica; or, A Dictionary of the Arts and Sciences, compiled upon a New Plan*, 3 vols. Edinburgh: Bell and Macfarquhar.

Stanley, Thomas (1701) *The History of Philosophy. Containing the Lives, Opinions and Actions, and Discourses of the Philosophers of Every Sect*, 3rd edition, London: W. Battersby.

Stauffer, D.A. (1930) *English Biography before 1700*. Cambridge: Cambridge University Press.

Stauffer, D.A. (1941) *The Art of Biography in Eighteenth-Century England*. New Jersey: Princeton University Press.

Stewart, D. (1980) [1796] *Account of the Life and Writings of Adam Smith* (ed. by I.S. Ross). In W.P.D. Wightman and J.C. Bryce (eds), *Adam Smith: Essays on Philosophical Subjects*. Oxford: Clarendon Press, pp. 265–352.

Stewart. L (1992) *The Rise of Public Science: Rhetoric, Technology and Natural Philosophy in Newtonian Britain, 1660–1750*, Cambridge: Cambridge University Press.

Strauss, G. (1966) A sixteenth-century encyclopaedia: Sebastian Munster's *Cosmography* and its editions. In C. Carter (ed.), *From the Renaissance to the Counter-Reformation*. London: J. Cape, pp. 145–63.

Tytler, J. (1778–84) *Encyclopaedia Britannica; or, A Dictionary of Arts, Sciences, etc. On a Plan Entirely New* 2nd edition, 10 vols. Edinburgh: Bell and Macfarquhar.

Wertz, S.K. (1993) Hume and the historiography of science. *Journal of the History of Ideas*, 54, 411–36.

Whiteley, S. (1992) The circle of learning: *Encyclopaedia Britannica*. In J. Rettic (ed.), *Distinguished Classics of Reference Publishing*. Phoenix, Arizona: Onyx Press, pp. 71–88.

Yeo, R. (1988) Genius, method and morality: images of Newton in Britain, 1760–1860. *Science in Context*, 2, 257–84.

Yeo, R. (1991) Reading encyclopaedias: science and the organisation of knowledge in British dictionaries of arts and sciences, 1730–1850. *Isis*, 82, 24–49.

Yeo, R. (1993) *Defining Science: William Whewell, Natural Knowledge and Public Debate in Early Victorian Britain*. Cambridge: Cambridge University Press.

Zedler, J.H. (ed.) (1732–50) *Grosses Vollstandiges Universal-Lexikon*, 64 vols. Leipzig and Halle.

XI

SCIENCE AND INTELLECTUAL AUTHORITY IN MID-NINETEENTH-CENTURY BRITAIN: ROBERT CHAMBERS AND *VESTIGES OF THE NATURAL HISTORY OF CREATION**

ROBERT CHAMBERS' ANONYMOUS WORK, *Vestiges of the Natural History of Creation*, is well known as a best selling statement of evolutionary speculation before Charles Darwin. Although mainly known as a publisher, Chambers was an amateur geologist and a voracious reader of scientific and philosophical literature. His book, which first appeared in 1844, offered a general theory about the origin and development of the natural world, and was presented as "the first attempt to connect the natural sciences into a history of creation."[1] The nature of Chambers' theory, its relation to Darwin's, and its social and intellectual impact on Victorian society have been studied by social historians and historians of science.[2] While drawing upon this scholarship, the present article attempts to consider the controversy provoked by Chambers' work in terms of the questions it raised about the image of science and the authority of the scientific community.

The *Vestiges*, as it came to be called, met with a deluge of criticism. The charges brought against the author were serious ones — lack of practical research, second-hand knowledge, and disregard of proper scientific methods. The extent of this attack led Chambers to

* I would like to thank the following people for their useful comments on drafts of this article: Charles S. Blinderman, Ludmilla Jordanova, David Oldroyd, and Barry Smith. A version of the article was read at the Wellcome Institute for the History of Medicine at Oxford, and I thank the participants for their discussion. Research was financially supported by the School of Humanities, Griffith University.

[1] [Robert Chambers], *Vestiges of the Natural History of Creation* (London: J. Churchill, 1844), p. 338. The authorship was revealed in the 12th edition of 1884, published after Chambers' death.

[2] The standard work is Milton Millhauser, *Just before Darwin: Robert Chambers and Vestiges* (Middletown, Connecticut: Wesleyan, 1959). For a more recent work which analyzes the differences between the projects of Chambers and Darwin, see M. J. S. Hodge, "The Universal Gestation of Nature: Chambers' *Vestiges* and *Explanations*," *Journal of the History of Biology*, 5 (1972), 127-151.

publish a sequel in which he claimed that his argument had been mis-understood,[3] and successive editions of the original work were also re-vised to take account of objections and to incorporate recent scientific material. This running battle with reviewers culminated with the tenth edition of 1853, which included an extensive appendix in which Chambers replied to criticism by showing that his authorities were also the authorities of contemporary science. He was able to quote experts in support of different parts of his book and cited passages from the works of Adam Sedgwick which suggested that the famous geologist once countenanced the general position he now condemned.[4] Chambers was also able to exploit tensions connected with the growing specialist divisions within science by indicating, for example, that Sedgwick was "neither an anatomist or naturalist" and could not be accepted as an authority in these areas (Chambers, *Vestiges*, 10th ed., p. xlix). He objected to the fact that the work of another critic, Hugh Miller — the Scottish writer and amateur geologist — had been "officially patronized and applauded" while his own had been castigated as "anti-scientific." Marshalling the kind of rhetoric which had been used against him, and again referring to the views of a well-known man of science, Chambers remarked that "twenty chapters from a mere working geologist and *litterateur* like Mr. Miller could not stand against" the authority of Professor Louis Agassiz (Chambers, *Vestiges*, 10th ed., p. xxxix). He suggested that the public should be "on their guard" against the duplicity of men of science and should decide which authorities to accept.

The way in which Chambers was able both to challenge and invoke authorities from the scientific community, to exploit tensions between them, and to appeal directly to public judgement, highlights the issues at stake in the controversy surrounding *Vestiges*. But before investigating this question of authority in science, it is important to place it in the context of a wider contemporary concern about the exercise of intellectual authority in a period of expanding knowledge and democratic reform.

I

In his *Democracy in America*, Alexis de Tocqueville suggested that the egalitarian ethos led to a rejection of the authority of learned

[3] [Robert Chambers], *Explanations: a Sequel to Vestiges of the Natural History of Creation by the author of that Work*, 2d ed. (London: J. Churchill, 1846).

[4] [Robert Chambers], *Vestiges of the Natural History of Creation, with extensive additions and emendations* (London: J. Churchill, 1853), pp. l-li, lviii.

elites.[5] This analysis found a receptive audience in Britain and was in fact anticipated by J. S. Mill who regarded the early decades of the century as an interregnum: traditional doctrinal authorities were in decline but no new intellectual authorities had appeared.[6] Instead, with the rise of the industrial middle classes and the emergence of public opinion as a political force, there was a danger that the criterion of popularity would come to decide the acceptability, and even the truth, of ideas.[7] He noticed a flowering of quackery and ephemeral literature all manipulated by the new "arts for attracting public attention."[8] With the growth of a mass market for information there was a multiplication of popular and elementary texts conveying knowledge in simple, accessible forms. Consequently, "the grand achievement of the present age" was the "diffusion of superficial knowledge" (Mill, "Spirit," p. 30). In itself, Mill did not view this as regrettable, but in a period characterized by the dogmatism of common sense, it created an indifference to theory and expertise. Against this attitude he urged the need for a clear distinction between common sense and informed judgement, between superficial and profound knowledge. He argued that the public would have to be convinced that the scope of individual judgement was limited, and that on some subjects it was necessary to "fall back on the authority of still more cultivated minds" (Mill, "Spirit," p. 44). This issue assumed a special urgency from the time of the First Reform Bill, when questions of intellectual and political authority were explicitly connected.

During the first half of the nineteenth century, the increasing specialization of knowledge made the locus of intellectual authority more difficult to define. Writing in 1849, in his *Essay on the Influence of Authority in Matters of Opinion*, George Cornewall Lewis declared that "there is no one body of persons who are competent to judge on all subjects, and who are qualified to guide all sorts of opinions; that there is no one intellectual aristocracy, separated from the rest of the community, and predominating over them indiscriminately. Every subject, in turn, has its own peculiar set of judges."[9] With wider dis-

[5] Alexis de Tocqueville, *Democracy in America*, 2 vols. (New York: A. Knopf, 1953), II, 3-12.

[6] John Stuart Mill, "The Spirit of the Age" in J. B. Schneewind, ed., *Mill's Essays on Literature and Society* (New York: Macmillan, 1965), pp. 33-38. See his favourable review of de Tocqueville, "Democracy in America," *Edinburgh Review* 72 (1840), 1-47.

[7] See William Mackinnon, *On the Rise, Progress and Present State of Public Opinion* (London: Saunders and Otley, 1828) for an early analysis.

[8] John Stuart Mill, "Civilization: Signs of the Times," *Westminster Review*, 27 (1836), 15.

[9] George Cornewall Lewis, *An Essay on the Influence of Authority in Matters of Opinion* (London: J. Parker, 1849), p. 167. This is a systematic treatment of the problem posed for individual judgement by the specialization of knowledge. On a related theme, see Ben Knights, *The Idea of the Clerisy in the Nineteenth Century* (Cambridge: Cambridge University Press, 1978).

semination of information through popular texts, the problem facing intellectual elites was one of ensuring that distinctions between expertise and common sense were maintained. As early as 1833, one writer stressed the need for what he called the "government of knowledge."[10]

A large part of this contemporary discussion dealt with social and political subjects in which the disposition to trust common-sense opinions was strongest. In the 1830s, for example, exponents of political economy such as Nassau Senior and Richard Whately contended that questions raised by this discipline required the unlearning of common-sense attitudes and the adoption of systematic scientific methods.[11] References were often made to the lack of certainty in the moral and social sciences in comparison with the strong consensus over fundamental principles in the natural sciences. But the question of authority was also relevant to natural science and it concerned the leaders of the scientific community. Lewis suggested one reason for this when he noted that the increasing popularity of science had encouraged the emergence of "mock sciences" such as mesmerism, homeopathy, and phrenology (Lewis, pp. 51-52): "No species of imposture is so captivating, so well-suited to the present time, and consequently so likely to meet with temporary success, as that which assumes the garb, and mimics the phraseology, of science" (Lewis, p. 55). Given this situation it was imperative that the scientific community assert its authority in public in order to distinguish between genuine and pseudo-science.[12] In order to appreciate the context of this problem, it is necessary to consider the position of science in early Victorian society.

Several recent authors have referred to the close relationship between "natural knowledge and the general culture" of the early Victorian period.[13] This situation is contrasted with the divorce between science and other forms of knowledge which occurred in the last half of

---

[10] Edward Bulwer Lytton, *England and the English*, 2 vols. (London: R. Bentley, 1833), II, 122.

[11] Richard Whately, *Introductory Lecture on Political Economy*, 3d ed. (London: B. Fellowes, 1847), pp. 55-64, 208, 215-17; Nassau Senior, *Introductory Lecture on Political Economy* (London: J. Mawman, 1827), pp. 1-2, 24-28. On the emergence of political economy as a discipline, see Maxine Berg, *The Machinery Question and the Making of Political Economy 1815-1848* (Cambridge: Cambridge University Press, 1980).

[12] On this problem of demarcation, see Roy Wallis, ed., *On the Margins of Science: The Social Construction of Rejected Knowledge* (Sociological Review Monograph, no. 27, University of Keele, 1979). On phrenology, see Roger Cooter, "Deploying 'Pseudoscience': Then and Now," in M. P. Hanen, M. J. Osler and R. G. Weyant, eds., *Science, Pseudo-Science and Society* (West Waterloo, Ontario: Wilfrid Laurier University Press, 1980), pp. 237-272.

[13] Arnold Thackray and Steven Shapin, "Prosopography as a Research Tool in the History of Science," *History of Science*, 12 (1974), 11. See also, Robert Young, "Natural Theology, Victorian Periodicals and the Fragmentation of a Common Context" in C. Chant and J. Fauvel, eds., *Darwin to*

the century, especially after 1870 when both scientific and non-scientific disciplines became more specialized and separated, by professional accreditation, from the generally educated public. Evidence of the integration of science with culture before this fragmentation can be found in the Victorian periodicals.[14] The major quarterlies and magazines treated science as part of a broad intellectual framework which included theology, literature, philosophy, and political economy. Reviewers often moved freely across these disciplines, and science was presented in a manner which assumed an informed readership. The great debates on geology, evolution, and their religious implications were conducted in these periodicals, and this meant that the meaning and status of science were matters of public discussion.

But while these debates reflected the interest in science amongst general readers, they also indicated the uncertain social position of science in early nineteenth-century Britain. Men of science were not differentiated from other educated groups by formal training, and the status of science was not secured by an institutionalized career structure such as that which characterized the legal, medical, and clerical professions.[15] This situation supported a general discussion of science but it also meant that men of science were compelled to defend the claims of science in a public forum against powerful opponents such as the clergy.

This close relationship between science and general cultural debate, together with the insecure status of the scientific community,[16] made the authority of science a significant issue. Scientists had to establish the domain of natural knowledge as their own, and monitor the boundaries between science and religion. But they also had to make distinctions between different kinds of scientific knowledge — general and specialist, empirical and theoretical, superficial and profound —

Einstein: Historical Studies on Science and Belief (New York: Longmans, 1980), pp. 69-107; Susan F. Cannon, *Science in Culture: The Early Victorian Period* (New York: Science History Publications, 1978), chaps. 1 and 9. Young refers to the place of science in a common intellectual context, Cannon to its position within a Victorian "Truth-Complex."

[14] For a useful general essay, see Walter E. Houghton, "Periodical Literature and the Articulate Classes" in J. Shattock and M. Wolff, eds., *The Victorian Periodical Press: Samplings and Soundings* (Leicester: Leicester University Press, 1982), pp. 3-82. Houghton's *Wellesley Index to Victorian Periodicals*, 3 vols (Toronto: University of Toronto Press, 1966-1976) has been used to identify authorship of the reviews referred to below.

[15] The contrast with the position of science in France is instructive. From the time of Napoleon, scientific institutions were financed by the state, and scientists were "professional" well before their counterparts in Britain. For an early notice of this, see Charles Babbage, *Reflections on the Decline of Science in England* (London: Fellowes, 1830), pp. 3-30; also Maurice Crosland, "The Development of a Professional Career in Science in France," in his edition of *The Emergence of Science in Europe* (London: Macmillan, 1975), pp. 140-141, 154-155.

[16] This is a problematic term in the context of early nineteenth-century Britain. See Susan Cannon, chaps. 5, 6. The debate over *Vestiges* reveals some of the difficulties surrounding terms such as "professional," "amateur," "expert," and "popular science."

10

which were produced by popularization and specialization. The following sections of this article investigate a prominent example of a public debate which involved these questions of demarcation.

II

One contemporary reviewer of *Vestiges* began by observing that "it has been warmly received by the public, and fiercely attacked by the physical philosophers."[17] The explicit target of this hostile reaction was Chambers' theory of Progressive Development which involved the concept of the transmutation of species and its application to man. In spite of his criticism of Jean Baptiste Lamarck, Chambers' work revived the prospect of a connection between science and materialism, an association which had political implications in the aftermath of the French Revolution. In Britain during this period, the spectre of French materialism and its radical connotations often shaped responses to scientific and philosophical thought. In the year of Peterloo, for example, the lectures of William Lawrence, which advocated a naturalistic approach to the study of the mind, were widely attacked not only for their theological consequences but because of their alleged affinity with French physiological theories, such as those of Xavier Bichat.[18] Lawrence was accused of both blasphemy and sedition, and while he proclaimed the internationalism of science, two major periodicals reminded him of the dangers of importing French ideas.[19]

Since 1819, although there was a continuing debate about the relationships between science and religion, the question of the materialist tendencies of science was not the central issue. Rather, most public discussion focussed on the new discipline of geology and its implications for Biblical literalism; and by the 1830s, various degrees of compromise had been reached.[20] The public relations of the British Association were designed to link science with a bland natural theology which stressed the moral and social benefits of natural knowledge.[21] In

---

[17] Francis Newman, "Explanations. A sequel to the Vestiges of the Natural History of Creation," *Prospective Review*, 2 (1846), 33.

[18] See Peter G. Mudford, "Lawrence's Natural History of Man (1819)," *Journal of the History of Ideas*, 29 (1968), 430-436; June Goodfield-Toulmin, "Some Aspects of English Physiology: 1780-1840," *Journal of the History of Biology*, 2 (1969), 307-320.

[19] William Lawrence, *Lectures on Physiology, Zoology, and the Natural History of Man* (London: J. Callow, 1819), p. 15; *Quarterly Review*, 22 (1819), 33; *British Critic*, 12 (1819), 95.

[20] R. M. Young, "The Impact of Darwin on Conventional Thought," in Anthony Symondson, ed., *The Victorian Crisis of Faith* (London: Society for Promoting Christian Knowledge, 1970), pp. 13-35.

[21] Jack Morrell and Arnold Thackray, *Gentlemen of Science: Early Years of the British Association for the Advancement of Science* (Oxford: Clarendon Press, 1981), pp. 224-229.

opening his review of *Vestiges*, David Brewster remarked that "we did not expect that this holy alliance would be disturbed either by the philosopher or the divine."[22] But by associating science with controversial materialist ideas in a work of popular circulation, Chambers threatened the rationale of this strategy and the harmony of science and religion it represented. In order to restore this alliance, and to preserve the religious and social respectability of science, it was crucial to show not only that the theories in the book were false and dangerous, but that the work itself could not be classified as "scientific."

It is possible to interpret the responses of leading scientists to Chambers' book as indicating a concern about the maintenance of this boundary between true and false science. Furthermore, it could be suggested that *Vestiges* appeared threatening because, in some respects, it was a difficult borderline case. For example, when the religious implications of the work were discussed, it was often noted that its natural theology, drawing on the evidence of general laws, was very close to that of leading men of science such as William Whewell, Adam Sedgwick, Baden Powell, John Herschel, and Charles Babbage.[23] Secondly, on the substantive question of its scientific content, even some of the critical reviewers remarked on the similarity between its doctrines and those of current European (if not British) science. One writer compared it with Alexander von Humboldt's *Kosmos*, thereby awarding it some respectability by association.[24]

In spite of its obvious flaws, *Vestiges* could not be easily dismissed as "pseudo-scientific" because it displayed too many positive links with mainstream science. The *Athenaeum* attempted to associate it with alchemy, astrology, mesmerism, phrenology, and "other kindred humbugs," alleging that like them, it masqueraded under "the form of true sciences."[25] But this equation was not effective because although mesmerism and phrenology were popular, philosophically controversial doctrines which claimed scientific status, they did not evince the dependence upon orthodox science which characterized

[22] [David Brewster], "Vestiges of the Natural History of Creation," *North British Review*, 3 (1845), 471.

[23] [W. H. Smith], "Vestiges of the Natural History of Creation," *Blackwood's Magazine*, 57 (1845), 449-451; "Natural History of Creation," *Westminster Review*, 48 (1847-48), 132-134. The additional problem here was that natural theology was a divided subject and, by linking it with evolutionary doctrine, Chambers exposed its weaknesses. On varieties of natural theology, see John H. Brooke, "Natural theology and the Plurality of Worlds: Observations on the Brewster-Whewell debate," *Annals of Science*, 34 (1977), 221-286 and Richard Yeo, "William Whewell, Natural Theology and the Philosophy of Science in Mid-Nineteenth-Century Britain," *Annals of Science*, 36 (1979), 495, 506-509.

[24] [J. Crosse], "The Vestiges," *Westminster Review*, 44 (1845), 195, 198; Newman, pp. 34-36.

[25] "Vestiges," *Athenaeum*, no. 897 (4 January 1845), 11.

Chambers' work; on the contrary, they often adopted a negative attitude towards scientific authorities. The author of *Vestiges*, however, employed evidence from recognized sciences, together with material from more dubious sources, in the service of a speculative generalization. Indeed, it was this synthetic feature of the book which made it difficult to dismiss *in toto*. David Brewster, for example, observed that it covered so many areas that individuals would always find something in it with which they could agree (Brewster, pp. 484-485). Writing to the anonymous author, Richard Owen, the leading comparative anatomist, said that "there are a few mistakes where you treat of my own department of science, easily rectified in your second edition."[26] Owen also remarked upon the "summary of the evidences from all the Natural Sciences bearing upon the origin of all Nature, by one who is evidently familiar with the principles of so extensive a range of human knowledge" (Owen, I, 249). Similarly, Darwin told Joseph Hooker that he had been "delighted with *Vestiges* for the multiplicity of parts he brings together though I do [not] agree with his conclusions at all."[27] Even a critical reviewer conceded that the author had "produced a book which may in some manner serve as an outline to the vast range of the natural sciences."[28]

Thus, although particular sections of the book could be shown to be inaccurate or misguided, it was difficult to represent the entire work as unscientific, because it deployed a large amount of recognized scientific thinking and data. The dilemma was exacerbated by Chambers' ability to benefit from criticism: in later editions he altered those sections which gave most offence without retreating from the central thesis. Having seen the fourth edition of 1845, James Forbes told Whewell that the author "has shown himself a very apt scholar, and has improved his knowledge and his argument so much since his first edition that his deformities no longer appear so disgusting."[29] In this way, Chambers continued, with some success, to implicate science, or sections of science, in his theory. This process was assisted by the fact that the work dealt with complex subjects in the organic

[26] Quoted in Richard Owen, *The Life of Richard Owen by his grandson*, 2 vols. (London: Murray, 1894), I, 251. For the question of Owen's attitude to Chambers' theory, see J. H. Brooke, "Richard Owen, William Whewell, and the Vestiges," *British Journal for the History of Science*, 10 (1977), 132-145.

[27] Quoted in Frank Egerton, "Refutation and Conjecture: Darwin's response to Sedgwick's Attack on Chambers," *Studies in History and Philosophy of Science*, 1 (1970), 178. See also p. 182 for Alfred Wallace's favourable reaction.

[28] "Explanations," *Athenaeum*, no. 946 (18 December 1845), 1191.

[29] James D. Forbes to Whewell, 8 January 1846 in J. C. Shairp, *Life and Letters of James David Forbes* (London: Macmillan, 1873), p. 178.

sciences about which there was no firm consensus. As one reviewer observed, "a false theory in natural philosophy is so much more easily corrected than one in physiology."[30] After the tenth edition of *Vestiges* had appeared, the young T. H. Huxley was sure that "in the popular mind the foolish fantasies of the 'Vestiges' are confounded with science, to the incalculable diminution of that reverence in which true philosophy should be held."[31] From this perspective it is possible to suggest that the massive critical effort expended on Chambers' book reflects not only the need to debunk its evolutionary theory and its materialism, but the need to clarify the meaning of science and scientific practice.

III

The popularity of the book — it reached a sixth edition by 1847 and sold 20,000 copies by 1860 — made this problem of demarcation at once more urgent and more difficult. Indeed, there is evidence that leading men of science regarded it as a case calling for delicate public relations. The correspondence of Whewell, Owen, and Forbes reveals an explicit discussion of strategies for dealing with *Vestiges*. As J. H. Brooke has shown, Whewell and Owen believed that it would be best not to reply directly since this would only give the book "an importance calculated to add greatly to its mischief."[32] Owen thought that it would be difficult to refute *Vestiges*, for the sake of those who could not already see its errors, without writing an essay which would inflate its importance. Furthermore, it is clear that these correspondents agreed that Sedgwick's eighty-five-page onslaught in the *Edinburgh Review* was ill-advised.[33] Whewell believed that *Vestiges* posed serious tactical problems because it offered bold answers to questions on which "men of real science" showed necessary restraint.[34] He later explained to Sedgwick that "the great difficulty in satisfying the mind of general

---

[30] "Vestiges of the Natural History of Creation," *British Quarterly Review*, 1 (1845), 490.

[31] Thomas Henry Huxley, "Vestiges of the Natural History of Creation," *British and Foreign Medico-chirurgical Review*, 13 (1854), 439.

[32] Owen to Whewell, 14 February 1844, quoted in Brooke, p. 138.

[33] Forbes to Whewell, 8 January 1846, Trinity College, Cambridge, Whewell Papers, Add. Mss.a. 204[70].

[34] William Whewell, *Indications of the Creator*, 2d ed. (London: J. Parker, 1846), p. 21. Whewell told his friend Richard Jones, the political economist, that no "really philosophical" work could have had the success of *Vestiges*. Whewell to Jones, 18 July 1845, in Isaac Todhunter, *William Whewell D.D., Master of Trinity College, Cambridge: An Account of His Writings, with Selections from his Literary and Scientific Correspondence*, 2 vols. (London: Macmillan, 1876), II, 326-327.

14

readers on such subjects is that you have to oppose to attractive positive generalizations nothing but negations and doubts."[35] Sedgwick, however, did not take this hint, and the fifth edition of his *Discourse on the Studies of the University of Cambridge* (1850) carried a monstrous preface which again did battle with the celebrated *Vestiges*.

These strategic considerations were posited upon inferences about the level of scientific understanding among the reading public. The danger of Chambers' book was said to lie in its appeal to those who were ill-informed on matters of science and logic; and by way of self-confirmation it was argued that the great popularity of the work proved the unscientific character of the public.[36] Reviewing the tenth edition of *Vestiges* in 1854, Huxley was depressed by the fact that a book of such transparent nonsense had not "sunk into its proper limbo." This testified, in his view, to the increasing "ignorance of the public mind as to the methods of science and the criterion of truth" (Huxley, p. 425). Addressing the readers of a professional medical journal, Huxley revealed a deep pessimism about public conceptions of science. He regarded the level of scientific understanding in *Vestiges* as that derived from the superficial information of "Chambers's [sic] Journal" or the "Penny Magazine" and compared its popularity with that of table-turning. This equation of spiritualism and "the science of a Mechanics' Institute" (Huxley, pp. 438-439) indicates a certain desperation about the problem of popularization, a matter which the phenomenon of *Vestiges* had highlighted.

In his book, *Keywords*, Raymond Williams has remarked that the notion of "popularization" still had favourable connotations during the nineteenth century.[37] This is largely true in the case of scientific knowledge, in that leaders of the scientific community believed that their interests would be served by the dissemination of scientific ideas and attitudes throughout society.[38] But some qualification is necessary, because not all scientists were sanguine about the benefits of popularization. Some of the major figures in the British Association for the Advancement of Science, for example, did not share the enthusiasm of utilitarian reformers such as Henry Brougham. At the fourth

---

[35] Whewell to Sedgwick, [ ] September 1849, Trinity College, Cambridge, 0.15.48[69].

[36] Whewell said that the book "has its hold on its readers . . . by *their* want of the apprehension of the difference of the nature of truth and falsehood in science and philosophy" (Whewell to F. Myers, 16 March 1845, in Janet M. Douglas, *The Life and Selections from the Correspondence of William Whewell D.D.* [London: Kegan Paul, 1882], pp. 317-318).

[37] Raymond Williams, *Keywords: A Vocabulary of Culture and Society* (London: Fontana, 1976), p. 199.

[38] For a section of those views, see Charles Gillispie, *Genesis and Geology* (New York: Harper and Row, 1959), chap. 7.

meeting of the Association in 1834, James Forbes warned that the diffusion of science might not be compatible with its advancement; and this theme was echoed by other important scientific writers such as William Swainson and Augustus De Morgan.[39] These anxieties were aggravated by the fact that much popular science appeared to convey an empirical and utilitarian image of science at the expense of its theoretical dimension.[40] There was a recognition that good popular works of science were necessary, but most leading scientists did not write at the level demanded for the widest audience. For reviewers, this was the lesson of *Vestiges:* "[I]f men competent to the task disdain to popularize science, the task will be attempted by men who are incompetent; popularized it will be" (Crosse, p. 153).

The task of enlisting leading scientists to write popular books was not new. In 1828, for example, Dionysious Lardner, the editor of the *Cabinet Cyclopaedia*, confided to Herschel that "my chief difficulty in Science is to find profound men who like yourself are able and willing to write a popular work."[41] Herschel wrote his famous *Preliminary Discourse on the Study of Natural Philosophy*, as well as a volume on astronomy, for Lardner's series. But although his efforts were loudly praised, there was some disquiet about the tension between the popularization of science and its advancement. One commentator doubted the value of employing the best scientists in the composition of popular texts, arguing that the progress of science would be better served if men such as Herschel concentrated on abstract research.[42]

By the 1840s, however, the problem of popularization was more complex than a question of authorship. The larger readership and the diversity of publications produced a more heterogenous audience for science, and the expansion of specialized knowledge exacerbated the difficulties of writing popular books. It was suggested that previous examples, such as the works of Mary Somerville which reached educated middle-class readers, were not appropriate: those might instruct serious students but they were useless to the general public. The *Westminster Review* diagnosed a need for broad surveys of science

[39] James D. Forbes, *British Association Reports* (London: Murray, 1835), pp. xii, xv; William Swainson, *Preliminary Discourse on the Study of Natural Philosophy* (London: Longman, 1834), pp. 426-427; Augustus De Morgan, "English Science," *British and Foreign Review*, 1 (1835), 157.

[40] [John Herschel], "Whewell on the Inductive Sciences," *Quarterly Review*, 68 (1841), 177-178, 185; [F. Whitwell], "Popular Science," *Quarterly Review*, 84 (1848-49), 322, 340-341; [A. J. Joyce], "The Progress of Mechanical Invention," *Edinburgh Review*, 39 (1849), 54.

[41] D. Lardner to Herschel, 28 July 1828, Library of the Royal Society, London, Herschel Papers, vol. 11, no. 108.

[42] [T. Galloway], "Sir John Herschel's Astronomy," *Edinburgh Review*, 58 (1833-34), 165.

XI

rather than elementary treatises on specific disciplines. What was needed was "something like a reasonable interpretation of the present state of science — something intelligible to at least the mass of thinking unscientific men" (Crosse, p. 153).[43]

To the great annoyance of leading scientists, Chambers' *Vestiges* apparently fitted this description. The fact that it met with popular success reinforced anxieties about the effects of popularization which had begun to surface in the early years of the British Association. During the 1840s, this disenchantment widened as various commentators drew attention to the public misconceptions about science, its theories, methods and functions.[44] These observations were made in contexts ranging from discussions of the role of scientific theory in technology to condemnations of mesmerism and spiritualism. Augustus De Morgan, for example, felt it necessary to examine several anti-Newtonian and astrological works in the *Dublin Review* of 1845, explaining that "it is desirable that, from time to time, the public should be guarded against the real evil consequences of ignorant speculation."[45] The manner in which Chambers gave roughly equal status to geological and phrenological theory in his first edition could only have confirmed these fears about the promiscuous nature of popular notions of science. Against this background, the success of *Vestiges* demanded that the scientific community assert its authority in order to clarify what was to count as "scientific."

IV

As the copies of *Vestiges* continued to flow from the press, the need for a strong scientific reply became evident. Whewell's *Indications of the Creator* was published early in 1845 but this was not seen as a scientific answer — Whewell himself regarded it as a collection of theological extracts from his earlier works. Thus, before Sedgwick's review appeared in July, Roderick Murchison wrote to Owen saying that "a real *man of armour* is required, and if you would undertake the concern you would do infinite service to *true* science and sincerely oblige your friends."[46] Murchison anticipated the difficulties con-

---

[43] There was some comment on the lack of suitable general surveys of science; Whewell's works were too sophisticated and Somerville's dealt only with the physical sciences; Herschel's *Discourse* was largely about method. In fact, most popular treatments of science dealt with particular disciplines.

[44] See for example, "Science in 1847," *Athenaeum* (12 January 1848), 60.

[45] Augustus De Morgan, "Speculators and Speculations," *Dublin Review*, 19 (1845), 100.

[46] Roderick Murchison to Owen, 2 April 1845 in Owen, *Life*, I, 254.

fronting a successful scientific rebuttal when he added that, "to be done at all it must be done by a master hand," (Murchison, p. 254). In order to appreciate those difficulties, it is necessary to look at the structure of *Vestiges*.

Chambers attempted to explain the origin and development of the natural world by reference to general laws of nature. For the purposes of this argument, he drew upon several areas of scientific evidence and theory. The book began with the notion of cosmical evolution by natural law, such as that suggested by the nebular hypothesis of contemporary astronomy. The second, and largest section, dealt with the case for a progressive development of plants and animals (including humans) from simple to complex forms by means of transmutation. Here Chambers marshalled evidence from the fossil record, comparative anatomy and embryology, ethnology, and phrenology.[47] Finally, in an attempt to document the possibility of transmutation from inorganic to organic forms, he offered cases of spontaneous generation obtained by experiments in electro-vitalism (Chambers, *Vestiges*, pp. 165-190). This meant — as contemporaries observed — that the book contained a confusing ensemble of material ranging from accepted scientific data and hypothesis to speculative theory and marginal doctrine.

All these facets of the book came under attack, but they raised different problems for the critics. Chambers' deployment of phrenology, his reference to the transmutation of oats into rye, and his acceptance of the experiments of Andrew Crosse and W. H. Weekes on spontaneous generation of simple organisms, met with either ridicule or confident dismissal.[48] However, where he dealt with more orthodox science, the reviewers had to enter into more extended argument.

The response to Chambers' use of the "nebular hypothesis" was out of proportion to its place in the book. Only the first chapter dealt with this, but the reviewers gave it considerable attention.[49] Chambers employed the ideas of William Herschel and Pierre Simon Laplace on

[47] One reviewer suggested that the references to foetal transformations were those which gave the author "his strongest hold on the popular mind, in favour of the development doctrines" ("Geology vs. Development," *Fraser's Magazine*, 42 [1850], 371). For the significance of Chambers' embryological argument, see Hodge, pp. 138, 142-146.

[48] The fact that the *acurus Crosii* was an insect, and not a more simple organism, made the disbelief especially spirited. But Chambers continued to stress the status of the evidence as experimental. See his *Explanations*, pp. 189-198. The spectre of Lamarck's evolutionary theory, which involved the notion of spontaneous generation, lay behind the scientific reaction in Britain. See also John Farley, *The Spontaneous Generation Controversy* (Baltimore: Johns Hopkins University Press, 1977), pp. 69-70, 94-97.

[49] For an account of Chambers' reaction to this criticism in successive editions, see M. B. Ogilvie, "Robert Chambers and the Nebular Hypothesis," *British Journal for the History of Science*, 8 (1975), 214-232.

the possible formation of solar systems from primeval clouds of gas in order to assert that this cosmic process had its analogue in the history of organic life on earth.[50] This speculative link had been made previously by John Nichol, an academic and writer of popular texts on astronomy. In his *Views of the Architecture of the Heavens* (1837), Nichol spoke of nebulae as "fossil relics . . . the germs, the elements of that Life, which in coming ages will bud and blossom"; he also stressed that "all things are in a state of change and PROGRESS."[51] Chambers reinforced these connections, arguing that astronomical theories of cosmic development were the prelude to a Law of Development in the organic creation (Chambers, *Vestiges*, pp. 27, 153-154, 359-360). To his critics, therefore, an attack upon the nebular hypothesis had the appeal of an effective dismissal of the entire work.

The case of the nebular hypothesis, however, illustrates the difficulties confronting a public, scientific refutation of *Vestiges*. As in other sections of the book, Chambers was using recognized scientific literature and was able to quote respectable authorities in physical astronomy who favoured this theory. As one reviewer of *Vestiges* acknowledged, "the nebular hypothesis . . . has assumed a shape and a consistency which forbids an entire rejection of it, which enforces our respect, and which, . . . habituates the imagination to regard our planetary system as having probably been evolved, under the will of Providence, by the long operation of the established laws of matter" (W. H. Smith, p. 448). Indeed, before the appearance of *Vestiges*, some of its most notable critics — Brewster, Whewell, and Herschel — had not been unfavourable towards the "nebular hypothesis." Whewell, who first used the phrase in his *Bridgewater Treatise*, argued that, if properly conceived, it was compatible with natural theology. In 1838 Brewster had given an account of Auguste Comte's version of the Laplacian hypothesis and found no necessary conflict with natural theology or Christian religion.[52]

Not long after the publication of *Vestiges*, William Parsons, Lord Rosse announced that his new telescope had resolved some nebu-

---

[50] On the relationship between cosmic and organic evolution, see Ronald L. Numbers, *Creation by Natural Law: Laplace's Nebular Hypothesis in American Thought* (Seattle: University of Washington Press, 1977); on Herschel, see Simon Schaffer, "Herschel in Bedlam: Natural History and Stellar Astronomy," *British Journal for the History of Science*, 13 (1980), 211-239.

[51] John P. Nichol, *Views of the Architecture of the Heavens* (Edinburgh: W. Tait, 1837), pp. 127, 206.

[52] William Whewell, *Astronomy and General Physics considered with Reference to Natural Theology* (London: W. Pickering, 1834), pp. 181-191; [David Brewster], "M. Comte's Course of Philosophy," *Edinburgh Review*, 67 (1838), 297-301. Herschel's popular *Treatise on Astronomy* (London: Longman, 1833), pp. 401-408, carried a discussion of the nebulae, but not the theory of stellar evolution. But for the view that Herschel presented it as a respectable theory, see Walter F. Cannon, "John Herschel and the Idea of Science," *Journal of the History of Ideas*, 22 (1961), 234.

lae, and this fact was used against Chambers. But no major writers took this evidence as sufficient grounds for the rejection of the nebular hypothesis as an account of how stars and planets were formed. Instead, it led to quite complex discussions which revealed the range of senses in which the hypothesis could be conceived. John Herschel was anxious to distinguish his father's views from those of Laplace, and Brewster tried to separate Comte's elaboration of Laplace from that of Chambers.[53] To an extent, the complexities of these discriminations blunted the attack upon Chambers' deployment of this concept. The different versions of the hypothesis,[54] and the fact that they were regularly conflated, made it hard to demonstrate to the public the "unscientific" character of the idea as it appeared in *Vestiges*. On the other hand, Chambers was able to present his construal of the nebular hypothesis as a legitimate interpretation of a controversial topic (Chambers, *Explanations*, pp. 5-25). Whewell summarized the situation, remarking that "the various discussions which have taken place respecting the Nebular Hypothesis have tended . . . to make it looked upon as far more probable than it really is" (Whewell, *Indications*, p. 27). There was also a lack of uniformity in the response of the scientific community: whereas geologists such as Sedgwick, Aggassiz and Hugh Miller were quick to oppose all forms of evolutionary theory in astronomy, the leading astronomers, such as Herschel and George Airy, were far more restrained.[55]

The most authoritative and detailed sections of Sedgwick's review focussed on the evidence from geology and paleontology which comprised the major part of the book. Chambers derived his Law of Development from the discoveries of these sciences and it was data which could not easily be set aside. In seeking to deprive Chambers of this support, Sedgwick argued that there were gaps in the fossil record and that there was no complete continuity from earlier simple to more recent complex species, such as the transmutation theory required.[56] The specialists, including Darwin, appreciated this reply, but in terms of a public response to *Vestiges*, it had certain drawbacks. For

---

[53] John Herschel, *British Association Reports* (London: Murray, 1846), pp. xxxvi-xxxviii; Brewster, "Vestiges," pp. 476-477.

[54] See Brooke, "Plurality," pp. 268-273.

[55] See Walter F. Cannon, "Herschel," pp. 234-235. Mary Somerville, the author of astronomy texts, was not hostile in her response. Thanking a friend for a copy of *Vestiges*, she wrote: "I think it a powerful production, and was highly pleased with it, but I can easily see that it will offend in some quarters" (Somerville to W. Grieg, 28 May 1845 in Martha Somerville, *Personal Recollections, from Early Life to Old Age, of Mary Somerville* [London: Murray, 1873], p. 278).

[56] [Adam Sedgwick], "Natural History of Creation," *Edinburgh Review*, 82 (1845), 29-64. Charles Lyell had made this point in criticizing Lamarck's theory in his *Principles of Geology*, 3 vols. (London: J. Murray, 1830-1833).

example, in trying to undermine Chambers' thesis of development by natural law, Sedgwick incurred the risk of seeming to reject the progressive character of geological history. Yet this perspective was important to his own natural theology, and he had endorsed it when criticizing Charles Lyell's Uniformitarianism in 1831 and again in his *Discourse* of 1850, where it was associated with a teleological view of creation.[57] Chambers forced Sedgwick to deny the *continuity* of progression and to stress the discontinuous aspects of organic creation because he had linked the idea of progress with that of transmutation.

But the rejection of this hypothesis begged the question of a *scientific* alternative. As Lyell understood, "the popularity of Vestiges arises from any theory being preferred to . . . a series of miracles."[58] Indeed, some reviewers were quick to acknowledge this by indicating that they had no *a priori* objections to explanations of the development of organic creation by natural law (W. H. Smith, pp. 449-451).[59] In his *Explanations*, Chambers was able to exploit this preference for natural explanation by shifting his focus from the substantive natural mechanism of evolution to a metaphysical commitment to the existence of general laws of nature (Chambers, *Explanations*, pp. 3-5, 147-149).[60] These two levels of proposition were not easily disentangled and, in attempting to oppose the doctrine of *Vestiges*, Chambers' scientific opponents ran the risk of seeming to prefer supernatural or miraculous to natural causes.[61]

In general, the range of its scientific content, its mixture of accredited evidence and marginal data, and its conflation of natural, moral, and metaphysical propositions made *Vestiges* a difficult target. Sound technical criticism could be levelled at specific sections, but often with the effect of granting them a degree of scientific status and with the larger result of leaving the general suggestion of the book only slightly damaged. While specialist critics could shatter the analogies which Chambers advanced, the reading public saw them as fascinating

[57] Adam Sedgwick, *Addresses delivered at the Anniversary Meetings of the Geological Society of London* (London: R. Taylor, 1831), 23-25; Sedgwick, *A Discourse on the studies of the University of Cambridge*, 5th ed. (London and Cambridge: Barker and Deighton, 1850), pp. ccxvii-ccxix.

[58] Quoted in Leonard G. Wilson, ed., *Sir Charles Lyell's Scientific Journals on the Species Question* (New Haven: Yale University Press, 1970), p. 84. For the status of the concept of "miracle" in scientific thought, see Walter F. Cannon, "The Problem of Miracles in the 1830s," *Victorian Studies*, 4 (1960), 5-32.

[59] For strong support of Chambers on this point, see Baden Powell, *Essays on the Spirit of the Inductive Philosophy: The Unity of Worlds and Philosophy of Creation* (London: Longman, 1855), p. vii.

[60] See also R. M. Young, "Darwin's Metaphor: Does Nature Select?" *Monist*, 55 (1971), 458-459.

[61] Chambers was also able to point to a contradiction between Herschel's opposition to natural explanations of species change in his Presidential Address of 1845, and his earlier acceptance of the possibility. See Chambers, *Explanations*, pp. 54, 141-142.

generalizations. If experts such as Darwin, Wallace, and Owen could express mild approval for the synthetic scope of the book, it is not surprising that, for non-scientists, including other intellectuals, this appeal survived the fire of the critics. One clear illustration of this is the response of Frederick Myers, the Cambridge academic, who confessed to Whewell his sympathy for the book:

I cannot but recognize a certain superiority in the book which will keep it alive, and am disposed to think that after all the scientific and other errors are deducted there will remain such a residuum of originality and truth as will demand and obtain for it a protracted consideration. The book is so far notable to me that it is the first sustained attempt to combine the highest results of astronomy, geology, physiology, and chemistry, into such a series as may serve to indicate, if not to demonstrate, the existence of a law of creation more general than any yet announced.[62]

Apart from attacking the content of the book, critics also made its methodology a major target. To some extent, this strategy of demarcation avoided the problem of explaining the technical details necessary for the rebuttal of Chambers' substantive scientific arguments. However, it was not without other difficulties, because the case of *Vestiges* illustrated the need to define proper scientific practice in more precise terms than those conveyed by the usual invocations of Baconianism.

Statements about scientific method carried strong normative connotations in the nineteenth century, because of intellectual traditions which associated the proper use of reason with moral virtue, and because spokesmen for the scientific community had presented method as the defining feature of the scientific enterprise.[63] Although there were methodological disputes among men of science, there was a uniformity in the way method was linked with the values of hard work, patience, and humility, when presented to the public.

Several critics had recourse to this rhetoric as a means of challenging the scientific credentials of *Vestiges*. Sedgwick, for example, declared that the author was "not only unacquainted with any of the severe lessons of inductive knowledge, but has a mind apparently incapable of comprehending them" (Sedgwick, *Discourse*, p. xx). The author had reversed the proper order of scientific research by looking for resemblances before fully ascertaining the *differences* among phenomena. This charge of methodological ignorance was made by

[62] Reverend Frederick Myers to Whewell, [ ] 1845, in Douglas, p. 317.
[63] On this theme, see Richard Yeo, "Scientific Method and the Image of Science, 1831-1891," in *The Parliament of Science: The British Association for the Advancement of Science, 1831-1981*, eds. Roy MacLeod and Peter Collins (Norwood, Pennsylvania: Science Reviews, 1981), pp. 65-88; and J. A. Schuster and R. R. Yeo, eds., *The Politics and Rhetoric of Scientific Method: Historical Studies* (Dordrecht: Reidel, forthcoming).

other reviewers: the book's mode of reasoning was almost "childish," its use of analogy deficient, and its reliance upon hypothesis excessive (W. H. Smith, pp. 449, 459). Both Sedgwick and Brewster also inferred a female authorship from the lack of disciplined attention to factual details: the inductive method demanded a masculine quality of mind capable of restraining the impulse to premature speculation (Sedgwick, "Natural History," pp. 3-4; Brewster, "Vestiges," p. 503).[64]

The normative sanctions of this rhetoric depended upon the non-specificity of statements about method beyond general references to induction and Bacon. But Chambers was able to exploit this vagueness by using the Baconian tradition against his critics. Thus, he argued that they had allowed theological dogma to determine their attitude to the question of transmutation so that they refused to recognize observations and experimental evidence. He castigated Whewell's idea of palaetiological sciences — those which studied origins but acknowledged a limit to natural explanations — as nothing but a mystification and a shackle on scientific research (Chambers, *Explanations*, pp. 126-130). Furthermore, Chambers was able to expose inconsistencies in statements by major writers about proper scientific methods, arguing, for example, that their criticism of his work as "hypothetical" contradicted the positive comments about the role of hypotheses made by authorities such as Herschel (Chambers, *Explanations*, pp. 179-180). In addition, the fact that Chambers did employ some respected scientific theories made it necessary for opponents to clarify the grounds of their methodolgical criticism in a way which revealed its connection with the issue of authority in science.

Millhauser has interpreted the methodological objections to *Vestiges* in terms of a distrust of hypothesis and speculation, or the "Baconian . . . particularizing spirit of most early-nineteenth-century science" (Millhauser, p. 144). This attitude was undoubtedly strong but the reaction to *Vestiges* was not simply a profession of empiricism: it was an attempt to restrict the privilege of theoretical speculation to a small circle of recognized researchers. The opposition to Chambers' thesis should not be referred to Baconianism, if this is meant to suggest a rejection of hypothesis in science.[65] The importance of hypothesis was

[64] Sedgwick told Lyell that women "have by nature a distaste for the dull realities of physical truth, and above all for the labour pains by which they are produced" (John W. Clark and Thomas M. Hughes, *Life and Letters of the Reverend Adam Sedgwick*, 2 vols. [Cambridge: Cambridge University Press, 1890], II, 85).

[65] The various meanings of Baconianism in the Victorian context would demand an extensive discussion which cannot be attempted here. What can be said is that there was opposition to the Baconian notion that theoretical activity could be divorced from practical empirical investigation.

not at stake: the issue was rather one of who was to be entrusted with theorizing. On this point, there was full agreement among those who might have differed on other questions of method: the qualifications for theoretical speculation had to be of the strictest kind. This position can be related to the division of scientific labour between amateur observers and the more elite group of theorists, a distinction that had been established in some sections of the British Association (Morrell and Thackray, pp. 268-275). One of the significant themes of the critical reviews was the complaint that *Vestiges* violated this principle: it was an attempt at grand speculation by an unrecognized amateur.

There is a sense in which the reaction to *Vestiges* was a criticism of the theorist as much as the theory. There are two dimensions to this claim. Firstly, in the case of speculations, such as the nebular hypothesis, the strongest attacks were not directed against the hypothesis itself, but against the uses to which it was put. Chambers was reprimanded for taking this hypothesis out of its proper context — physical astronomy — and extrapolating it to the realm of organic nature. As Brewster remarked, the danger of the nebular hypothesis was magnified by the "use which has been made of it as a basis for other errors" (Brewster, "Vestiges," p. 479). Closely connected with this objection was the complaint that Chambers had taken a tentative hypothesis from the safe circle of scientific debate and proclaimed it "*to the world*" as a dogmatic doctrine (Whewell, p. 25).

There was, secondly, a denial of Chambers' right to theorize. Chambers presented himself in the role of a theorist standing above the data of the various sciences, drawing them together into large generalizations. One recent Darwinian scholar has suggested that the British scientific community, especially its geological and natural history sections, did not recognize a role for the "theorist" distinct from other participants of the scientific enterprise, and that this affected the way Darwin presented his theories to scientific audiences.[66] If the posture of the theorist was problematic in the case of a recognized researcher, it was clearly an improper role for Chambers because, as he frequently acknowledged, his qualifications were those of "a general student," not those of an expert (Chambers, *Explanations*, p. 103). For this reason, as Brewster explained, his theory was "not entitled to the privilege which we concede to the original inquirer" (Brewster, "Vestiges," p. 474).

---

[66] Sandra Herbert, "The Place of Man in the Development of Darwin's Theory of Transmutation, Part II," *Journal of the History of Biology*, 10 (1977), 157-171.

Most critics branded the author of *Vestiges* as a novice and dismissed the work as being solely dependent on second-hand knowledge of science. Thus, one reviewer predicted that if the author had "performed one single chemical experiment, and endeavoured to understand its import . . . he would never have presumed to write this book" ("Vestiges," *British Quarterly Review*, 1 [1845], 507). Similarly, Huxley could sympathize with Sedgwick's indignation when confronted with a grand hypothesis from a man "who would have been an astronomer — but for sitting up at night; a geologist — but for dirtying his fingers" (Huxley, p. 438). These criticisms established practical scientific research as one of the preconditions for theoretical speculation. From this perspective, the doctrine of *Vestiges* could be denounced as rash conjecture because its author did not evince the skills, knowledge, and reputation which derived from serious scientific research.

It is therefore possible to see the reaction to Chambers as the affirmation of a philosophy of science which gave authority in matters of theory to those whose work had been sanctioned by scientific institutions. On this view, theoretical speculation was dangerous if it did not proceed from thorough acquaintance with empirical data and if it was not guided by proper methodology. Licence to theorize was denied to those who had not accomplished significant scientific work in a specialized area. Brewster underlined one of the key issues when, in effect, he advised Chambers to specialize: to concentrate on some "little department of science" before seeking to "enjoy the luxury of generalization."[67] But this emphasis on detailed knowledge of a particular field meant that theories embracing material from several disciplines would become increasingly rare as specialization intensified. It was this question of specialization, and its implications for the participation of laymen in science, which became the centre of Chambers' reply to his critics.

V

In his *Explanations*, published in 1845 as a response to the criticism of *Vestiges*, Chambers launched an attack on the authority of "the scientific class." In effect, he refused to accept that the judgement of "scientific men" on particular parts of the book was sufficient to dis-

---

[67] [David Brewster], "Explanations of Vestiges of the Natural History of Creation," *North British Review*, 4 (1846), 504.

credit his main theory. He feared that this piecemeal criticism had distracted public attention from the general scope of the work. But he then argued that it was precisely these general views which the "scientific class" were unable to appreciate because their research had become too specialized:

They are, almost without exception, engaged, each in his own little department of science, and able to give little or no attention to other parts of that vast field. From year to year, and from age to age, we see them at work, adding no doubt much to the known, and advancing many important interests, but, at the same time, doing little for the establishment of comprehensive views of nature. Experiments in however narrow a walk, facts of whatever minuteness, make reputations in scientific societies; all beyond is regarded with suspicion and distrust.

(Chambers, *Explanations*, pp. 175-176).

It is important to recognize that Chambers was not alone in expressing dissatisfaction with the current image of science. In fact, some of the less critical reviews used the case of his book to comment on tendencies of scientific research and their effects. Thus, Edward Forbes, the naturalist, remarked that "for some time back we have been so immersed in the *facts* of science, that to read a volume of speculations is like a breath of fresh air to the workman in a crowded factory." Although critical of some of its scientific arguments, he concluded that "the great merit of the 'Vestiges' is, the attempt made to show the mutual bearing of sciences, at present too often regarded as far apart."[68] The *Westminster Review*, which carried articles on both *Vestiges* and *Explanations*, continued this theme. It referred to the accumulation of unorganized facts produced by specialization and welcomed *Vestiges* as the "most skilful generalization" available about the natural world.[69] One of its writers concluded that "the error of the author of the '*Vestiges*' has been too general an acquaintance with loose scientific literature; the error of his antagonists has been too limited a knowledge of grand scientific views" (Crosse, p. 194).

The problem of specialization and its implications for the public image of science was recognized by the leaders of the scientific community. It had been noticed at the early meetings of the British Association in the context of discussions about the need to foster communication between researchers in different fields of science. The Association was meant to facilitate this process and to give institutional

[68] [Edward Forbes], "Vestiges of the Natural History of Creation," *Lancet*, 2 (23 November 1844), 265-266.

[69] "Natural History of Creation," *Westminster Review*, 48 (1847-48), 130. For another criticism of "the specialty of most scientific men," see George Henry Lewes, *Comte's Philosophy of the Sciences* (London: H. Bohn, 1853), p. 13.

expression to the unity of science. But the existence of an expanding number of sections devoted to particular groups of sciences indicated the pressure of specialization. Men of science committed to the ideal of general knowledge were sensitive to this development, and Whewell coined the word "scientist" in the context of these fears about the fragmentation of scientific knowledge.[70] By the 1840s, this specialization was firmly established, and its consequences frequently noted.[71]

The implications of this tendency were thought to be twofold: firstly, that potentially fruitful interaction between different branches of science would not occur; and secondly, that it would become increasingly difficult for writers to present a general picture of all science to the lay public. G. C. Lewis, in his discussion of the exercise and recognition of scientific authority, warned that it was therefore "necessary that men of comprehensive minds should survey the whole circle of the sciences, should understand their mutual relations, . . . to avoid that narrowing influence which is produced by restricting the mind to the exclusive contemplation of one subject" (Lewis, p. 112). Chambers was able to portray himself as one who had taken up this challenge by transcending the specialist horizons of the experts. As several reviewers suggested, this was the appeal of *Vestiges*: it offered a synthesis of recent ideas from several sciences and presented them under seductive generalizations such as the Law of Development. The frustration this produced in scientific authorities was epitomized by Whewell's reaction to positive comments about this feature of Chambers' book: "If the mere combining chemistry, geology, physiology, and the like, into a nominal system, while you violate the principles of each at every step of your hypothesis, be held to be a philosophical merit, because the speculator is seeking a wider law than gravitation, I do not see what we, whose admiration of the discovery of gravitation arises from its truth, and the soundness of every step to the truth, have to do, except seek another audience."[72]

Chambers was indeed seeking another audience. In replying to the attack upon his amateur standing, he argued that the scientific specialists were too narrow in outlook to offer an unbiased opinion on a

[70] [William Whewell], "On the Connexion of the Physical Sciences," *Quarterly Review*, 51 (1834), 59. On specialization, see Yeo, "Scientific Method," pp. 68-70.

[71] Apart from capitalizing on the issue of specialization in *Explanations*, Chambers altered a passage in the "Note Conclusory" of the original edition of *Vestiges* in order to emphasize his role as a synthesizer attempting "to weave a great generalization out of the truths already established" (*Vestiges*, 4th ed. [London: Churchill, 1845], p. 404).

[72] Whewell to Myers, 16 March 1845, in Douglas, p. 318.

general theory. He concluded that "it must be before another tribunal,
that this new philosophy is to be truly and righteously judged"
(Chambers, *Explanations*, p. 179). Confronted by "these extraordinary
opinions," Brewster ridiculed the prospect of a work about the natural
world, refusing the judgement of men of science and their learned
institutions. "If they be disqualified for the task," he asked, "is it from
the Royal Society of Literature — from the Antiquaries . . . or from a
committee of Blue-stockings, that he is to summon the judges who are
to preside at the grand inquest on this new philosophy?" (Brewster,
"Explanations," pp. 488-489). This was indeed a pertinent question
because, having rejected the authority of the scientific community,
Chambers was appealing to "ordinary readers" (Chambers,
Explanations, p. 2).

In his attack upon the scientific establishment, Chambers constructed a dichotomy between a narrow, pragmatic physical science
limited to specialists, and a broad, speculative natural philosophy in
which all people could participate.[73] He condemned the "chilling
repression of all saliency in investigation, which characterizes the
scientific men of our country and age" (Chambers, *Explanations*, p.
179). In a description of the scientific ethos which many men of science
would not have recognized, Chambers claimed that the values of
science were becoming too pragmatic and utilitarian. The consequence
was that the fundamental questions which people had about the
natural world and their relationship to its Creator were not being
addressed. In this context, he appealed to the non-scientific public for a
judgement on *Vestiges* and urged the need for a broad natural philosophy closer to the demands of society than the limited physical science
of learned institutions. In this way, Chambers challenged the right of
scientific authorities to restrict the scope of conjecture in science and
proclaimed the right of the lay person to speculate in the field of
natural knowledge. Against this view, scientific spokesmen were
adamant that certain questions required specialist knowledge and
could not be answered on the basis of general reading. Referring to the
fossil record, Brewster asserted that "geologists alone are entitled to
adjudicate upon such a question" (Brewster, "Vestiges," p. 485).

As we have seen, this assertion of the authority of practising
scientists in their areas of expertise was a major theme in the response

---

[73] For a contrast between popular science and the image of pure science, which refers to the *Vestiges*
dispute, see R. G. A. Dolby, "On the Autonomy of Pure Science: The Construction and
Maintenance of Barriers between Scientific Establishments and Popular Culture," in *Scientific
Establishments and Hierarchies*, eds. N. Elias and H. Martins (Dordrecht: Reidel, 1982), pp.
267-292.

28

to *Vestiges*. But the debate also revealed the problems which confronted a definition of expertise in terms of first-hand knowledge, such as field research or experimentation. Chambers was able to reply to the charge that he lacked this experience by showing that it was a simplistic conception of scientific knowledge. Thus, he admonished Sedgwick for proclaiming that he had *"seen* what he speaks of" — namely, geological evidence against the development hypothesis — as if this "personal and direct observation of facts" immediately settled the question (Chambers, *Explanations*, pp. 101-102). Such an attitude ignored the way in which scientific knowledge depended upon a dialogue between different interpretations of reported observations. This response was possible because the increasing division of labour in scientific research was making the distinction between first and second-hand knowledge an inappropriate criterion of scientific authority. In this particular debate, it did not discriminate between Chambers and, say, Whewell, who had done little practical geology, or Brewster, who worked on optics. Chambers, like Whewell, was asserting the right to hypothesize and generalize about the results of scientific discoveries in various fields. But in addition to this, he claimed that the general public had a right to opinions in matters of science.

Apart from Chambers, the most forceful exponent of this position was Francis Newman, who reviewed both *Vestiges* and *Explanations* for the *Prospective Review*. Newman also resented the manner in which Sedgwick scorned the notion of second-hand knowledge. He agreed that the non-scientific public was dependent upon "physical professors" for facts and details, but stressed that "they must not be allowed to monopolize the functions of thought or philosophy" (Newman, pp. 33-34). He was indignant about the implication that "practical geologists are to dictate concerning the Laws of Evidence; as if nobody but they could properly know what 'Induction' meant" (Newman, p. 34). The proofs of science were not strictly demonstrative but were of the kind addressed to the "ordinary understanding," and debate about scientific theories should therefore contain all views. Newman realized that, in practice, this wider participation was limited and that "the public at large have the *unanimity of men of science* as the great guarantee of truth." But he stressed that the public confidence in this consensus was based on the assumption that "no authority [had been] allowed to domineer."[74] Like Chambers, he

---

[74] [Francis Newman], "Vestiges of the Natural History of Creation," *Prospective Review*, 1 (1845), 50.

insisted that natural scientists should not be allowed to define the limits of debate or restrict the number of participants, even in the area of natural knowledge.

It is interesting that Chambers' concern with the place of authority in science did not end with the controversy over *Vestiges*. In an essay of 1859, his target was the "scientific scepticism" which rejected the everyday evidence of the senses whenever these conflicted with general laws of nature. Although not clearly specified, it is likely that the occasion of this paper was the debunking of psychic phenomena by Michael Faraday and other prominent scientists. Chambers understood Faraday to be claiming that the testimony of the senses could not be trusted and that there was "no safety but in a knowledge of the laws of nature."[75]

In Chambers' opinion, this attitude crippled scientific progress because it prevented the recognition of new facts. Reproducing his earlier critique, he suggested that the outlook of sceptical scientists arose from "habits of mind induced by exclusive dealing with one class of facts": physicists, for example, because they relied upon mathematical proofs, denigrated other forms of evidence. He also contended that natural laws were themselves based on a complex set of sense testimony and inference and were therefore not necessarily more certain than well-documented testimonies about specific occurrences (Chambers, *Testimony*, pp. 2-9, 15, 22-24).

Given his earlier stress upon the efficacy of the laws of nature, this was a curious position to take. There is, however, an element of continuity between these two episodes: namely, his reluctance to allow scientific authorities to restrict speculation about the natural world. In the defence of *Vestiges*, there was an appeal to the judgement of "ordinary readers" over the particular objections of specialists; in the essay relating to psychic research, there was a claim for the value of human testimony against the methodological standards of prominent scientists. By affirming the status of common observation in competition with the experimental methods of experts, Chambers was reserving a place for the untrained public in debates about natural knowledge.

This article has suggested that the controversy over *Vestiges* involved not only scientific and theological issues, but questions concerning the social relations of science. It has shown how this debate

[75] [Robert Chambers], *Testimony: Its Posture in the Scientific World* (London and Edinburgh: Ward R. Chambers, 1859), pp. 1-2.

was, in part, constituted by the related problems of popularization and specialization. These developments comprised the social and intellectual conditions which made the authority of science a significant issue in the mid-nineteenth century; and the rhetoric of both the scientific spokesmen and their opponents employed these concepts to advance alternative notions of authority in science. The great popularity of *Vestiges* reinforced existing fears among leaders of the scientific community about public understanding of science and about the ways in which a variety of doctrines were represented as "scientific."

The task of distinguishing between science and pseudo-science in the case of *Vestiges* was complicated by its diverse content and by the problem of explaining its technical deficiencies to the lay public. The synthetic character of the book was attractive as a counter to the increasing specialization of science which, in turn, made the task of effective popularization more difficult. Apart from attacking the substantive content of Chambers' book, the critics condemned it on methodological grounds. In doing so, they attempted to establish criteria such as practical research and specialist knowledge which invested authority in full-time experts. But the debate exposed difficulties in simple equations between first-hand knowledge and scientific authority and underlined the problem of formulating general methodological prescriptions for different disciplines.

Chambers was able to reply to both substantive and methodological criticisms, exploiting differences of opinion and perspective among scientific authorities. Furthermore, in responding to those who wished to circumscribe the role of amateurs in scientific debate, he questioned the specialization of the scientific enterprise, arguing that it was not fulfilling its higher moral and cultural purpose. This response constructed a contrast between natural philosophy and physical science in order to assert the right of the lay public to speculate on general questions about the natural world; and, it rejected any exclusive claims to authority on the part of the scientific community. The arguments over Chambers' work therefore involved not only scientific data and theory, but rival conceptions of the scope and practice of science.

The controversy over *Vestiges* also reminds us of the historical nature of the language in which the social relations of Victorian science are often discussed. Although, in retrospect, it is tempting to explain this episode as a simple confrontation between "professional" and "amateur" science, it is necessary to recognize that these are problematic terms. In the mid-nineteenth century, science was still part of a general intellectual culture, and institutions such as the British Association catered for a broad audience. This meant that the limits of scientific debate were not narrowly defined — in addition to the technical content of a scientific work, its style, accessibility, and social implications were considered appropriate criteria. In this context, terms such as "professional" and "amateur," "specialist," and "popular" did not refer to readily accepted categories but were themselves an integral part of disputes about the proper form of natural knowledge. This article has attempted to show that the relationships of authority which these dichotomies reflect were still being negotiated in the period in which Chambers' book appeared.[76] It was not until about the 1870s, when the institutional and educational status of science became more secure, that the boundaries between an accredited scientific community and the lay public became clearly delineated. This did not mean that the problem of authority was resolved; rather, it assumed different forms. The ascendency of experts was established and the need to confront amateur speculation in public was not so urgent, but there were clashes between experts from different disciplines.[77] And with this greater specialization, the problem of responding to synthetic theories (illustrated by the case of *Vestiges*) became a significant issue in scientific debate.[78] Finally, there was an intensification of the more general question of claims to cultural authority beyond the sphere of natural knowledge, as some sections of the scientific community sought to make British culture more "scientific."[79]

[76] For an analysis of a famous mid-nineteenth-century debate that raises interesting comparisons, see J. R. Lucas, "Wilberforce and Huxley: A Legendary Encounter," *Historical Journal*, 22 (1979), 313-330.

[77] For the clash between geologists and physicists, see J. D. Burchfield, *Lord Kelvin and the Age of the Earth* (New Haven: Yale University Press, 1975), chap. 3.

[78] The theory of continental drift, proposed by Alfred Wegener in the early twentieth century, posed this kind of problem because it cut across several disciplines. The work of Immanuel Velikovsky, which draws upon archaeology and mythology as well as physics and geology, is another example.

[79] See Frank M. Turner, "Public Science in Britain, 1880-1919," *Isis*, 71 (1980), 580-608.

INDEX

INDEX